Advanced Signal Processing and Digital Noise Reduction

Advanced Signal Processing and Digital Noise Reduction

Saeed V. Vaseghi
Queen's University of Belfast
UK

⊛WILEY ⊞TEUBNER

A Partnership between John Wiley & Sons and B. G. Teubner Publishers

Chichester · New York · Brisbane · Toronto · Singapore · Stuttgart · Leipzig

Copyright © 1996 jointly by John Wiley & Sons Ltd. and B.G. Teubner

John Wiley & Sons Ltd
Baffins Lane
Chichester
West Sussex
PO19 1UD
England

B.G. Teubner
Industriestraße 15
70565 Stuttgart (Vaihingen)
Postfach 80 10 69
70510 Stuttgart
Germany

National Chichester 01243 779777
International   (+44) 1243 779777

National Stuttgart (0711) 789010
International   +49 711 789010

Other Wiley Editorial Offices

John Wiley & Sons, Inc., 605 Third Avenue
New York, NY 10158-0012, USA

Brisbane • Toronto • Singapore

Other Teubner Editorial Offices

B.G. Teubner, Verlagsgesellschaft mbH, Johannisgaße 16
D-04103 Leipzig, Germany

Die Deutsche Bibliotheck - CIP-Einheitsaufnahme
Vaseghi, Saeed V.
Advanced signal processing and digital noise reduction / Saeed
V. Vaseghi. -Stuttgart ; Leipzig ; Teubner ; Chichester ; New York ;
Brisbane ; Toronto ; Singapore :Wiley, 1996
    ISBN 3 519 06451 0 (Teubner)
    ISBN 0 471 95875 1 (Wiley)

WG: 37	DBN 94.687092.6	96.02.16
8058	fm	

British Library Cataloguing in Publication Data

A catalogue record for this book is available from the British Library

ISBN Wiley 0 471 95875 1
ISBN Teubner 3 519 06451 0

Produced from camera-ready copy supplied by the authors using MacWord 5.1.
Printed and bound in Great Britain by Bookcraft (Bath) Ltd.
This book is printed on acid-free paper responsibly manufactured from sustainable foresta-
tion, for which at least two trees are planted for each one used for paper production.

To my Parents

with thanks to Peter Rayner and Ben Milner

Contents

Preface xvi

1 Introduction 1
    1.1 Signals and Information 2
    1.2 Signal Processing Methods 3
        1.2.1 Non-parametric Signal Processing 3
        1.2.2 Model-based Signal Processing 3
        1.2.3 Bayesian Statistical Signal Processing 4
        1.2.4 Neural Networks 4
    1.3 Applications of Digital Signal Processing 4
        1.3.1 Adaptive Noise Cancellation and Noise Reduction 5
        1.3.2 Blind Channel Equalisation 7
        1.3.3 Signal Classification and Pattern Recognition 8
        1.3.4 Linear Prediction Modelling of Speech 9
        1.3.5 Digital Coding of Audio Signals 11
        1.3.6 Detection of Signals in Noise 13
        1.3.7 Directional Reception of Waves: Beamforming 14
    1.4 Sampling and Analog to Digital Conversion 16
        1.4.1 Time-Domain Sampling and Reconstruction of Analog Signals ... 17
        1.4.2 Quantisation 20
    Bibliography 21

2 Stochastic Processes 23
    2.1 Random Signals and Stochastic Processes 24
        2.1.1 Stochastic Processes 25
        2.1.2 The Space or Ensemble of a Random Process 26
    2.2 Probabilistic Models of a Random Process 27
    2.3 Stationary and Nonstationary Random Processes 31
        2.3.1 Strict Sense Stationary Processes 33
        2.3.2 Wide Sense Stationary Processes 34
        2.3.3 Nonstationary Processes 34
    2.4 Expected Values of a Stochastic Process 35
        2.4.1 The Mean Value 35
        2.4.2 Autocorrelation 36
        2.4.3 Autocovariance 37
        2.4.4 Power Spectral Density 38
        2.4.5 Joint Statistical Averages of Two Random Processes 40
        2.4.6 Cross Correlation and Cross Covariance 40
        2.4.7 Cross Power Spectral Density and Coherence 42
        2.4.8 Ergodic Processes and Time-averaged Statistics 42

2.4.9 Mean-ergodic Processes... 43
2.4.10 Correlation-ergodic Processes... 44
2.5 Some Useful Classes of Random Processes 45
2.5.1 Gaussian (Normal) Process 45
2.5.2 Multi-variate Gaussian Process... 47
2.5.3 Mixture Gaussian Process 48
2.5.4 A Binary-state Gaussian Process 49
2.5.5 Poisson Process 50
2.5.6 Shot Noise 52
2.5.7 Poisson-Gaussian Model for Clutters and Impulsive Noise 53
2.5.8 Markov Processes 54
2.6 Transformation of a Random Process 57
2.6.1 Monotonic Transformation of Random Signals....................... 58
2.6.2 Many-to-one Mapping of Random Signals............................. 60
Summary 62
Bibliography... 63

3 Bayesian Estimation and Classification.................................... 65
3.1 Estimation Theory: Basic Definitions 66
3.1.1 Predictive and Statistical Models in Estimation 66
3.1.2 Parameter Space 67
3.1.3 Parameter Estimation and Signal Restoration....................... 68
3.1.4 Performance Measures 69
3.1.5 Prior, and Posterior Spaces and Distributions....................... 71
3.2 Bayesian Estimation.. 74
3.2.1 Maximum a Posterior Estimation 75
3.2.2 Maximum Likelihood Estimation 76
3.2.3 Minimum Mean Squared Error Estimation............................ 79
3.2.4 Minimum Mean Absolute Value of Error Estimation 81
3.2.5 Equivalence of MAP, ML, MMSE and MAVE........................ 82
3.2.6 Influence of the Prior on Estimation Bias and Variance............ 82
3.2.7 The Relative Importance of the Prior and the Observation 86
3.3 Estimate-Maximise (EM) Method 90
3.3.1 Convergence of the EM algorithm 91
3.4 Cramer-Rao Bound on the Minimum Estimator Variance 93
3.4.1 Cramer-Rao Bound for Random Parameters 95
3.4.2 Cramer-Rao Bound for a Vector Parameter.......................... 95
3.5 Bayesian Classification 96
3.5.1 Classification of Discrete-valued Parameters 96
3.5.2 Maximum a Posterior Classification 98
3.5.3 Maximum Likelihood Classification 98
3.5.4 Minimum Mean Squared Error Classification 99
3.5.5 Bayesian Classification of Finite State Processes.................... 99
3.5.6 Bayesian Estimation of the Most Likely State Sequence 101
3.6 Modelling the Space of a Random Signal.................................. 102
3.6.1 Vector Quantisation of a Random Process 103
3.6.2 Design of a Vector Quantiser: K-Means Algorithm 103

    3.6.3 Design of a Mixture Gaussian Model 104
    3.6.4 The EM Algorithm for Estimation of Mixture Gaussian
          Densities 105
  Summary 108
  Bibliography 109

4 Hidden Markov Models 111
  4.1 Statistical Models for Nonstationary Processes 112
  4.2 Hidden Markov Models 114
    4.2.1 A Physical Interpretation of Hidden Markov Models 115
    4.2.2 Hidden Markov Model As a Bayesian Method 116
    4.2.3 Parameters of a Hidden Markov Model 117
    4.2.4 State Observation Models 118
    4.2.5 State Transition Probabilities 119
    4.2.6 State-Time Trellis Diagram 120
  4.3 Training Hidden Markov Models 121
    4.3.1 Forward-Backward Probability Computation 122
    4.3.2 Baum-Welch Model Re-Estimation 124
    4.3.3 Training Discrete Observation Density HMMs 125
    4.3.4 HMMs with Continuous Observation PDFs 127
    4.3.5 HMMs with Mixture Gaussian pdfs 128
  4.4 Decoding of Signals Using Hidden Markov Models 129
    4.4.1 Viterbi Decoding Algorithm 131
  4.5 HMM-based Estimation of Signals in Noise 133
    4.5.1 HMM-based Wiener Filters 135
    4.5.2 Modelling Noise Characteristics 136
  Summary 137
  Bibliography 138

5 Wiener Filters 140
  5.1 Wiener Filters: Least Squared Error Estimation 141
  5.2 Block-data Formulation of the Wiener Filter 145
  5.3 Vector Space Interpretation of Wiener Filters 148
  5.4 Analysis of the Least Mean Squared Error Signal 150
  5.5 Formulation of Wiener Filter in Frequency Domain 151
  5.6 Some Applications of Wiener Filters 152
    5.6.1 Wiener filter for Additive Noise Reduction 153
    5.6.2 Wiener Filter and Separability of Signal and Noise 155
    5.6.3 Squared Root Wiener Filter 156
    5.6.4 Wiener Channel Equaliser 157
    5.6.5 Time-alignment of Signals 158
    5.6.6 Implementation of Wiener Filters 159
  Summary 161
  Bibliography 162

6 Kalman and Adaptive Least Squared Error Filters 164
  6.1 State-space Kalman Filters 165
  6.2 Sample Adaptive Filters 171
  6.3 Recursive Least Squares (RLS) Adaptive Filters 172

6.4 The Steepest Descent Method 177
6.5 The LMS Adaptation Method 181
Summary 182
Bibliography 183

7 Linear Prediction Models 185
7.1 Linear Prediction Coding 186
    7.1.1 Least Mean Squared Error Predictor 189
    7.1.2 The Inverse Filter: Spectral Whitening 191
    7.1.3 The Prediction Error Signal 193
7.2 Forward, Backward and Lattice Predictors 193
    7.2.1 Augmented Equations for Forward and Backward Predictors 195
    7.2.2 Levinson-Durbin Recursive Solution 196
    7.2.3 Lattice Predictors 198
    7.2.4 Alternative Formulations of Least Squared Error Predictors 200
    7.2.5 Model Order Selection 201
7.3 Short-term and Long-term Predictors 202
7.4 MAP Estimation of Predictor Coefficients 204
7.5 Signal Restoration Using Linear Prediction Models 207
    7.5.1 Frequency Domain Signal Restoration 209
Summary 212
Bibliography 212

8 Power Spectrum Estimation 214
8.1 Fourier Transform, Power Spectrum and Correlation 215
    8.1.1 Fourier Transform 215
    8.1.2 Discrete Fourier Transform (DFT) 217
    8.1.3 Frequency Resolution and Spectral Smoothing 217
    8.1.4 Energy Spectral Density and Power Spectral Density 218
8.2 Non-parametric Power Spectrum Estimation 220
    8.2.1 The Mean and Variance of Periodograms 221
    8.2.2 Averaging Periodograms (Bartlett Method) 221
    8.2.3 Welch Method :Averaging Periodograms from Overlapped and
          Windowed Segments 222
    8.2.4 Blackman-Tukey Method 224
    8.2.5 Power Spectrum Estimation from Autocorrelation of Overlapped
          Segments 225
8.3 Model-based Power Spectrum Estimation 225
    8.3.1 Maximum Entropy Spectral Estimation 227
    8.3.2 Autoregressive Power Spectrum Estimation 229
    8.3.3 Moving Average Power Spectral Estimation 230
    8.3.4 Autoregressive Moving Average Power Spectral Estimation 231
8.4 High Resolution Spectral Estimation Based on Subspace Eigen Analysis .. 232
    8.4.1 Pisarenko Harmonic Decomposition 232
    8.4.2 Multiple Signal Classification (MUSIC) Spectral Estimation 235
    8.4.3 Estimation of Signal Parameters via Rotational Invariance

     Techniques (ESPRIT) 238
  Summary 240
  Bibliography... 240

9 Spectral Subtraction 242
  9.1 Spectral Subtraction 243
     9.1.1 Power Spectrum Subtraction...................................... 246
     9.1.2 Magnitude Spectrum Subtraction................................ 247
     9.1.3 Spectral Subtraction Filter: Relation to Wiener Filters.............. 247
  9.2 Processing Distortions.. 248
     9.2.1 Effect of Spectral Subtraction on Signal Distribution 250
     9.2.2 Reducing the Noise Variance...................................... 251
     9.2.3 Filtering Out the Processing Distortions 251
  9.3 Non-linear Spectral Subtraction.. 252
  9.4 Implementation of Spectral Subtraction 255
     9.4.1 Application to Speech Restoration and Recognition 257
  Summary 259
  Bibliography... 259

10 Interpolation 261
  10.1 Introduction 262
     10.1.1 Interpolation of a Sampled Signal............................. 262
     10.1.2 Digital Interpolation by a Factor of I 263
     10.1.3 Interpolation of a Sequence of Lost Samples............ 265
     10.1.4 Factors that Affect Interpolation 267
  10.2 Polynomial Interpolation... 268
     10.2.1 Lagrange Polynomial Interpolation 269
     10.2.2 Newton Interpolation Polynomial 270
     10.2.3 Hermite Interpolation Polynomials 273
     10.2.4 Cubic Spline Interpolation 273
  10.3 Statistical Interpolation 276
     10.3.1 Maximum a Posterior Interpolation 277
     10.3.2 Least Squared Error Autoregressive Interpolation..................... 279
     10.3.3 Interpolation Based on a Short-term Prediction Model 279
     10.3.4 Interpolation Based on Long-term and Short-term Correlations 282
     10.3.5 LSAR Interpolation Error 285
     10.3.6 Interpolation in Frequency-Time Domain 287
     10.3.7 Interpolation using Adaptive Code Books 289
     10.3.8 Interpolation Through Signal Substitution 289
  Summary 291
  Bibliography... 292

11 Impulsive Noise 294
  11.1 Impulsive Noise 295
     11.1.1 Autocorrelation and Power Spectrum of Impulsive Noise 297
  11.2 Stochastic Models for Impulsive Noise 298
     11.2.1 Bernoulli-Gaussian Model of Impulsive Noise 299
     11.2.2 Poisson-Gaussian Model of Impulsive Noise 299

11.2.3 A Binary State Model of Impulsive Noise 300
11.2.4 Signal to Impulsive Noise Ratio 302
11.3 Median Filters 302
11.4 Impulsive Noise Removal Using Linear Prediction Models 304
11.4.1 Impulsive Noise Detection 304
11.4.2 Analysis of Improvement in Noise Detectability 306
11.4.3 Two-sided Predictor 308
11.4.4 Interpolation of Discarded Samples 308
11.5 Robust Parameter Estimation 309
11.6 Restoration of Archived Gramophone Records 311
Summary 312
Bibliography 312

12 Transient Noise 314
12.1 Transient Noise Waveforms 315
12.2 Transient Noise Pulse Models 316
12.2.1 Noise Pulse Templates 317
12.2.2 Autoregressive Model of Transient Noise 317
12.2.3 Hidden Markov Model of a Noise Pulse Process 318
12.3 Detection of Noise Pulses 319
12.3.1 Matched Filter 320
12.3.2 Noise Detection Based on Inverse Filtering 321
12.3.3 Noise Detection Based on HMM 322
12.4 Removal of Noise Pulse Distortions 323
12.4.1 Adaptive Subtraction of Noise pulses 323
12.4.2 AR-based Restoration of Signals Distorted by Noise Pulses 324
Summary 327
Bibliography 327

13 Echo Cancellation 328
13.1 Telephone Line Echoes 329
13.1.1 Telephone Line Echo Suppression 330
13.2 Adaptive Echo Cancellation 331
13.2.1 Convergence of Line Echo Canceller 333
13.2.2 Echo Cancellation for Digital Data Transmission over
Subscriber's Loop 334
13.3 Acoustic Feedback Coupling 335
13.4 Sub-band Acoustic Echo Cancellation 339
Summary 341
Bibliography 341

14 Blind Deconvolution and Channel Equalisation 343
14.1 Introduction 344
14.1.1 The Ideal Inverse Channel Filter 345
14.1.2 Equalisation Error, Convolutional Noise 346
14.1.3 Blind Equalisation 347
14.1.4 Minimum and Maximum Phase Channels 349

14.1.5 Wiener Equaliser 350
14.2 Blind Equalisation Using Channel Input Power Spectrum 352
14.2.1 Homomorphic Equalisation 354
14.2.2 Homomorphic Equalisation using a Bank of High Pass Filters .. 356
14.3 Equalisation Based on Linear Prediction Models 356
14.3.1 Blind Equalisation Through Model Factorisation 358
14.4 Bayesian Blind Deconvolution and Equalisation 360
14.4.1 Conditional Mean Channel Estimation 360
14.4.2 Maximum Likelihood Channel Estimation 361
14.4.3 Maximum a Posterior Channel Estimation 361
14.4.4 Channel Equalisation Based on Hidden Markov Models 362
14.4.5 MAP Channel Estimate Based on HMMs 365
14.4.6 Implementations of HMM-Based Deconvolution 366
14.5 Blind Equalisation for Digital Communication Channels 369
14.6 Equalisation Based on Higher-Order Statistics 375
14.6.1 Higher-Order Moments 376
14.6.2 Higher Order Spectra of Linear Time-Invariant Systems 379
14.6.3 Blind Equalisation Based on Higher Order Cepstrum 379
Summary 385
Bibliography 385

Frequently used Symbols and Abbreviations 388

Index 391

Preface

Stochastic signal processing plays a central role in telecommunication and information processing systems, and has a wide range of applications in speech technology, audio signal processing, channel equalisation, radar signal processing, pattern analysis, data forecasting, decision making systems etc. The theory and application of signal processing is concerned with the identification, modelling, and utilisation of patterns and structures in a signal process. The observation signals are often distorted, incomplete and noisy. Hence, noise reduction and the removal of channel distortions is an important part of a signal processing system. The aim of this book is to provide a coherent and structured presentation of the theory and applications of stochastic signal processing and noise reduction methods. This book is organised in fourteen chapters.

Chapter 1 begins with an introduction to signal processing, and provides a brief review of the signal processing methodologies and applications. The basic operations of sampling and quantisation are reviewed in this chapter.

Chapter 2 provides an introduction to the theory and applications of stochastic signal processing. The chapter begins with an introduction to random signals, stochastic processes, probabilistic models and statistical measures. The concepts of stationary, non-stationary and ergodic processes are introduced in this chapter, and some important classes of random processes such as Gaussian, mixture Gaussian, Markov chains, and Poisson processes are considered. The effects of transformation of a signal on its distribution are considered.

Chapter 3 is on Bayesian estimation/classification. In this chapter the estimation and classification problems are formulated within the general framework of the Bayesian inference. The chapter includes Bayesian theory, classical estimators, estimate-maximise method, Cramer-Rao bound on the minimum variance estimate, Bayesian classification, and the modelling of the space of a random signal. This chapter provides a number of examples on Bayesian estimation of signals observed in noise.

Chapter 4 considers hidden Markov models for nonstationary signals. The chapter begins with an introduction to the modelling of nonstationary signals and then concentrates on the theory and applications of hidden Markov models (HMMs). HMM is introduced as a Bayesian model and the methods of training HMMs, and using HMMs for decoding and classification are considered. The chapter also includes the application of HMMs in noise reduction.

Chapter 5 considers Wiener Filters. The least squared error filter is formulated first through minimisation of the expectation of the squared error function over the space of the error signal. Then a block-signal formulation of Wiener filters, and a vector space interpretation of Wiener filters, are considered. The frequency response of the Wiener filter is derived through minimisation of mean squared error in the frequency domain. Some applications of the Wiener filter are considered, and a case study of the Wiener filter for removal of additive noise, provides useful insight into the operation of the filter.

Chapter 6 considers the state-space Kalman filters and the sample-adaptive least squared error filters. The chapter begins with the state-space equation for Kalman filters. The optimal filter coefficients are derived using the principle of orthogonality of the innovation signal. The recursive least squared (RLS) filter which is an exact sample-adaptive implementation of the Wiener filter is derived in this chapter. Then the steepest descent search method for the optimal filter is introduced. The chapter concludes with a study of the LMS adaptive filters.

Chapter 7 considers linear prediction models. Forward prediction, backward prediction and lattice predictors are studied. This chapter introduces a modified predictor for the modelling of the short term and the pitch period correlation structures. A maximum a posterior (MAP) estimate of a predictor model which includes prior probability density function of the predictor is introduced. This chapter concludes with application of linear prediction models in signal restoration.

Chapter 8 consider frequency analysis and power spectrum estimation. The chapter begins with an introduction to Fourier transform, and the role of power spectrum in identification of patterns and structures in a signal process. The chapter considers nonparametric spectral estimation, model-based spectral estimation, maximum entropy method, and high resolution spectral estimation based on eigen analysis.

Chapter 9 considers spectral subtraction. A general form of spectral subtraction is formulated and the processing distortions that result form spectral subtraction are considered. The effects of processing distortions on the distribution of a signal is illustrated. The chapter considers methods for removal of the distortions and also nonlinear methods of spectral subtraction. This chapter concludes with an implementation of spectral subtraction for signal restoration.

Chapter 10 considers interpolation of a sequence of unknown samples. This chapter begins with a study of the ideal interpolation of a band limited signal, a simple model for the effects of a number of missing samples, and the factors that effect interpolation. Interpolators are divided into two categories of polynomial and statistical interpolators. A general form of polynomial interpolation, and its special forms Lagrange, Newton, Hermite, and cubic spline interpolators are considered. Statistical interpolators in this chapter include maximum a posterior interpolation, least squared error interpolation based on an autoregressive model, time-frequency interpolation, and interpolation through search of an adaptive codebook for the best signal.

Chapters 11 and 12 cover the modelling detection and removal of impulsive noise and transient noise pulses. In chapter 11 impulsive noise is modelled as a binary state nonstationary processes and several stochastic models for impulsive noise are considered. For removal of impulsive noise, the median filters, and a method based on a linear prediction model of the signal process are considered. The materials in chapter 12 closely follow chapter 11. In this chapter a template-based method, an HMM-based method, and AR model-based method for removal of transient noise are considered.

Chapter 13 covers echo cancellation. The chapter begins with introduction to telephone line echoes, and consider line echo suppression and adaptive line echo cancellation. Then the problem of acoustic echoes and acoustic coupling between loudspeaker and microphone systems are considered. The chapter concludes with a study of a sub-band echo cancellation system

Chapter 14 is on blind deconvolution and channel equalisation. This chapter begins with an introduction to channel distortion models and the ideal channel equaliser. Then the Wiener equaliser, blind equalisation using the channel input power spectrum, blind deconvolution based on linear predictive models, Bayesian channel equalisation, and blind equalisation for digital communication channels are considered. The chapter concludes with equalisation of maximum phase channels using the higher-order statistics.

Saeed Vaseghi
December 1995

1

Introduction

1.1 Signals and Information
1.2 Signal Processing Methods
1.3 Some Applications of Stochastic Signal Processing
1.4 Sampling and Analog to Digital Conversion

Signal processing is concerned with the identification, modelling, and utilisation of patterns and structures in a signal process. Signal processing methods plays a central role in information technology and digital telecommunication, in the efficient and optimal transmission, reception, and extraction of information. Stochastic signal processing theory provides the foundations for modelling the signals and the environments in which the signals propagate. Stochastic models are applied in signal processing, and in decision making systems, for extracting information from an observation signal which may be noisy, distorted or incomplete. This chapter begins with a definition of signals, and a brief introduction to various signal processing methodologies. We consider several key applications of digital signal processing in adaptive noise reduction, channel equalisation, pattern classification/recognition, audio signal coding, signal detection, and spatial processing for directional reception of signals. The chapter concludes with a study of the basic processes of sampling and digitisation of analog signals.

1.1 Signals and Information

A signal can be defined as the variation of a quantity by which information is conveyed on the state, the characteristics, the composition, the trajectory, the course of action or the intention of the signal source. *A signal can be regarded primarily as a means to convey information.* The information conveyed in a signal may be used by humans or machines for communication, forecasting, decision making, control, exploration etc. Figure 1.1 illustrates an information source followed by; a system for signalling the information, a communication channel for propagation of the signal, and a signal processing unit at the receiver for extraction of the information from the signal. In general, there is a translation or mapping between the information $I(t)$ and the signal $x(t)$ that carries the information, this mapping function may be denoted as $T[.]$ and expressed as

$$x(t) = T[I(t)] \tag{1.1}$$

For example, in human speech communication, the voice generating mechanism provides a means for the talker to map each word into a distinct acoustic speech signal that can propagate to the listener. To communicate a word w, the talker generates an acoustic signal realisation of the word, this acoustic signal $x(t)$ may be contaminated by ambient noise and/or distorted by a communication channel, or impaired by the speaking disabilities of the talker and received as the noisy and distorted signal $y(t)$. In addition to conveying the spoken word, the acoustic speech signal has the capacity to convey information on the speaking characteristics, accents, and the emotional state of the talker. The listener extracts these information by processing the signal $y(t)$.

In the past few decades, the theory and applications of digital signal processing have evolved to play a key role in the development of telecommunication and information processing technology. Signal processing methods are central to the development of intelligent man/machine interface in such areas as speech and visual pattern recognition systems. In general, digital signal processing is concerned with two broad areas of information theory : (a) efficient and reliable transmission, reception, storage and representation of signals in telecommunication systems, and (b) extraction of information from noisy signals for pattern recognition, forecasting, decision making, signal enhancement, control etc. In the next section we consider four broad approaches to signal processing problems.

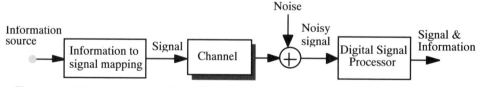

Figure 1.1 Illustration of an information communication and signal processing system.

1.2 Signal Processing Methods

Signal processing methods have evolved in algorithmic complexity aiming for optimal utilisation of the information in order to achieve the best performance. In general the computational requirement of signal processing methods increases, often exponentially, with the algorithmic complexity. However, the implementation cost of advanced signal processing methods has been offset and made affordable by the continuing decrease in the cost of signal processing hardware in recent years.

Depending on the method used, digital signal processing algorithms can be categorised into one or a combination of four broad categories. These are non-parametric signal processing, model-based signal processing, Bayesian statistical signal processing, and neural networks. These methods are briefly described in the followings.

1.2.1 Non-parametric Signal Processing

Nonparametric methods as the name implies do *not* utilise a parametric model of the signal generation or a model of the statistical distribution of the signal. The signal is processed as a waveform or a sequence of digits. Nonparametric methods are not specialised to any particular class of signals, they are broadly applicable methods that can be applied to any signal regardless of the characteristics or the source of the signal. The drawback of these methods is that they do not utilise the distinct characteristics of the signal process which may lead to substantial improvement in performance. Examples of nonparametric methods include digital filtering, and transform-based signal processing such as the Fourier analysis/synthesis relations. Some nonparametric methods of power spectrum estimation and signal restoration are described in Chapters 8 and 9.

1.2.2 Model-based Signal Processing

Model-based signal processing methods utilise a parametric model of the signal generation process. The parametric model normally describes the predictable structures and the observable patterns in the signal process, and can be used to forecast the future values of a signal from its past trajectory. Model-based methods normally outperform nonparametric methods as they utilise more information in the form of a model of the signal process. However, they can be sensitive to deviations of the input from the class of signals characterised by the model. The most widely used parametric model is the linear prediction model, Chapter 7. Linear prediction models have facilitated the development of advanced signal processing methods for a range of applications such as low bit rate speech coding in cellular telephony, high resolution spectral analysis, and speech recognition.

1.2.3   Bayesian Statistical Signal Processing

The fluctuations of a random signal, or the distribution of a class of random signals, can not be modelled by a predictive equation, but can be described in terms of the statistical average values, and the distribution of the signal in a multi-dimensional signal space. For example, a particular acoustic realisation of a spoken word may be characterised by a linear prediction model driven by a random signal as described in Chapter 7. However, the random input of the linear prediction model, and the variations in the characteristics of different acoustic realisations of the same word across the speaking population, can only be described in statistical terms and in terms of probability density functions. Bayesian inference theory provides a generalised framework for statistical processing of random signals, and for formulating and solving estimation and decision making problems. Chapter 3 describes Bayesian estimation and classification of random processes observed in noise.

1.2.4   Neural Networks

Neural networks are combinations of relatively simple nonlinear adaptive processing units, arranged to have a resemblance to the transmission and processing of signals in biological neurones. In a neural network several layers of parallel processing elements are interconnected with a hierarchically structured connection network, and trained to perform a signal processing function. Neural networks are particularly useful in nonlinear partitioning of a signal space, in feature extraction and pattern recognition, and in decision making systems. Since the main objective of this book is to provide a coherent presentation of the theory and applications of stochastic signal processing, neural networks are not included in the scope of this book.

1.3 Applications of Digital Signal Processing

Since the 1980's, the development and commercial availability of increasingly powerful and affordable digital computers has been accompanied by the development of advanced digital signal processing algorithms for a wide variety of applications such as noise reduction, telecommunication, radar, sonar, image-based control systems, video and audio signal processing, pattern recognition, geophysics explorations, data forecasting, and the processing of a large data base for the identification and extraction of unknown underlying structures and patterns.

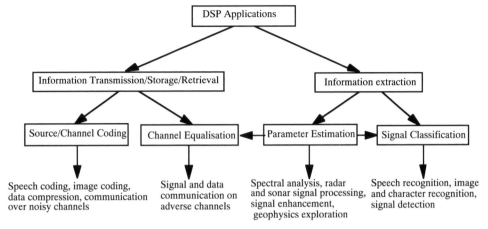

Figure 1.2 shows a possible categorisation of DSP applications. This section provides a review of several applications of digital signal processing methods.

Figure 1.2 A classification of the applications of digital signal processing.

1.3.1   Adaptive Noise Cancellation and Noise Reduction

In communication from a noisy acoustic environment, or over a noisy telephone channel, and often in signal measurements, the information bearing signal is observed in an additive random noise. The noisy observation $y(m)$ can be modelled as

$$y(m) = x(m) + n(m) \tag{1.2}$$

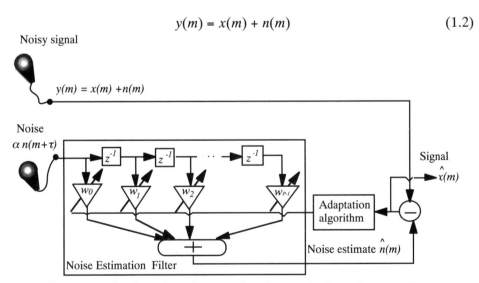

Figure 1.3 Configuration of a two-microphone adaptive noise canceller.

where $x(m)$ and $n(m)$ are the signal and the noise, and the integer variable m is the discrete-time index. In some situations, for example when using a radio-phone in a car or in an aircraft cockpit, it may be possible to measure and estimate the instantaneous amplitude of the ambient noise using a directional microphone. The signal $x(m)$ may then be recovered by subtraction of the noise estimate from the noisy signal.

Figure 1.3 shows a two-input adaptive noise cancellation system, developed for enhancement of noisy speech. In this system a directional microphone takes as input the noisy signal $x(m) + n(m)$, and a second directional microphone, positioned some distance away, measures the noise $\alpha n(m + \tau)$. The attenuation factor α and the time-delay τ provide a rather over-simplified model of the effects of the propagation of noise to different positions in the space where the microphones are placed. The noise from the second microphone, is processed by an adaptive digital filter to make it equal to the noise contaminating the signal, and then subtracted from the noisy signal to cancel out the noise. The adaptive noise canceller is more effective in cancelling out the low frequency part of the noise, but generally suffers from the nonstationary character of the signals and from the assumption that the diffusion and propagation of the noise sound in the space can be modelled by a linear filter.

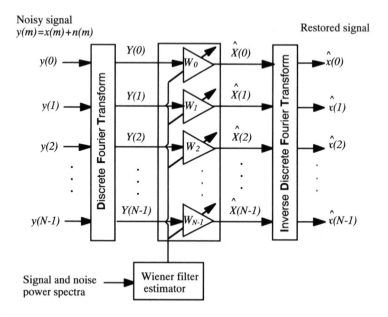

Figure 1.4 A frequency domain Wiener filter for reducing additive noise.

In many applications, for example at the receiver of a telecommunication system, there is no access to the instantaneous value of the contaminating noise, and only the noisy signal is available. In such cases the noise can not be cancelled out, but it may be reduced, in an average sense, using the statistics of the signal and the noise process. Figure 1.4 shows a bank of Wiener filters for reducing additive noise when only the noisy signal is available. The filter bank coefficients attenuate each noisy signal frequency in an inverse proportion to the signal to noise ratio at that frequency. The Wiener filter-bank coefficients, derived in Chapter 5, are calculated from estimates of the power spectra of the signal and the noise.

1.3.2   Blind Channel Equalisation

Channel equalisation is the recovery of a signal distorted in transmission through a communication channel with a nonflat magnitude or a nonlinear phase response. When the channel response is unknown the process of signal recovery is called blind equalisation. Blind equalisation has a wide range of applications, for example in digital telecommunications for removal of inter-symbol interference, in speech recognition for removal of the effects of the microphones and the channels, in correction of distorted images, analysis of seismic data, de-reverberation of acoustic recordings etc.

In practice, blind equalisation is feasible only if some useful statistics of the channel input are available. The success of a blind equalisation method depends on how much is known about the characteristics of the input signal, and how useful this knowledge can be in the channel identification and equalisation process. Figure 1.5 illustrates the configuration of a decision-directed equaliser. This blind channel equaliser is composed of two distinct sections : an adaptive equaliser that removes a large part of the channel distortion, followed by a decision device for an improved estimate of the

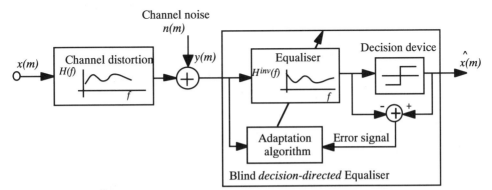

Figure 1.5 Configuration of a decision-directed blind channel equaliser.

channel input. The output of the decision device is the final estimate of the channel input, and it is used as the desired signal *to direct* the equaliser adaptation. Blind equalisation is covered in detail in Chapter 14.

1.3.3   Signal Classification and Pattern Recognition

Signal classification is used in detection, pattern recognition and decision making systems. For example, a simple binary-state classifier can act as the detector of the presence, or the absence, of a known waveform in noise. In signal classification, the aim is to design a system for *labelling* a signal with one of a number of likely categories.

To design a classifier, a set of models are trained for the classes of signals which are of interest in the application. The simplest form that the models can assume is a bank of waveforms each representing the prototype for one class of signals. A more complete model for each class of signals takes the form of a probability distribution function.

In the classification phase, a signal is labelled with the nearest or the most likely class. For example, in communication of a binary bit stream over a bandpass channel, the binary phase shift keying (BPSK) scheme signals the bit "1" using the waveform $A_c \sin \omega_c t$ and the bit "0" using $-A_c \sin \omega_c t$. At the receiver, the decoder has to classify and label the received noisy signal as a "1" or a "0". Figure 1.6 illustrates a correlation receiver for a BPSK signalling scheme. The receiver has two correlators, each programmed with one of the two symbols representing the binary states "1" and "0". The decoder correlates the unlabelled signal with each of the two candidate symbols and selects the candidate which has a higher correlation with the input. Figure 1.7 illustrates the use of a classifier in a limited vocabulary, isolated-word,

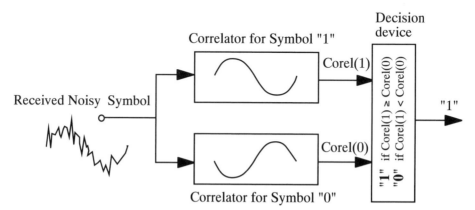

Figure 1.6 A block diagram illustration of the classifier in a binary phase shift keying demodulation.

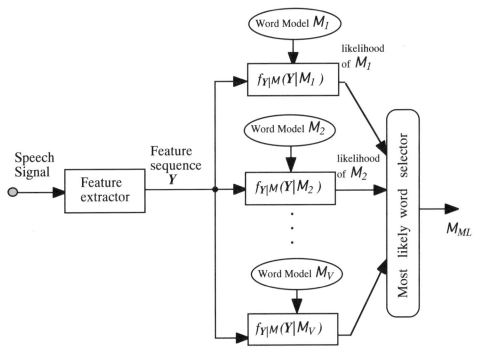

Figure 1.7 Configuration of speech recognition system, $f(Y|M_i)$ is the likelihood of the model M_i given an observation sequence Y.

speech recognition system. Assume there are V words in the vocabulary. For each word a model is trained, on many different examples of the spoken word, to capture the average characteristics and the statistical variations of the word. The classifier has access to a bank of V models, one for each word in the vocabulary. In the speech recognition phase, the task is to label an acoustic speech feature sequence, representing an unlabelled spoken word, as one of the V likely words. For each candidate word the classifier calculates a probability score and selects the word with the highest score.

1.3.4 Linear Prediction Modelling of Speech

Linear predictive models are widely used in speech processing applications such as low bit rate speech coding in cellular telephony, speech enhancement and speech recognition. Speech is produced by inhaling air into the lungs and exhaling it through a vibrating glottis and the vocal tract. The random, noise-like, air flow from the lungs is spectrally shaped and amplified by the vibrations of the glottal cords and the resonance of the vocal tract. The effect of the glottal cords and the vocal tract is to introduce a measure of correlation and predictability on the random variations of the air from the lungs. Figure 1.8 illustrates a model for speech production. The source models the lung and emits a random excitation signal which is filtered, first by a pitch filter model of the glottal cords, and then by a model of the vocal tract.

The main source of correlation in speech is the vocal tract which can be modelled by a linear predictor. A linear predictor forecasts the amplitude of the signal at time m, $x(m)$, using a linear combination of P previous samples $[x(m-1),\cdots,x(m-P)]$ as

$$\hat{x}(m) = \sum_{k=1}^{P} a_k x(m-k) \qquad (1.3)$$

Where $\hat{x}(m)$ is the prediction of the signal $x(m)$, and the vector $\mathbf{a}^T = [a_1,\cdots,a_P]$ are the coefficients of a predictor of order P. The prediction error $e(m)$, the difference between the actual sample $x(m)$ and its predicted value $\hat{x}(m)$, is defined as

$$e(m) = x(m) - \sum_{k=1}^{P} a_k x(m-k) \qquad (1.4)$$

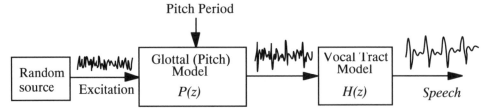

Figure 1.8 Linear predictive model of speech.

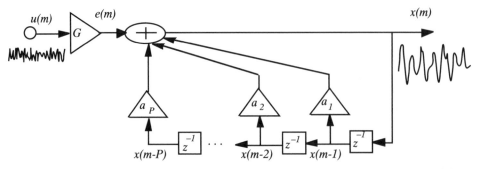

Figure 1.9 Illustration of a signal generated by an all-pole, linear prediction model.

The prediction error $e(m)$ may also be interpreted as the random excitation or the information content of $x(m)$. From Eq. (1.4) a signal generated by a linear predictor can be synthesised as

$$x(m) = \sum_{k=1}^{P} a_k x(m-k) + e(m) \qquad (1.5)$$

Eq. (1.5) describes a model for speech synthesis illustrated in Figure 1.9.

1.3.5 Digital Coding of Audio Signals

In digital audio, the memory required to record a signal, or the bandwidth required for signal transmission, and the signal to quantisation noise ratio, are directly proportional to the number of bits per sample. The objective in the design of a coder is to achieve high fidelity with as few bits per sample as possible, and at an affordable implementation cost. Audio signal coding schemes utilise the statistical structure of the signal, and a model of the signal generation, together with information on the psycho-acoustics of hearing. In general there are two main categories of audio coders. These are model-based coders, used for low bit rate speech coding in applications such as cellular telephony, and transform-based coders used in high quality coding of speech and digital hi-fi audio.

Figure 1.10 shows a simplified block diagram configuration of a speech coder-

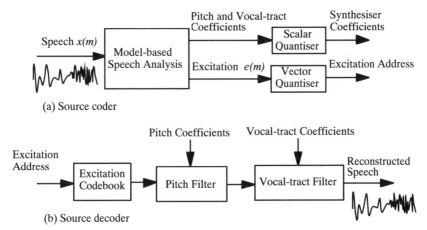

Figure 1.10 Block diagram configuration of a model-based speech coder.

synthesiser of the type used in digital cellular telephone. The speech signal is modelled as the output of a filter excited by a random signal. The random excitation models the air exhaled through the lung, and the filter models the vibrations of the glottal cords and the vocal tract. At the transmitter, speech is segmented into blocks of about 30 ms duration, and each block of speech samples is analysed to extract and transmit a set of excitation and filter parameters that can be used to synthesis the speech. At the receiver, the model parameters and the excitation are used to reconstruct the speech.

A transform-based coder is shown in Figure 1.11. The aim of transformation is to convert the signal into a form where it lends itself to a more convenient and useful interpretation and manipulation. In Figure 1.11 the input signal can be transformed to the frequency domain using a filterbank, a discrete Fourier transform, or a discrete cosine transform. Three main advantages of coding a signal in the frequency domain are the following : (a) the frequency spectrum of a signal has a relatively well defined structure, for example most of the signal power is usually concentrated in the lower regions of the spectrum, (b) a relatively low amplitude frequency would be masked in the near vicinity of a large amplitude frequency and can therefore be coarsely encoded without any audible degradation, and (c) the frequency samples are orthogonal and can be coded independently with different precision. The number of bits assigned to each frequency is a variable that reflects its contribution to the reproduction of a perceptually high quality signal. In an adaptive coder, the allocation of bits to different frequencies is made to vary with the time-variations of the power spectrum of the signal.

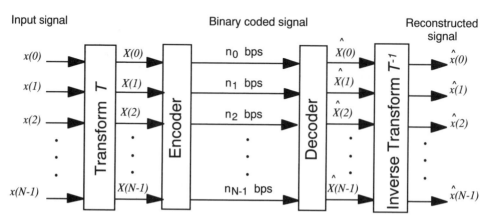

Figure 1.11 Illustration of a transform-based coder.

1.3.6 Detection of Signals in Noise

In the detection of signals in noise, the aim is to determine if the observation consists of noise alone, or if it contains an information bearing signal. The noisy observation $y(m)$ can be modelled as

$$y(m) = b(m)x(m) + n(m) \qquad (1.6)$$

where $x(m)$ is the signal to be detected, $n(m)$ is the noise and $b(m)$ is a binary-valued state sequence, such that $b(m) = 1$ indicates the presence of the signal $x(m)$, and $b(m) = 0$ indicates that the signal is absent. If the signal $x(m)$ has a known shape, then a correlator or a matched filter can be used to detect the signal. The impulse response $h(m)$ of the matched filter for detection of a signal $x(m)$ is the time reversed $x(m)$ given by

$$h(m) = x(N\text{-}1\text{-}m) \qquad 0 \le m \le N\text{-}1 \qquad (1.7)$$

where N is the length of $x(m)$. The output of the matched filter is given by

$$z(m) = \sum_{m=0}^{N-1} h(m - k)y(m) \qquad (1.8)$$

The matched filter output is compared to a threshold and a binary decision is made as

$$\hat{b}(m) = \begin{cases} 1 & \text{if } z(m) \ge Threshold \\ 0 & otherwise \end{cases} \qquad (1.9)$$

where $\hat{b}(m)$ is an estimate of $b(m)$, and it may be erroneous in particular if the signal to noise ratio is low. Table.1 lists four possible outcomes that together $b(m)$ and its estimate $\hat{b}(m)$ can assume. The choice of the threshold level affects the sensitivity of the detector. The higher the threshold, the less the likelihood that noise would be classified as signal, so the false alarm rate falls, but the probability of miss-classification of signal as noise increases.

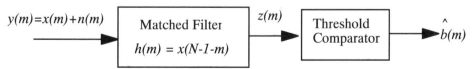

Figure 1.12 Configuration of a matched filter followed by a threshold comparator for detection of signals in noise.

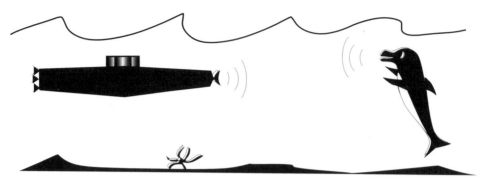

Figure 1.13 Sonar : Detection of objects using the intensity and time delay of reflected sound waves.

$\hat{b}(m)$	$b(m)$	Detector decision
0	0	signal absent $\checkmark$
0	1	signal absent *(Missed)*
1	0	signal present *(False alarm)*
1	1	signal present $\checkmark$

Table-1 Four possible outcomes in the signal detection problem.

The risk in choosing a threshold θ can be expressed as

$$\mathcal{R}\,(Threshold = \theta) = P_{False\,Alarm}(\theta) + P_{Miss}(\theta) \tag{1.10}$$

The choice of the threshold reflects a trade-off between the miss rate $P_{Miss}(\theta)$ and the false alarm rate $P_{False\,Alarm}(\theta)$.

1.3.7   Directional Reception of Waves: Beamforming

Beamforming is the spatial processing of plane waves received by an array of sensors such that the waves incident at a particular spatial angle are passed through, whereas those arriving from other directions are attenuated. Beamforming is used in radar and sonar signal processing to steer the reception of signals towards a desired direction, and in speech processing for reducing the effects of ambient noise.

To explain the process of beamforming consider a uniform linear array of sensors as illustrated in Figure 1.14. The term *linear array* implies that the array of sensors is spatially arranged in a straight line and with equal spacing d between the sensors. Consider a sinusoidal, far field, plane wave with a frequency f_0 propagating towards the sensors at an incidence angle of θ as illustrated in Figure 1.14. The array of sensors samples the incoming wave as it propagates in the space. The time delay for the wave to travel a distance of d between two adjacent sensors is given by

$$\tau = \frac{d \sin \theta}{c} \tag{1.11}$$

where c is the speed of propagation of the wave in the medium. The phase difference corresponding to a delay of τ is given by

$$\varphi = 2\pi \frac{\tau}{T_0} = 2\pi f_0 \frac{d \sin \theta}{c} \tag{1.12}$$

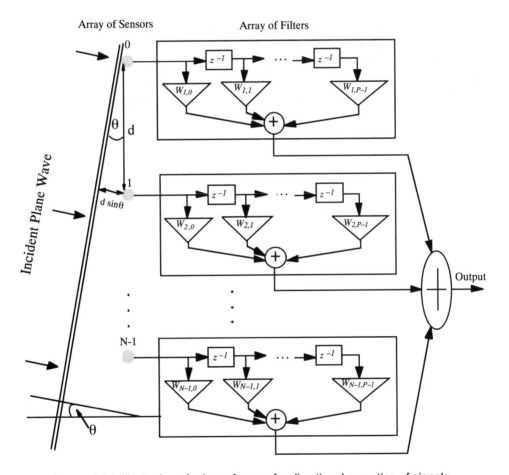

Figure 1.14 Illustration of a beamformer, for directional reception of signals.

where T_0 is the period of the sine wave. By inserting appropriate corrective time delays in the path of the samples at each sensor, and then averaging the outputs of the sensors, the signals arriving from the direction θ will be time-aligned and coherently combined, whereas those arriving from other directions will suffer some cancellations and attenuations. Figure 1.14 illustrates a beamformer which is an array of digital filters arranged in space. The filter array acts as a two dimensional space-time signal processing system. The space filtering allows the beamformer to be steered towards a desired direction for example towards the direction along which the incoming signal has maximum intensity. The phase of each filter controls the time delay, and can be adjusted to coherently combine the signals. The magnitude frequency response of each filter can remove the out of band noise.

1.4 Sampling and Analog to Digital Conversion

A digital signal is a sequence of real or complex valued numbers, representing the fluctuations of an information bearing quantity with time, space, or some other variable. The *basic* digital signal is the unit-sample signal $\delta(m)$ defined as

$$\delta(m) = \begin{cases} 1 & m = 0 \\ 0 & m \neq 0 \end{cases} \tag{1.13}$$

where m is the discrete time index. A digital signal $x(m)$ can be expressed as the sum of a number of scaled and shifted unit samples as

$$x(m) = \sum_{k=-\infty}^{\infty} x(k)\delta(m-k) \tag{1.14}$$

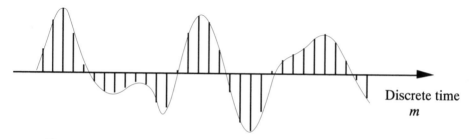

Figure 1.15 A discrete-time signal and its envelope of variation with time.

Many random processes, such as speech, music, radar, sonar etc. generate signals which are continuous in time and continuous in amplitude. Continuos signals are termed analog because their fluctuations with time is analogous to the variations of the signal source. For digital processing, analog signals are sampled, and each sample is converted into an n-bit digit. The digitisation process should be such that the original signal can be recovered from its digital version with no loss of information, and with as high a fidelity as is required in an application. Figure 1.16 illustrates a block diagram configuration of a digital signal processor with an analog input. The lowpass filter removes out of band signal frequencies above a preselected range. The sample and hold (S/H) unit periodically samples the signal to convert the continuous- time signal into a discrete-time signal. The analog to digital converter (ADC) maps each continuous amplitude sample into an n-bit digit. After processing, the digital output of the processor can be converted back into an analog signal using a digital to analog converter (DAC) and a lowpass filter as illustrated in Figure 1.16.

1.4.1 Time-Domain Sampling and Reconstruction of Analog Signals

The conversion of an analog signal to a sequence of n-bit digits consists of two basic steps of sampling and quantisation. The sampling process, when performed with sufficiently high speed, can capture the fastest fluctuations of the signal, and can be a lossless operation, in that the analog signal can be recovered from the sampled sequence as described in Chapter 10. The quantisation of each sample into an n-bit digit, involves some irrevocable error and possible loss of information. However, in practice the quantisation error can be made negligible by using an appropriately high number of bits as in a digital audio hi-fi.

A sampled signal can be modelled as the product of a continuous-time signal $x(t)$ and a periodic impulse train $p(t)$ as

$$x_{sampled}(t) = x(t)p(t)$$

$$= \sum_{m=-\infty}^{\infty} x(t)\delta(t - mT_s) \tag{1.15}$$

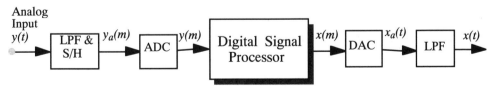

Figure 1.16 Configuration of a digital signal processing system.

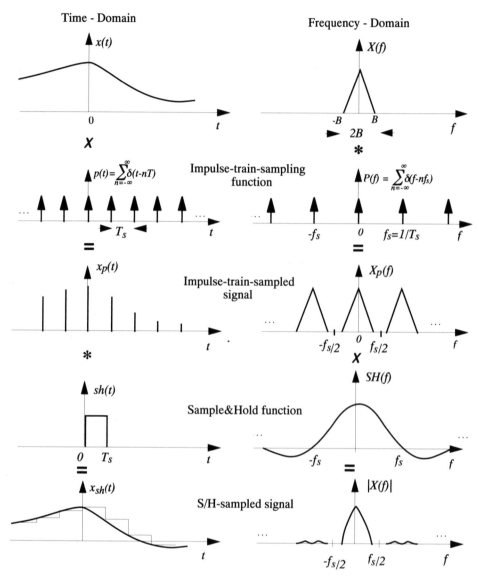

Figure 1.17 Sample&Hold signal modelled as impulse-train sampling followed by convolution with a rectangular pulse.

where T_s is the sampling interval and the sampling function $p(t)$ is defined as

$$p(t) = \sum_{m=-\infty}^{\infty} \delta(t - mT_s) \qquad (1.16)$$

The spectrum $P(f)$ of the sampling function $p(t)$ is also a periodic impulse train given by

$$P(f) = \sum_{k=-\infty}^{\infty} \delta(f - kf_s)$$ (1.17)

where $f_s = 1/T_s$ is the sampling frequency. Since multiplication of two time-domain signals is equivalent to the convolution of their frequency spectra we have

$$X_{sampled}(f) = FT[x(t) \cdot p(t)] = X(f)*P(f) = \sum_{k=-\infty}^{\infty} X(f - kf_s)$$ (1.18)

where the operator *FT[]* denotes the Fourier transform. In Eq. (1.18) the convolution of a signal spectrum $X(f)$ with each impulse $\delta(f - kf_s)$, shifts $X(f)$ and centres it on kf_s. *Hence as expressed in Eq. (1.18) the sampling of a signal x(t) results in a periodic repetition of its spectrum X(f) centred on frequencies* $0, \pm f_s, \pm 2f_s, \cdots$. When the sampling frequency is higher than twice the maximum frequency content of the signal, then the repetitions of the signal spectra are separated as shown in Figure 1.1. In this case, the analog signal can be recovered if the sampled signal is passed through an analog lowpass filter with a cutoff frequency of f_s. If the sampling frequency is less than $2f_s$, then the adjacent repetitions of the spectrum overlap and the original spectrum can not be recovered. The distortion, due to an insufficiently high sampling rate, is irrevocable and is known as *aliasing* This observation is the basis of the *Nyquist sampling theorem :* a bandlimited continuous-time signal, with a highest frequency content (bandwidth) of B Hz, can be uniquely recovered from its samples provided that the sampling speed $f_s > 2B$ samples per second.

In practice sampling is achieved using an electronic switch that allows a capacitor to charge up or down to the level of the input voltage once every T_s seconds as illustrated in Figure 1.18. The sample-and-hold signal can be modelled as the output of a filter with a rectangular impulse response, and with the impulse-train sampled signal as the input as illustrated in Figure 1.17.

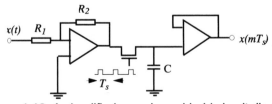

Figure 1.18 A simplified sample-and-hold circuit diagram.

1.4.2 Quantisation

For digital signal processing, continuous amplitude samples from the sample-and-hold are quantised and mapped into n-bit binary digits. For quantisation to n bits, the amplitude range of the signal is divided into 2^n discrete levels, and each sample is quantised to the nearest quantisation level, and then mapped to the binary code assigned to that level. Figure 1.19 illustrates the quantisation of a signal into 4 discrete levels. Quantisation is a many to one mapping, in that all the values that fall within the continuum of a quantisation band are mapped to the centre of the band. The mapping between an analog sample $x_a(m)$ and its quantised value $x(m)$ can be expressed as

$$x(m) = Q[x_a(m)] \tag{1.19}$$

where $Q[.]$ is the quantising function.
The performance of a quantiser is measured by signal to quantisation noise ratio SQNR per bit. The quantisation noise is defined as

$$e(m) = x(m) - x_a(m) \tag{1.20}$$

Now consider an n-bit quantiser with an amplitude range of $\pm V$ volts. The quantisation step size is $\Delta = V/2^n$. Assuming that the quantisation noise is a zero mean uniform process with an amplitude range of $-\Delta/2$ and $\Delta/2$ we can express the noise power as

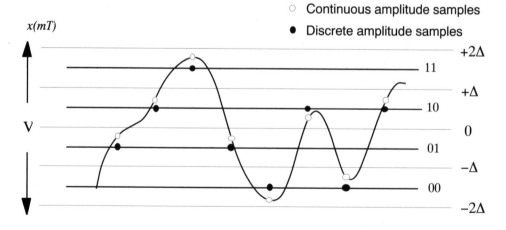

Figure 1.19 Offset binary scalar Quantisation

$$\mathcal{E}[e^2(m)] = \int_{-\Delta/2}^{\Delta/2} f_E(e(m))e^2(m)\, de(m) = \frac{1}{\Delta} \int_{-\Delta/2}^{\Delta/2} e^2(m)\, de(m)$$

$$= \frac{\Delta^2}{12} = \frac{V^2 2^{-2n}}{12}$$

(1.21)

where $f_E(e(m)) = 1/\Delta$ is the uniform probability density function of noise. Similarly assuming that the signal $x(m)$ is uniformly distributed the signal power is given by

$$\mathcal{E}[x^2(m)] = \frac{V^2}{3}$$

(1.22)

From Eq. (1.21) and (1.22) the signal to quantisation noise ratio is given by

$$SQNR = 10 \log_{10}\left(\frac{\mathcal{E}[x^2(m)]}{\mathcal{E}[e^2(m)]}\right) = 10 \log_{10}\left(\frac{V^2/3}{V^2 2^{-2n}/12}\right)$$

$$= 6.02 + 6.02n \quad dB$$

(1.23)

Therefore, from Eq. (1.23) every additional bit in an analog to digital converter results in a 6 dB improvement in signal to quantisation ratio.

Bibliography

ALEXANDER S.T. (1986), Adaptive Signal Processing Theory and Applications. Springer-Verlag, New York.

DAVENPORT W. B., ROOT W. L. (1958), An Introduction to the Theory of Random Signals and Noise. McGraw-Hill, New York.

EPHRAIM Y., (1992),Statistical Model Based Speech Enhancement Systems, Proc. IEEE, Vol. 80, No. 10, Pages 1526-1555.

GAUSS K. G. (1963), Theory of Motion of Heavenly Bodies, Dover, New York.

GALLAGER R. G. (1968), Information Theory and Reliable Communication, Wiley, New York.

HAYKIN S. (1991), Adaptive Filter Theory, Prentice-Hall, Englewood Cliffs, N. J.

HAYKIN S. (1985), Array Signal Processing, Prentice Hall, Englewood Cliffs, N. J.

KAILATH T. (1980), Linear Systems, Prentice Hall, Englewood Cliffs, N. J.

KALMAN R. E. (1960), A New Approach to Linear Filtering and Prediction Problems, Trans. of the ASME, Series D, Journal of Basic Engineering, Vol. 82 Pages 35-45.

KAY S. M. (1993), Fundamentals of Statistical Signal Processing, Estimation Theory Prentice-Hall, Englewood Cliffs, N. J.

LIM J. S. (1983), Speech Enhancement, Prentice Hall, Englewood Cliffs, N. J.

LUCKY R.W., SALZ J., WELDON E.J (1968), Principles of Data Communications McGraw-Hill.

KUNG S. Y.(1993), Digital Neural Networks, Prentice-Hall, Englewood Cliffs, N. J.

MARPLE S. L. (1987), Digital Spectral Analysis with Applications. Prentice Hall, Englewood Cliffs, N. J.

OPPENHEIM A. V., SCHAFER R. W. (1989), Discrete-Time Signal Processing, Prentice-Hall, Englewood Cliffs, N. J.

PROAKIS J. G., RADER C. M., LING F., NIKIAS C. L. (1992), Advanced Signal Processing, Macmillan.

RABINER L. R., GOLD B. (1975), Theory and Applications of Digital Processing, Prentice-Hall, Englewood Cliffs, N. J.

RABINER L. R., SCHAFER R. W. (1978), Digital Processing of Speech Signals, Prentice-Hall, Englewood Cliffs, N. J.

SCHARF L.L. (1991), Statistical Signal Processing : Detection, Estimation, and Time Series Analysis, Addison Wesley, Reading, Massachusetts.

THERRIEN C. W.(1992), Discrete Random Signals and Statistical Signal Processing, Prentice-Hall, Englewood Cliffs, N. J.

VAN-TREES H. L. (1971), Detection, Estimation and Modulation Theory, Parts I., II and III., Wiley, New York.

SHANNON C. E. (1948),A Mathematical Theory of Communication, Bell Systems Tech. J., Vol. 27, pages 379-423, 623-656.

WILSKY A. S. (1979), Digital Signal Processing, Control and Estimation Theory : Poins of Tangency, Areas of Intersection and Parallel Directions, MIT Press.

WIDROW B. (1975),Adaptive Noise Cancelling : Principles and Applications, Proc. IEEE, Vol. 63, Pages 1692-1716.

WIENER N. (1948), Extrapolation, Interpolation and Smoothing of Stationary Time Series, MIT Press Cambridge, Mass.

WIENER N. (1949), Cybernetics, MIT Press Cambridge, Mass.

ZADEH L. A., DESOER C. A. (1963), Linear System Theory : The State-Space Approach, McGraw-Hill.

2

Stochastic Processes

2.1 Random Signals and Stochastic Processes
2.2 Probabilistic Models of a Stochastic Process
2.3 Stationary and Nonstationary Processes
2.4 Expected Values of a Stochastic Process
2.5 Some Useful Classes of Random Processes
2.6 Transformation of a Random Process

Stochastic processes are classes of signals whose fluctuations in time are partially or completely random. Examples of signals that can be modelled by a stochastic process are speech, music, image, time-varying channels, noise, and any information bearing function of time. Stochastic signals are completely described in terms of a probability model, but they can also be characterised with relatively simple statistics, such as the mean, the correlation and the power spectrum. This chapter begins with a study of the basic concepts of random signals and stochastic processes, and the models that are used for characterisation of random processes. We study the important concept of ergodic stationary processes in which time-averages obtained from a single realisation of a stochastic process can be used instead of the ensemble averages. We consider some useful and widely used classes of random signals, and study the effect of filtering or transformation of a signals on its probability distribution.

2.1 Random Signals and Stochastic Processes

Signals, in terms of one of their fundamental characteristics, may be classified into two broad categories : *deterministic* signals and *random (or stochastic)* signals. In each class a signal may be continuous or discrete in time, and may have continuous-valued or discrete-valued amplitudes.

A deterministic signal can be defined as one that traverses a predetermined trajectory in time and space. The fluctuations of a deterministic signal can be completely described in terms of a function of time, and the exact value of the signal at any time is predictable from the functional description and the past history of the signal. For example, a sine wave $x(t)$ can be modelled, and accurately predicted either by a second order linear predictive equation, or by the more familiar equation $x(t) = A \sin(2\pi f t + \phi)$.

Random signals have unpredictable fluctuations; it is not possible to formulate an equation that can predict the *exact* value of a random signal from its past history. Most signals such as speech and noise are at least in part random. The concept of randomness is closely associated with the concepts of information and noise. In fact much of the work on the processing of random signals is concerned with the extraction of information from noisy observations. If a signal is to have a capacity to convey information, it must have a degree of randomness; a predictable signal conveys no information. Therefore the random part of a signal is either the information content of the signal, or noise, or a mixture of both information and noise. Although a random signal is not completely predictable, it often exhibits a set of well defined statistical characteristics such as the maximum, the minimum, the mean, the median, the variance, the power spectrum etc. A random process is described in terms of its statistics, and most completely in terms of a probability density function, from which all its statistics can be calculated.

Example 2.1 Figure 2.1(a) shows a block diagram model of a deterministic digital signal. The model generates an output signal $x(m)$ from P past samples as

$$x(m) = h_1(x(m-1), x(m-2), ..., x(m-P)) \qquad (2.1)$$

where the function h_1 can be a linear or a nonlinear model. A knowledge of the model h_1 and the P initial samples is all that is required to predict the future values of the signal $x(m)$. For a sinusoidal signal generator Eq. (2.1) becomes

$$x(m) = a\, x(m-1) - x(m-2) \qquad (2.2)$$

where the choice of the parameter $a = 2\cos(2\pi f_o / f_s)$ determines the oscillation frequency f_o of the sinusoid, at a sampling frequency of f_s.

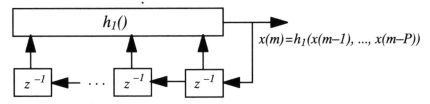

$$x(m)=h_1(x(m-1), ..., x(m-P))$$

(a) A deterministic signal model.

Random
Input $e(m)$

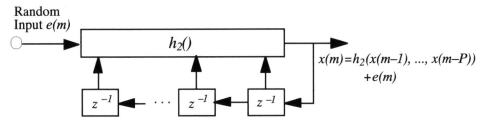

$$x(m)=h_2(x(m-1), ..., x(m-P)) +e(m)$$

(b) A stochastic signal model.

Figure 2.1 Illustration of deterministic and stochastic signal models.

Figure 2.1(b) is a model for a stochastic process given by

$$x(m)= h_2(x(m-1),x(m-2),...,x(m-P))+ e(m) \tag{2.3}$$

where the random input $e(m)$ models the unpredictable part of the signal $x(m)$, and the function h_2 models the part of the signal that is correlated with the past samples. For example, a narrow-band, second order autoregressive process can be modelled as

$$x(m)= a_1 x(m-1) + a_2 x(m-2)+e(m) \tag{2.4}$$

where the choice of a_1 and a_2 determine the centre frequency and the bandwidth of the process.

2.1.1 Stochastic Processes

The term stochastic process, is broadly used to describe a random process that generates sequential signals such speech and noise. In signal processing terminology, a stochastic process is a probability model of a class of random signals, e.g. Gaussian process, Markov process, Poisson process, *etc*. The classical example of a stochastic process is the so called Brownian motion of particles in a fluid. Particles in the

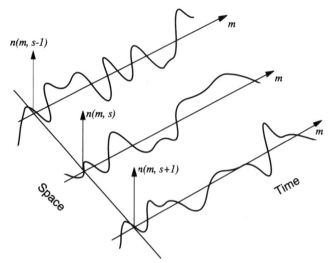

Figure 2.2 Illustration of three realisations in the space of a random noise $N(m)$.

space of a fluid move randomly due to the bombardment by the fluid molecules. The erratic motion of each particle, is a single realisation of a stochastic process. The motion of all particles in the fluid form the collection, or the space, of different realisations of the process.

In this chapter we are mainly concerned with discrete time random processes that may occur naturally or may be obtained by sampling a continuous-time band limited. random process. The term 'discrete-time stochastic process' refers to a class of discrete-time random signals, $X(m)$, that can be characterised by a probabilistic model. Each realisation of a discrete stochastic process $X(m)$ may be indexed in time and space as $x(m,s)$, where m is the discrete time index, and s is an integer variable that designates a space index to each realisation of the process.

2.1.2 The Space or Ensemble of a Random Process

The collection of all realisations of a random process is known as the ensemble, or the space, of the process. For illustration, consider a random noise process over a telecommunication network as shown in Figure 2.2. The noise on each line fluctuates randomly with time, and may be denoted as $n(m,s)$, where m is the discrete time index, and s denotes the line number. The collection of noise on different lines form the ensemble (or the space) of the noise process denoted by $N(m)=\{n(m,s)\}$, where $n(m,s)$ denotes a realisation of the noise process $N(m)$ on the line s. The "true" statistics of a random process are obtained from averages taken over the ensemble of different realisations of the process. However, in many practical cases only one realisation of a process is available. In Section 2.4 we consider ergodic processes in which time-averaged statistics, from a single realisation of a process, may be used instead of the ensemble averaged statistics.

Notation : The following notation is used in this chapter : $X(m)$ denotes a random process, the signal $x(m,s)$ is a particular realisation of the process $X(m)$, the random signal $x(m)$ is any realisations of $X(m)$, and the collection of all realisations of $X(m)$ denoted as $\{x(m,s)\}$ form the ensemble of the random process $X(m)$.

2.2 Probabilistic Models of a Random Process

Probability models provide the most complete description of a random process. For a fixed time instant m, the sample realisations of a random process $\{x(m,s)\}$, is a random variable that takes on values across the space, s, of the process. The main difference between a random variable and a random process is that the latter generates a time series. Therefore, the probability models used for random variables may also be applied to random processes. We start this section with definitions of the probability functions for a random variable.

The space of a random variable is the collection of all the values, or outcomes, that the variable can assume. The space of a variable can be partitioned, according to some criteria, into a number of subspaces. A subspace is a collection of signal values with a common attribute, such as a cluster of closely spaced samples, or the collection of samples with an amplitude within a given band of values. Each subspace is called an event, and the probability of an event A, $P(A)$, is the ratio of the number of observed outcomes from the space of A , N_A, divided by the total number of observations

$$P(A) = \frac{N_A}{\displaystyle\sum_{All\,events\,i} N_i} \qquad (2.5)$$

From Eq. (2.5) the sum of the probabilities of all events in an experiment is unity.

Example 2.2 The space of two discrete numbers obtained as outcomes of throwing a pair of dice is shown in Figure 2.3. This space can be partitioned in different ways, for example the two subspaces shown in Figure 2.3 are associated with the pair of numbers that add up to less than or equal to 8, and to greater than 8. In this example, assuming that all numbers are equally likely, the probability of each event is proportional to the total number of outcomes in the space of the event.

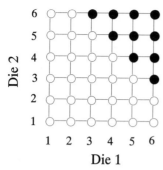

● Outcome from event A : die1+die2 > 8

○ Outcome from event B : die1+die2 ≤ 8

$$P_A = \frac{10}{36} \qquad P_B = \frac{26}{36}$$

Figure 2.3 A two-dimensional representation of the outcomes of two dice, and the subspaces associated with the events corresponding to the sum of the dice being greater than 8 or, less than or equal to 8.

For a random variable X, that can only assume discrete values from a finite set of numbers $\{x_1, x_2, ..., x_N\}$, each outcome x_i may be considered an event and assigned a probability of occurrence. The probability that a discrete-valued random variable X takes a value x_i, $P(X=x_i)$, is called the *probability mass function (pmf)*. For two such random variables, X and Y, the probability of an outcome in which X takes a value x_i and Y takes a values y_j $P(X=x_i, Y=y_j)$ is called the joint probability mass function. The joint pmf can be described in terms of the conditional and marginal probability mass functions as

$$
\begin{aligned}
P_{X,Y}(x_i, y_j) &= P_{Y|X}(y_j|x_i)P_X(x_i) \\
&= P_{X|Y}(x_i|y_j)P_Y(y_j)
\end{aligned}
\tag{2.6}
$$

where $P_{Y|X}(y_j|x_i)$ is the probability of the random variable Y taking a value y_j conditioned on X having taken a value of x_i, and the so called marginal pmf of X is obtained as

$$
\begin{aligned}
P_X(x_i) &= \sum_{j=1}^{M} P_{X,Y}(x_i, y_j) \\
&= \sum_{j=1}^{M} P_{X|Y}(x_i|y_j)P_Y(y_j)
\end{aligned}
\tag{2.7}
$$

where M is the number of values, or outcomes, in the space of the random variable Y. From Eqs. (2.6) and (2.7) we have the Bayes rule for conditional probability mass function, given by

$$P_{X|Y}(x_i|y_j) = \frac{1}{P_Y(y_j)} P_{Y|X}(y_j|x_i) P_X(x_i)$$

$$= \frac{P_{Y|X}(y_j|x_i) P_X(x_i)}{\sum_{i=1}^{M} P_{Y|X}(y_j|x_i) P_X(x_i)} \tag{2.8}$$

Now consider a continuous-valued random variable. A continuous-valued variable can assume an infinite number of values, and hence, the probability that it takes on any one particular value vanishes to zero. For continuous-valued random variables the cumulative distribution function is defined as the probability that the outcome is less than x as

$$F_X(x) = Prob(X \leq x) \tag{2.9}$$

Where *Prob*() denotes probability. The probability that a random variable X takes on a value within a band of width Δ, centred on x, can be expressed as

$$\frac{1}{\Delta} Prob(x - \Delta/2 \leq X \leq x + \Delta/2) = \frac{1}{\Delta}[Prob(X \leq x + \Delta/2) - Prob(X \leq x - \Delta/2)]$$

$$= \frac{1}{\Delta}[F_X(x + \Delta/2) - F_X(x - \Delta/2)] \tag{2.10}$$

As Δ tends to zero we obtain the *probability density function (pdf)* as

$$f_X(x) = \lim_{\Delta \to 0} \frac{1}{\Delta}[F_X(x + \Delta/2) - F_X(x - \Delta/2)]$$

$$= \frac{\partial F_X(x)}{\partial x} \tag{2.11}$$

Since $F_X(x)$ increases with x, the pdf which is the rate of change of $F_X(x)$ is a nonnegative-valued function; $f_X(x) \geq 0$. The integral of the pdf of a random variable x in the range $\pm\infty$ is unity

$$\int_{-\infty}^{\infty} f_X(x) dx = 1 \tag{2.12}$$

The conditional and marginal probability functions and the Bayes rule, of Eqs.(2.6) to (2.8), also apply to probability density functions of continuous-valued variables. Now, the probability models for random variables may also be applied to random processes. For a continuous-valued random process $X(m)$, the simplest probabilistic model is the uni-variate pdf $f_{X(m)}(x)$, which is the probability density function that a sample from the random process $X(m)$ takes the value x. A bi-variate pdf $f_{X(m)X(m+n)}(x_1, x_2)$ describes the pdf that the samples of the process at time instances m and $m+n$ take the values x_1 and x_2 respectively. In general, an M-variate pdf $f_{X(m_1)X(m_2)\cdots X(m_M)}(x_1, x_2, ..., x_M)$ describes the pdf of M samples of a random process taking specific values at specific time instances. For an M-variate pdf we can write

$$\int\limits_{-\infty}^{\infty} f_{X(m_1),...,X(m_M)}(x_1, \ ...,x_M) \ dx_M = f_{X(m_1),\cdots X(m_{M-1})}(x_1, ...,x_{M-1}) \quad (2.13)$$

and the sum of the pdfs of all possible realisations of a random process is unity i.e.

$$\int\limits_{-\infty}^{\infty} \cdots \int\limits_{-\infty}^{\infty} f_{X(m_1),...X(m_M)}(x_1, \ ...,x_M) \ dx_1,...dx_M \ = 1 \quad (2.14)$$

The probability of a realisation of a random process at specified time instances may be conditioned on given values of the process at other time instances, and expressed in the form of a conditional probability density function as

$$f_{X(m)|X(n)}(x_m|\, x_n) = \frac{f_{X(n)|X(m)}(x_n|x_m) f_{X(m)}(x_m)}{f_{X(n)}(x_n)} \quad (2.15)$$

If the realisation of a random process at any time is independent of its realisations at other time instances, then the random process is uncorrelated. For an uncorrelated process a multi-variate pdf can be written in terms of products of uni-variate pdfs as

$$f_{[X(m_1)\cdots X(m_M)|X(n_1)\cdots X(n_N)]}\big(x_{m_1}, ..., x_{m_M} \big| x_{n_1}, ..., x_{n_N}\big) = \prod_{i=1}^{M} f_{X(m_i)}(x_{m_i}) \quad (2.16)$$

Discrete-valued stochastic processes can only assume values from a finite set of allowable numbers $[x_1, x_2, ..., x_n]$. An example is the output of a binary message coder which generates a sequence of 1's and 0's. Discrete-time, discrete-valued, stochastic processes are characterised by multi-variate probability mass functions (pmf) denoted as

$$P_{[x(m_1), ..., x(m_M)]}\big(x(m_1) = x_i, ..., x(m_M) = x_k\big) \quad (2.17)$$

The probability that a discrete random process $X(m)$ takes on a value of x_m at time instant m can be conditioned on the process taking on a value x_n at some other time instant n, and expressed in the form of a conditional pmf as

$$P_{X(m)|X(n)}(x_m | x_n) = \frac{P_{X(n)|X(m)}(x_n|x_m)P_{X(m)}(x_m)}{P_{X(n)}(x_n)} \tag{2.18}$$

and for a statistically independent process

$$P_{[X(m_1)\cdots X(m_M)|X(n_1)\cdots X(n_N)]}(x_{m_1}, \ldots, x_{m_M} | x_{n_1}, \ldots, x_{n_N}) = \prod_{i=1}^{M} P_{X(m_i)}(X(m_i) = x_{m_i})$$

$$\tag{2.19}$$

2.3 Stationary and Nonstationary Random Processes

Although the amplitude of a signal $x(m)$ fluctuates with time m, the characteristics of the process that generates the signal may be time-invariant (stationary), or time-varying (non-stationary). An example of a nonstationary process is speech whose loudness and spectral composition changes continuously as the speaker generates various sounds. A process is stationary if the parameters of the probability model of the process are time-invariant, otherwise the process is nonstationary. Stationarity implies that all the parameters, such as the mean, the variance, the power spectral composition, and the higher order moments of the process are time-invariant. In practice there are various degrees of stationarity, it may be that one set of the statistics of a process is stationary, whereas another set is time-varying. For example a random process may have a time-invariant mean, but a time-varying power.

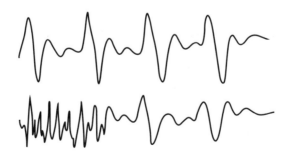

Figure 2.4 Examples of a quasi-stationary and a non-stationary speech segments.

Example 2.3 In this example we consider the *time-averaged* values of the mean and the power of a stationary signal $A\sin\omega t$ and a transient signal $Ae^{-\alpha t}$: The mean and power of the sinusoid are

$$Mean(A\sin\omega t) = \frac{1}{T}\int_T A\sin\omega t\, dt = 0 \qquad \text{constant} \qquad (2.20)$$

$$Power(A\sin\omega t) = \frac{1}{T}\int_T A^2\sin^2\omega t\, dt = \frac{A^2}{2} \qquad \text{constant} \qquad (2.21)$$

where T is the period of the sine wave. The mean and the power of the transient signal are

$$Mean(Ae^{-\alpha t}) = \frac{1}{T}\int_t^{t+T} Ae^{-\alpha t}dt = \frac{A}{\alpha T}(1 - e^{-\alpha T})e^{-\alpha t} \qquad \text{time-varying} \qquad (2.22)$$

$$Power(Ae^{-\alpha t}) = \frac{1}{T}\int_t^{t+T} A^2 e^{-2\alpha\tau}\, d\tau = \frac{A^2}{2\alpha T}(1 - e^{-2\alpha T})e^{-2\alpha t}$$

$$\text{time-varying} \qquad (2.23)$$

In Eqs. (2.22) and (2.23) the signal mean and power are exponentially decaying functions of the time variable t.

Example 2.4 Consider a nonstationary signal $y(m)$ generated by a binary-state random process and described by the following equation

$$y(m) = \bar{s}(m)x_0(m) + s(m)x_1(m) \qquad (2.24)$$

where $s(m)$ is a binary-valued state variable, and $\bar{s}(m)$ denotes the binary complement of $s(m)$. From Eq. (2.24) we have

$$y(m) = \begin{cases} x_0(m) & \text{if } s(m) = 0 \\ x_1(m) & \text{if } s(m) = 1 \end{cases} \qquad (2.25)$$

Let μ_{x_0} and P_{x_0} denote the mean and the power of the signal $x_0(m)$, and μ_{x_1} and P_{x_1} the mean and the power of $x_1(m)$ respectively. The expectation of $y(m)$, given the state $s(m)$, is obtained as

$$\mathcal{E}[y(m)|s(m)] = \bar{s}(m)\mathcal{E}[x_0(m)] + s(m)\mathcal{E}[x_1(m)]$$
$$= \bar{s}(m)\mu_{x_0} + s(m)\mu_{x_1} \tag{2.26}$$

In Eq. (2.26) the mean of $y(m)$ is expressed as a function of the state of the process at time m. The power of $y(m)$ is given by

$$\mathcal{E}[y^2(m)|s(m)] = \bar{s}(m)\mathcal{E}[x_0^2(m)] + s(m)\mathcal{E}[x_1^2(m)]$$
$$= \bar{s}(m)P_{x_0} + s(m)P_{x_1} \tag{2.27}$$

Although many signals are nonstationary, the concept of stationarity has played an important role in the development of signal processing methods. Furthermore even nonstationary signals such as speech can often be considered as approximately stationary for a short period of time. In stochastic signal theory two classes of stationary processes are defined as : (a) strict sense stationary processes, and (b) wide side sense stationary processes which is a less strict form of stationarity in that it only requires that the first and second order statistics of the process should be time-invariant.

2.3.1   Strict Sense Stationary Processes

A random process $X(m)$ is stationary in a strict sense, if all its distributions and statistical parameters are time-invariant. The strict sense stationarity implies that the nth order distribution is translation-invariant for all $n=1, 2,3 \ldots$

$$Prob[x(m_1) \le x_1, x(m_2) \le x_2, \ldots, x(m_n) \le x_n)]$$
$$= Prob[x(m_1 + \tau) \le x_1, x(m_2 + \tau) \le x_2, \ldots, x(m_n + \tau) \le x_n)] \tag{2.28}$$

From Eq. (2.28) the statistics of a strict sense stationary process including the mean, the correlation, and the power spectrum are time-invariant, therefore we have

$$\mathcal{E}[x(m)] = \mu_x \tag{2.29}$$

$$\mathcal{E}[x(m)x(m+k)] = r_{xx}(k) \tag{2.30}$$

and

$$\mathcal{E}\left[|X(f,m)|^2\right] = \mathcal{E}\left[|X(f)|^2\right] = P_{XX}(f) \tag{2.31}$$

where μ_x, $r_{xx}(m)$, and $P_{XX}(f)$ are the mean value, the autocorrelation and the power spectrum of the signal $x(m)$ respectively, and $X(f,m)$ denotes the frequency-time spectrum of $x(m)$.

2.3.2 Wide Sense Stationary Processes

The strict sense stationarity condition requires that all of the statistics of the process should be time-invariant. A less restrictive form of a stationary process is the so called wide sense stationarity. A process is said to be wide sense stationary if the mean and the autocorrelation functions of the process are time invariant;

$$\mathcal{E}[x(m)] = \mu_x \tag{2.32}$$

$$\mathcal{E}[x(m)x(m+k)] = r_{xx}(k) \tag{2.33}$$

From the definitions of strict and wide sense stationary processes, it is clear that a strict sense stationary process is also wide sense stationary, whereas the reverse is not necessarily true.

2.3.3 Nonstationary Processes

A random process is said to be nonstationary if its distributions or statistics vary with time. Most stochastic processes such as video signals, audio signals, financial data

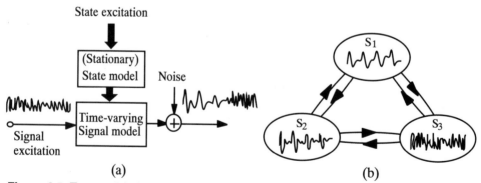

Figure 2.5 Two models for a nonstationary processes: (a) a stationary process drives the parameters of a continuously time-varying model, (b) a finite state model with each state having a different set of statistics.

meteorological data, biomedical signals etc. are nonstationary, because they are generated by systems whose environments and parameters vary over time.

For example, speech is a non-stationary process generated by a time-varying articulatory system. The loudness and the frequency composition of speech changes over time, and sometimes the change can be quite abrupt. Time-varying processes may be modelled by a combination of stationary random models as illustrated in Figure 2.5. In Figure 2.5(a), a non-stationary process is modelled as the output of a time-varying system whose parameters are controlled by a stationary process. In Figure 2.5(b) a time-varying process is modelled with a chain of time-invariant states, with each state having a different set of statistics or probability distributions. Finite state statistical models for time-varying processes are discussed in detail in Chapter 4.

2.4 Expected Values of a Stochastic Process

Expected values of a process play a central role in the modelling, and processing, of stochastic signals. Furthermore, the probability models of a random process are usually expressed as functions of the expected values. For example, a Gaussian pdf is defined as a function of the mean and the covariance of the process, and a Poisson pdf is defined in terms of the mean of the process. Often in signal processing applications, we may have a suitable statistical model of the process, e.g. a Gaussian pdf, and to complete the model we need the expected value parameters. Furthermore in many signal processing algorithms, such as spectral subtraction for noise reduction in Chapter 9, or linear prediction in Chapter 7, what we essentially need is an estimate of the mean or the correlation function of the process. The expected value of a function, $h(X(m_1), X(m_2), ..., X(m_M))$. of a random process is defined as

$$\mathcal{E}\left[h(X(m_1),...,X(m_M))\right] = \int\limits_{-\infty}^{\infty} ... \int\limits_{-\infty}^{\infty} h(x_1,...,x_M)\, f_{X(m_1),...,X(m_M)}(x_1,...,x_M)\, dx_1...dx_M$$

$$(2.34)$$

The most important, and widely used, expected values are the mean value, the correlation, the covariance, and the power spectrum.

2.4.1 The Mean Value

The mean value of a process plays an important part in signal processing and parameter estimation from noisy observations. For example, in Chapter 3 it is shown that the optimal linear estimate of a signal, from a noisy observation, is an interpolation between the mean value and the observed value of the noisy signal. The mean value of a random vector $[X(m_1), ..., X(m_M)]$ is its average value across the ensemble of the process defined as

$$\mathcal{E}[X(m_1), ..., X(m_M)] = \int_{-\infty}^{\infty} \cdots \int_{-\infty}^{\infty} (x_1, ..., x_M) f_{X(m_1), ..., X(m_M)}(x_1, ..., x_M)\, dx_1 ... dx_M$$

$$(2.35)$$

2.4.2 Autocorrelation

The correlation function, and its Fourier transform the power spectral density, can be used to describe and identify patterns and structures in a signal process. Correlators play a central role in signal processing and telecommunication systems, including predictive coders, equalisers, digital receivers, delay estimators, classifiers, and signal restoration systems. The autocorrelation function of a random process $X(m)$, denoted by $r_{xx}(m_1, m_2)$, is defined as

$$r_{xx}(m_1, m_2) = \mathcal{E}[x(m_1)x(m_2)]$$

$$= \int_{-\infty}^{\infty}\int_{-\infty}^{\infty} x(m_1)x(m_2) f_{X(m_1), X(m_1)}(x(m_1), x(m_2))\, dx(m_1)\, dx(m_2) \quad (2.36)$$

The autocorrelation function $r_{xx}(m_1, m_2)$ is a measure of similarity, or mutual relation, of the outcomes of the process at time instances m_1 and m_2 If the outcome of a random process at time m_1 bears no relation to that at time m_2 then $X(m_1)$ and $X(m_2)$ are said to be independent or uncorrelated and $r_{xx}(m_1, m_2) = 0$. For a wide sense stationary process the autocorrelation function is time-invariant and depends on the time difference $m = m_1 - m_2$

$$r_{xx}(m_1 + \tau, m_2 + \tau) = r_{xx}(m_1, m_2) = r_{xx}(m_1 - m_2) = r_{xx}(m) \quad (2.37)$$

where $m = m_1 - m_2$. The autocorrelation function of a real-valued wide sense stationary process is a symmetric function with the following properties

$$r_{xx}(-m) = r_{xx}(m) \quad (2.38)$$

$$r_{xx}(m) \leq r_{xx}(0) \quad (2.39)$$

For a zero mean signal is $r_{xx}(0)$ the signal power.

Example 2.5 Autocorrelation of the Output of a Linear Time-Invariant (LTI) System.
Let $y(m)$ and $h(m)$ denote the input and output of a LTI system, with impulse response $h(m)$, respectively. The input output relation is given by

$$y(m) = \sum_k h_k x(m - k) \tag{2.40}$$

The autocorrelation function of the output $y(m)$ can be related to the autocorrelation of the input signal $x(m)$ as

$$
\begin{aligned}
r_{yy}(k) &= E[y(m)y(m + k)] \\
&= \sum_i \sum_j h_i h_j E[x(m - i)x(m + k - j)] \\
&= \sum_i \sum_j h_i h_j r_{xx}(k + i - j)
\end{aligned}
\tag{2.41}
$$

When the input is an uncorrelated random signal with unit variance, then Eq. (2.41) becomes

$$r_{yy}(k) = \sum_i h_i h_{k+i} \tag{2.42}$$

2.4.3   Autocovariance

The autocovariance function $c_{xx}(m_1, m_2)$ of a random process $X(m)$ is measure of the average similarity of the zero-mean samples $[x(m_1)-\mu_x(m_1)]$ and $[x(m_2)-\mu_x(m_2)]$ and is defined as

$$
\begin{aligned}
c_{xx}(m_1, m_2) &= E\big[(x(m_1) - \mu_x(m_1))(x(m_2) - \mu_x(m_2))\big] \\
&= r_{xx}(m_1, m_2) - \mu_x(m_1)\mu_x(m_2)
\end{aligned}
\tag{2.43}
$$

where $\mu_x(m)$ is the mean of $X(m)$. Note that for a zero mean process the autocorrelation and the autocovariance functions are identical. Note also that $c_{xx}(m_1, m_1)$ is the variance of the process. For a stationary process the autocovariance function of Eq. (2.43) becomes

$$c_{xx}(m_1, m_2) = c_{xx}(m_1 - m_2) = r_{xx}(m_1 - m_2) - \mu_x^2 \tag{2.44}$$

2.4.4 Power Spectral Density

The power spectral density (PSD) function, also called the power spectrum, of a random process gives the spectrum of the distribution of the power among the individual frequency contents of the process. The power spectrum of a wide sense stationary process $X(m)$ is defined, by the Wiener-Kinchin theorem in Chapter 8, as the Fourier transform of the autocorrelation function

$$
\begin{aligned}
P_{XX}(f) &= \mathcal{E}[X(f)X^*(f)] \\
&= \sum_{m=-\infty}^{\infty} r_{xx}(k)e^{-j2\pi fm}
\end{aligned}
\tag{2.45}
$$

where $r_{xx}(m)$ and $P_{XX}(f)$ are the autocorrelation and power spectrum of $x(m)$ respectively and f is the frequency variable. For a real-valued stationary signal the autocorrelation function is symmetric and the power spectrum may be written in terms of a cosine transform as

$$
P_{XX}(f) = r_{xx}(0) + \sum_{m=1}^{\infty} 2r_{xx}(m)\cos(2\pi fm)
\tag{2.46}
$$

The power spectral density is a real and non-negative function, expressed in units of watts per hertz. From Eq. (2.45), the autocorrelation sequence of a random process may be obtained as the inverse Fourier transform of the power spectrum as

$$
r_{xx}(m) = \int_{-1/2}^{1/2} P_{XX}(f)\, e^{j2\pi fm}\, df
\tag{2.47}
$$

Note that the autocorrelation and the power spectrum represent the second order statistics of a process in the time and the frequency domains respectively.

Example 2.6 Power Spectrum and Autocorrelation of White Noise

A noise process with uncorrelated, independent, samples is called a white noise process. The autocorrelation of a stationary white noise $n(m)$ is defined as

$$
r_{nn}(k) = \mathcal{E}[n(m)n(m+k)] = \begin{cases} Noise\,power & k = 0 \\ 0 & k \neq 0 \end{cases}
\tag{2.48}
$$

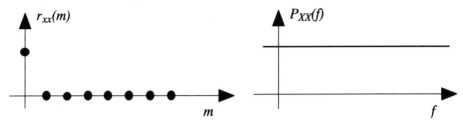

Figure 2.6 Autocorrelation and power spectrum of white noise.

Eq. (2.48) is a mathematical statement of the definition of an uncorrelated (white) noise process. The equivalent description in the frequency domain is derived by taking Fourier transform of $r_{nn}(k)$ as

$$P_{NN}(f) = \sum_{k=-\infty}^{\infty} r_{nn}(k)e^{-j2\pi fk} = r_{nn}(0) = Noise \ \ power \tag{2.49}$$

The power spectrum of a white noise process is spread equally across all time instances and across all frequency bins. White noise is one of the most difficult types of noise to remove, because it does not have a localised structure either in the time domain or in the frequency domain.

Example 2.7 Autocorrelation and Power Spectrum of Impulsive Noise.

Impulsive noise is a random, binary-state ("on/off") sequence of impulses of random amplitudes an random time of occurrences. In Chapter 11 a random impulsive noise sequence $n_i(m)$ is modelled as an amplitude-modulated random binary sequence as

$$n_i(m) = n(m) \ b(m) \tag{2.50}$$

where $b(m)$ is a binary-state random sequence that indicates the presence or the absence of an impulse, and $n(m)$ is a random noise process. Assuming that impulsive noise is an uncorrelated process, the autocorrelation of impulsive noise can be defined as a binary-state process as

$$r_{nn}(k, m) = \mathcal{E}[n_i(m)n_i(m+k)] = \sigma_n^2 \ \delta(k)b(m) \tag{2.51}$$

Where σ_n^2 is the noise variance. Note that in Eq. (2.51) the autocorrelation is expressed as a binary-state function that depends on the on/off state of impulsive noise at time $m..$ The power spectrum of an impulsive noise sequence is obtained, by taking the Fourier transform of the autocorrelation function, as

$$P_{NN}(f,m) = \sigma_n^2\, b(m) \qquad\qquad (2.52)$$

2.4.5 Joint Statistical Averages of Two Random Processes

In many signal processing problems, for example in processing the outputs of an array of sensors, we deal with more than one random process. Joint statistics, and joint distributions, are used to describe the statistical inter-relationship between two or more random processes. For two discrete-time random processes $x(m)$ and $y(m)$, the joint pdf is denoted as

$$f_{X(m_1),\ldots,X(m_M),Y(n_1),\ldots,Y(n_N)}(x_1, \ldots,\, x_M,\; y_1,\ldots,y_N) \qquad\qquad (2.53)$$

When two random processes, $X(m)$ and $Y(m)$ are uncorrelated, the joint pdf can be expressed as product of the pdfs of each process as

$$
\begin{aligned}
&f_{X(m_1),\ldots,X(m_M),Y(n_1),\ldots,Y(n_N)}(x_1, \ldots,\, x_M,\; y_1,\ldots,y_N) \\
&= f_{X(m_1),\ldots,X(m_M)}(x_1, \ldots,\, x_M)\, f_{Y(n_1),\ldots,Y(n_N)}(y_1,\ldots,y_N)
\end{aligned}
\qquad (2.54)
$$

2.4.6 Cross Correlation and Cross Covariance

The cross correlation function of two random process $x(m)$ and $y(m)$ is defined as

$$
\begin{aligned}
r_{xy}(m_1,m_2) &= \mathcal{E}[x(m_1)y(m_2)] \\
&= \int\limits_{-\infty}^{\infty}\int\limits_{-\infty}^{\infty} x(m_1)y(m_2)\, f_{X(m_1)Y(m_2)}(x(m_1),y(m_2))\; dx(m_1)\; dy(m_2)
\end{aligned}
\qquad (2.55)
$$

For wide sense stationary signals, the cross correlation function $r_{xy}(m_1,m_2)$ depends only on the time difference $m=m_1-m_2$ as

$$r_{xy}(m_1 + \tau, m_2 + \tau) = r_{xy}(m_1,m_2) = r_{xy}(m_1 - m_2) = r_{xy}(m) \qquad (2.56)$$

The cross covariance function is defined as

$$c_{xy}(m_1,m_2) = \mathcal{E}\left[(x(m_1)-\mu_x(m_1))(y(m_2)-\mu_y(m_2))\right]$$
$$= r_{xy}(m_1,m_2) - \mu_x(m_1)\mu_y(m_2) \tag{2.57}$$

Note that for zero mean processes the cross correlation function and the cross covariance function are the same. For a wide sense stationary process the cross-covariance function of Eq. (2.57) becomes

$$c_{xy}(m_1,m_2) = c_{xy}(m_1-m_2) = r_{xy}(m_1-m_2) - \mu_x\mu_y \tag{2.58}$$

Example 2.8 Time Delay Estimation : Consider two signals $y_1(m)$ and $y_2(m)$ each composed of an information bearing signal $x(m)$ plus an additive noise, and given by

$$y_1(m) = x(m) + n_1(m) \tag{2.59}$$

$$y_2(m) = A x(m-D) + n_2(m) \tag{2.60}$$

where A is an amplitude factor, and D is a time delay variable. Cross correlation of the signals $y_1(m)$ and $y_2(m)$ yields

$$\begin{aligned}
r_{y_1 y_2}(k) &= \mathcal{E}[y_1(m)y_2(m+k)] \\
&= \mathcal{E}\left[(x(m)+n_1(m))(Ax(m-D+k)+n_2(m+k))\right] \\
&= Ar_{xx}(k-D) + r_{xn_2}(k) + Ar_{xn_1}(k-D) + r_{n_1 n_2}(k)
\end{aligned} \tag{2.61}$$

Assuming that the signal and noise are uncorrelated we have $r_{y_1 y_2}(k) = Ar_{xx}(k-D)$. As shown in Figure 2.7 the cross correlation function has its maximum at the lag D.

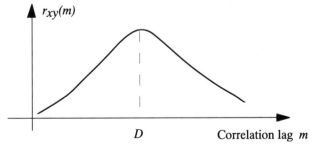

Figure 2.7 The peak of the cross correlation of two delayed signals can be used to estimate the time delay D.

2.4.7 Cross Power Spectral Density and Coherence

The cross power spectral density of two random processes $X(m)$ and $Y(m)$ is defined as the Fourier transform of their cross correlation function as

$$P_{XY}(f) = \mathcal{E}[X(f)Y^*(f)]$$

$$= \sum_{m=-\infty}^{\infty} r_{xy}(m)\, e^{-j2\pi fm} \tag{2.62}$$

The cross power spectral density of two processes is a measure of similarity, or coherence of their power spectra. The coherence, or spectral coherence, of two random processes is a normalised form of the cross power spectral density, and is defined as

$$C_{XY}(f) = \frac{P_{XY}(f)}{\sqrt{P_{XX}(f)P_{YY}(f)}} \tag{2.63}$$

The coherence function is used in applications such as time delay estimation and signal to noise ratio measurements.

2.4.8 Ergodic Processes and Time-averaged Statistics

In many signal processing problems there is only one single realisation of a random process from which its statistical parameters, such as the mean, the correlation, and the power spectrum can be estimated. In such cases time-averaged statistics, obtained from averages along the time dimension of a single realisation of the process, are used instead of the "true" ensemble averages obtained across the space of different realisations of the process. This section considers ergodic random processes for which time-averages can be used instead of the ensemble-averages. *A stationary stochastic process is said to be ergodic if it exhibits the same statistical characteristics along the time dimension of a single realisation, as across the space (or ensemble) of the different realisations of the process.* Over a very long time, a single realisation of an ergodic process takes on all the values, the characteristics and the configurations exhibited across the entire space of the process. For an ergodic process $\{x(m,s)\}$ we have

$$\underset{\text{Along time } m}{Statistical\,averages[x(m,s)]} = \underset{\text{Across space } s}{Statistical\,averages[x(m,s)]} \tag{2.64}$$

Where the function *statistical averages*[.] refers to any statistical operation such as the mean, the variance, the power spectrum etc.

2.4.9   Mean-ergodic Processes

The time-averaged estimate of the mean of a signal $x(m)$ obtained from N samples is given by

$$\hat{\mu}_X = \frac{1}{N} \sum_{m=0}^{N-1} x(m) \qquad (2.65)$$

A stationary process is said to be mean-ergodic if the time-averaged value of an infinitely long realisation of the process is the same as the ensemble-mean taken across the space of the process. Therefore for a mean-ergodic process we have

$$\lim_{N \to \infty} \mathcal{E}[\hat{\mu}_X] = \mu_X \qquad (2.66)$$

$$\lim_{N \to \infty} Var[\hat{\mu}_X] = 0 \qquad (2.67)$$

Condition (2.67) is also referred to as mean-ergodicity in the mean squared error (or minimum variance of error) sense.

The time-averaged estimate of the mean of a signal, obtained from a random realisation of the process, is itself a random variable, with is own mean, variance, and probability density function. If the number of observation samples N is relatively large, then from the central limit theorem the probability density function of the estimate $\hat{\mu}_X$ is Gaussian. The expectation of $\hat{\mu}_X$ is given by

$$\mathcal{E}[\hat{\mu}_x] = \mathcal{E}\left(\frac{1}{N} \sum_{m=0}^{N-1} x(m)\right) = \frac{1}{N} \sum_{m=0}^{N-1} \mathcal{E}[x(m)] = \frac{1}{N} \sum_{m=0}^{N-1} \mu_x = \mu_x \qquad (2.68)$$

Form Eq. (2.68), the time-averaged estimate of the mean is unbiased. The variance of $\hat{\mu}_X$ is given by

$$\begin{aligned} Var[\hat{\mu}_x] &= \mathcal{E}[\hat{\mu}_x^2] - \mathcal{E}^2[\hat{\mu}_x] \\ &= \mathcal{E}[\hat{\mu}_x^2] - \mu_x^2 \end{aligned} \qquad (2.69)$$

Now the term $\mathcal{E}[\hat{\mu}_x^2]$ in Eq. (2.69) may be expressed as

$$\begin{aligned} \mathcal{E}[\hat{\mu}_x^2] &= \mathcal{E}\left(\left(\frac{1}{N} \sum_{m=0}^{N-1} x(m)\right)\left(\frac{1}{N} \sum_{k=0}^{N-1} x(k)\right)\right) \\ &= \frac{1}{N} \sum_{m=-(N-1)}^{N-1} \left(1 - \frac{|m|}{N}\right) r_{xx}(m) \end{aligned} \qquad (2.70)$$

Substitution of Eq. (2.70) in Eq. (2.69) yields

$$
\begin{aligned}
\mathrm{Var}[\hat{\mu}_x^2] &= \frac{1}{N} \sum_{m=-(N-1)}^{N-1} \left(1 - \frac{|m|}{N}\right) r_{xx}(m) - \mu_x^2 \\
&= \frac{1}{N} \sum_{m=-(N-1)}^{N-1} \left(1 - \frac{|m|}{N}\right) c_{xx}(m)
\end{aligned}
\tag{2.71}
$$

Therefore the condition for a process to be mean-ergodic, in the mean squared error sense, is

$$
\lim_{N \to \infty} \frac{1}{N} \sum_{m=-(N-1)}^{N-1} \left(1 - \frac{|m|}{N}\right) c_{xx}(m) = 0
\tag{2.72}
$$

2.4.10   Correlation-ergodic Processes

The time-averaged estimate of the autocorrelation of a random process, estimated from N samples is given by

$$
\hat{r}_{xx}(m) = \frac{1}{N} \sum_{k=0}^{N-1} x(k)x(k+m)
\tag{2.73}
$$

A process is correlation-ergodic, in the mean squared error sense, if

$$
\lim_{N \to \infty} \mathcal{E}[\hat{r}_{xx}(m)] = r_{xx}(m)
\tag{2.74}
$$

$$
\lim_{N \to \infty} \mathrm{Var}[\hat{r}_{xx}(m)] = 0
\tag{2.75}
$$

where $r_{xx}(m)$ is the ensemble-averaged autocorrelation. Taking the expectation of $\hat{r}_{xx}(m)$ shows that it is an unbiased estimate

$$
\mathcal{E}[\hat{r}_{xx}(m)] = \mathcal{E}\left(\frac{1}{N} \sum_{k=0}^{N-1} x(k)x(k+m)\right) = \frac{1}{N} \sum_{k=0}^{N-1} \mathcal{E}[x(k)x(k+m)] = r_{xx}(m)
\tag{2.76}
$$

The variance of $\hat{r}_{xx}(m)$ is given by

$$
\mathrm{Var}[\hat{r}_{xx}(m)] = \mathcal{E}[\hat{r}_{xx}^2(m)] - r_{xx}^2(m)
\tag{2.77}
$$

The term $\mathcal{E}[\hat{r}_{xx}^2(m)]$ in Eq. (2.77) may be expressed as

$$\mathcal{E}[\hat{r}_{xx}^2(m)] = \frac{1}{N^2} \sum_{k=0}^{N-1} \sum_{j=0}^{N-1} \mathcal{E}[x(k)x(k+m)x(j)x(j+m)]$$

$$= \frac{1}{N^2} \sum_{k=0}^{N-1} \sum_{j=0}^{N-1} \mathcal{E}[z(k,m)z(j,m)] \tag{2.78}$$

$$= \frac{1}{N} \sum_{k=-N+1}^{N-1} \left(1 - \frac{|k|}{N}\right) r_{zz}(k,m)$$

where $z(i,m)=x(i)x(i+m)$. Therefore the condition for correlation ergodicity in mean squared error sense is

$$\lim_{N\to\infty} \left(\frac{1}{N} \sum_{k=-N+1}^{N-1} \left(1 - \frac{|k|}{N}\right) r_{zz}(k,m) - r_{xx}^2(m) \right) = 0 \tag{2.79}$$

2.5 Some Useful Classes of Random Processes

In this section we consider some important classes of random processes that are extensively used in the modelling of signals and noise in statistical signal processing applications.

2.5.1 Gaussian (Normal) Process

The Gaussian process, also called the normal process, is perhaps the most widely applied of all stochastic models. Some advantages of the Gaussian probability models are the following:

(a) Many physical phenomena, including some important classes of signal and noise, can be approximated by a Gaussian process.
(b) Many non-Gaussian random processes can be approximated with a weighted combination (i.e. a mixture) of a number of Gaussian densities of appropriate means and variances.
(c) Optimal estimation methods based on Gaussian models often result in linear and mathematically tractable solutions.
(d) The sum of many independent random processes has a Gaussian distribution. This phenomenon is known as the central limit theorem.

A scalar Gaussian random variable is described by the following probability density function

$$f_X(x) = \frac{1}{\sqrt{2\pi}\,\sigma_x} \exp\left(-\frac{(x-\mu_x)^2}{2\sigma_x^2}\right)$$ (2.80)

where μ_x and σ_x^2 are the mean and the variance of the random variable x. The Gaussian process of Eq. (2.80) is also denoted as $N(x, \mu_x, \sigma_x^2)$. The maximum of a Gaussian pdf occurs at the mean μ_x and is given by

$$f_X(\mu_x) = \frac{1}{\sqrt{2\pi}\,\sigma_x}$$ (2.81)

From Eq. (2.80), the Gaussian pdf of x decreases exponentially with the increasing distance of x from the mean value μ_x. The distribution function $F(x)$ is given as

$$F_X(x) = \frac{1}{\sqrt{2\pi}\,\sigma_x} \int_{-\infty}^{x} \exp\left(-\frac{(\chi-\mu_x)^2}{2\sigma_x^2}\right) d\chi$$ (2.82)

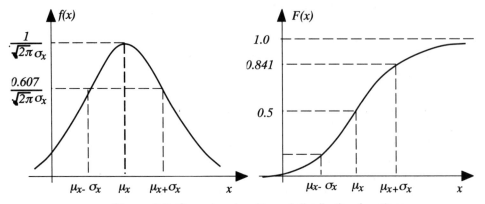

Figure 2.8 Gaussian density and distribution functions.

2.5.2 Multi-variate Gaussian Process

Multi-variate densities model vector-valued processes. Consider an P-variate Gaussian vector process $\{x=[x(m_0), x(m_1), \ldots, x(m_{P-1})]^T\}$ of mean vector μ_x, and covariance matrix Σ_{xx}. The multi-variate Gaussian pdf of x is given by

$$f_X(x) = \frac{1}{(2\pi)^{P/2}|\Sigma_{xx}|^{1/2}} \exp\left(-\frac{1}{2}(x-\mu_x)^T \Sigma_{xx}^{-1}(x-\mu_x)\right) \qquad (2.83)$$

where the mean vector μ_x is defined as

$$\mu_x = \begin{pmatrix} \mathcal{E}[x(m_0)] \\ \mathcal{E}[x(m_2)] \\ \vdots \\ \mathcal{E}[x(m_{P-1})] \end{pmatrix} \qquad (2.84)$$

and the covariance matrix Σ_{xx} is

$$\Sigma_{xx} = \begin{pmatrix} c_{xx}(m_0,m_0) & c_{xx}(m_0,m_1) & \cdots & c_{xx}(m_0,m_{P-1}) \\ c_{xx}(m_1,m_0) & c_{xx}(m_1,m_1) & \cdots & c_{xx}(m_1,m_{P-1}) \\ \vdots & \vdots & \ddots & \vdots \\ c_{xx}(m_{P-1},m_0) & c_{xx}(m_{P-1},m_1) & \cdots & c_{xx}(m_{P-1},m_{P-1}) \end{pmatrix} \qquad (2.85)$$

The Gaussian process of Eq. (2.83) is also denoted as $N(x, \mu_x, \Sigma_{xx})$. If the elements of a vector process are uncorrelated then the covariance matrix is a diagonal matrix with zeros in the off-diagonal elements. In this case the multi-variate pdf may be described as the product of the pdfs of the individual elements of the vector as

$$f_X\left(x = [x(m_0), \ldots, x(m_{P-1})]^T\right) = \prod_{i=0}^{P-1} \frac{1}{\sqrt{2\pi}\,\sigma_{xi}} \exp\left(-\frac{(x(m_i)-\mu_{xi})^2}{2\sigma_{xi}^2}\right) \qquad (2.86)$$

Example 2.9 Conditional multi-variate Gaussian probability density function. Consider two vector realisations $x(m)$ and $y(m+k)$ from the two vector-valued correlated stationary Gaussian processes $N(x, \mu_x, \Sigma_{xx})$ and $N(y, \mu_y, \Sigma_{yy})$. The joint probability density function of $x(m)$ and $y(m+k)$ is a multi-variate Gaussian density $N([x(m),y(m+k)], \mu_{xy} \Sigma_{xy})$ with its mean vector and covariance matrix given by

$$\mu_{(x,y)} = \begin{bmatrix} \mu_x \\ \mu_y \end{bmatrix} \tag{2.87}$$

$$\Sigma_{(x,y)} = \begin{bmatrix} \Sigma_{xx} & \Sigma_{xy} \\ \Sigma_{xy} & \Sigma_{yy} \end{bmatrix} \tag{2.88}$$

The conditional density of $x(m)$ given $y(m+k)$ is given from Bayes rule as

$$f_{X|Y}(x(m)|\,y(m+k)) = \frac{f_{X,Y}(x(m),y(m+k))}{f_Y(y(m+k))} \tag{2.89}$$

It can be shown that conditional density is also a multi-variate Gaussian with its mean vector and covariance matrix given by

$$\begin{aligned} \mu_{(x|y)} &= \mathcal{E}\big[x(m)|\,y(m+k)\big] \\ &= \mu_x + \Sigma_{xy}\Sigma_{yy}^{-1}(y-\mu_y) \end{aligned} \tag{2.90}$$

$$\Sigma_{(x|y)} = \Sigma_{xx} - \Sigma_{xy}\Sigma_{yy}^{-1}\Sigma_{yx} \tag{2.91}$$

2.5.3   Mixture Gaussian Process

Probability density functions of many processes, such as speech, are non-Gaussian. A non-Gaussian pdf may be approximated by a weighted sum (i.e. a mixture) of a number of Gaussian densities of appropriate mean vectors and covariance matrices. An M-mixture Gaussian density is defined as

$$f_X(x) = \sum_{i=1}^{M} P_i \mathcal{N}_i(x,\mu_{x_i},\Sigma_{xx_i}) \tag{2.92}$$

where $\mathcal{N}_i(x,\mu_{x_i},\Sigma_{xx_i})$ is a multi-variate Gaussian density of mean vector μ_{x_i} and covariance matrix Σ_{xx_i}, and P_i are the mixing coefficients. The parameter P_i is the prior probability of it mixture component given by

$$P_i = \frac{N_i}{\sum_{j=1}^{M} N_j} \tag{2.93}$$

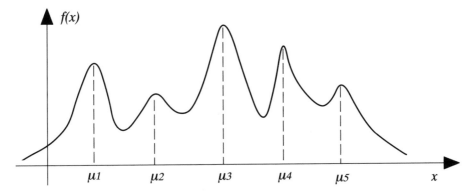

Figure 2.9 A mixture Gaussian density.

Where is the number of observations associated with the mixture i. Figure 2.9 shows a non-Gaussian pdf modelled as a mixture of five Gaussian densities. Algorithms developed for Gaussian processes can be extended to mixture Gaussian densities.

2.5.4 A Binary-state Gaussian Process

Consider a random process $x(m)$ with two statistical states; such that in the state S_0 the process has a Gaussian pdf with a mean of $\mu_{x,0}$ and a variance of $\sigma_{x,0}^2$, and in the state S_1 the process is also Gaussian with a mean of $\mu_{x,1}$ and a variance of $\sigma_{x,1}^2$. The state-dependent pdf of $x(m)$ can be expressed as

$$f_{X|S}\big(x(m)|s_i\big)= \frac{1}{\sqrt{2\pi}\,\sigma_{x,i}}\,\exp\!\left(-\frac{1}{2\sigma_{x,i}^2}\big(x(m)-\mu_{x,i}\big)^2\right) \qquad i=0,1 \quad (2.94)$$

The joint probability distribution of the binary-valued state *is* and the continuous-valued signal $x(m)$ can be expressed as

$$f_{X,S}\big(x(m),s_i\big) = f_{X|S}\big(x(m)|s_i\big)P_S\big(s_i\big)$$

$$= \frac{1}{\sqrt{2\pi}\,\sigma_{x,i}}\,\exp\!\left(-\frac{1}{2\sigma_{x,i}^2}\big(x(m)-\mu_{x,i}\big)^2\right) \times P_S\big(s_i\big) \qquad (2.95)$$

where $P_S(s_i)$ is the state probability. For a multi-state process we have the following probabilistic relations between the joint and the marginal probabilities :

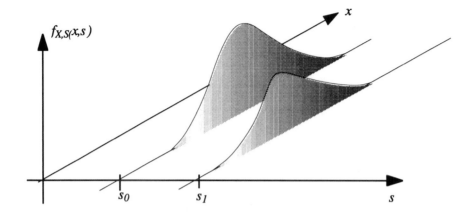

$f_{X,S}(x,s)$

s_0 s_1 s

Figure 2.10 Illustration of a binary-state Gaussian process

$$\sum_S f_{X,S}(x(m),s_i) = f_X(x(m)) \tag{2.96}$$

$$\int_X f_{X,S}(x(m),s_i)\,dx = P_S(s_i) \tag{2.97}$$

and

$$\sum_S \int_X f_{X,S}(x(m),s_i)\,dx = 1 \tag{2.98}$$

Note that in a multi-state model the statistical parameters of the process *switch* between a number of different states, whereas in a single-state mixture pdf a *weighted* combination of a number of pdfs models the process. In chapter 4 on hidden Markov models we consider multi-state models with a mixture pdf per state.

2.5.5 Poisson Process

The Poisson process is a continuous time, integer-valued counting process, used for modelling the occurrences of a random event in various time intervals. An important area of application of the Poisson process is in the queuing theory for the analysis and modelling of the distributions of demand on a service facility such as a telephone exchange, a shared computer system, a financial service, a petrol station etc. Other applications of Poisson distributions include the counting of the number of emissions in particle physics, the number of times that a component may fail in a system, and modelling of clutter in radar, shot noise, and impulsive noise.

Consider an event counting process $X(t)$, in which the probability of occurrence of the event is governed by a rate function $\lambda(t)$, such that the probability that an event occurs in a small time interval Δt is

$$Prob(1 \;\; occurrence \;\; in \;\; the \;\; interval(t, t + \Delta t)) \; = \; \lambda(t)\Delta t \qquad (2.99)$$

Assuming that, in the small interval Δt, no more than one occurrence of the event is possible, the probability of no occurrence is given by

$$Prob(0 \;\; occurrence \;\; in \;\; the \;\; interval(t, t + \Delta t)) = 1 - \lambda(t)\Delta t \qquad (2.100)$$

When the parameter $\lambda(t)$ is independent of time $\lambda(t)=\lambda$, the process is called a homogeneous Poisson process. Now, for a homogeneous Poisson process, consider the probability of k occurrence of an event in a time interval of $t+\Delta t$ denoted by $P(k, (0, t+\Delta t))$

$$
\begin{aligned}
P(k, (0, t + \Delta t)) \; &= \; P(k, (0, t))P(0, (t, t + \Delta t)) + P(k - 1, (0, t))P(1, (t, t + \Delta t)) \\
&= P(k, (0, t))(1 - \lambda\Delta t) + P(k - 1, (0, t))\lambda\Delta t
\end{aligned} \qquad (2.101)
$$

Rearranging Eq. (2.101), and letting Δt tend to zero, we obtain the following linear differential equation

$$\frac{dP(k, t)}{dt} \; = \; -\lambda P(k, t) + \lambda P(k - 1, t) \qquad (2.102)$$

Where $P(k,t)=P(k,(0, t))$. The solution of this differential equation is given by

$$P(k, t) \; = \; \lambda e^{-\lambda t} \int_0^t P(k - 1, \tau)\lambda\, e^{-\lambda\tau}\, d\tau \qquad (2.103)$$

This equation can be solved recursively : starting with $P(0,t)=e^{-\lambda t}$, $P(1,t)=\lambda t\, e^{-\lambda t}$, we obtain the Poisson density

$$P(k, t) \; = \; \frac{(\lambda t)^k}{k!}\, e^{-\lambda t} \qquad (2.104)$$

From Eq. (2.104) it is easy to show that for a homogenous Poisson process the probability of k occurrences of an event in a time interval (t_1, t_2) is given by

$$P[k,(t_1,t_2)] \quad = \quad \frac{[\lambda(t_2 - t_1)]^k}{k!} e^{-\lambda(t_2 - t_1)} \tag{2.105}$$

A Poisson process $X(t)$ is incremented by one every time the event occurs. From Eq. (2.104), the mean and the variance of a Poisson process $X(t)$ are

$$\mathcal{E}[X(t)] = \lambda t \tag{2.106}$$

$$r_{XX}(t_1,t_2) = \mathcal{E}[X(t_1)X(t_2)] = \lambda^2 t_1 t_2 + \lambda \min(t_1,t_2) \tag{2.107}$$

$$\mathrm{Var}[X(t)] = \mathcal{E}[X^2(t)] - \mathcal{E}^2[X(t)] = \lambda t \tag{2.108}$$

2.5.6 Shot Noise

Shot noise is a Poisson process for modelling a sequence of short duration pulses emitted at random time instances. For example, in a potodetection circuit a current pulse is generated every time a photoelectron is emitted by the light falling on the cathode. The current pulse sequence can be modelled as the response of a linear filter excited by a Poisson distributed binary impulse input sequence.

Consider a Poisson distributed binary-valued impulse process $x(t)$. Divide the time axis into uniform short intervals of Δt such that only one occurrence of an impulse is possible within each time interval. Let $x(m\Delta t)$ be "1" if an impulse is present in the interval $m\Delta t$ to $(m+1)\Delta t$, and "0" otherwise. For $x(m\Delta t)$ we have

$$\mathcal{E}[x(m\Delta t)] = 1 \times P(x(m\Delta t) = 1) + 0 \times P(x(m\Delta t) = 0) = \lambda \Delta t \tag{2.109}$$

and

$$\mathcal{E}[x(m\Delta t)x(n\Delta t)] = \begin{cases} 1 \times P(x(m\Delta t) = 1) = \lambda \Delta t & m = n \\ 1 \times P(x(m\Delta t) = 1)) \times P(x(n\Delta t) = 1) = (\lambda \Delta t)^2 & m \neq n \end{cases} \tag{2.110}$$

A shot noise process $y(m)$ is defined as the output of a linear system with impulse response $h(t)$, excited by a Poisson-distributed binary impulse input $x(t)$ as

$$y(t) = \int_{-\infty}^{\infty} x(\tau)h(t - \tau)d\tau$$

$$= \sum_{k=-\infty}^{\infty} x(m\Delta t)h(t - m\Delta t) \tag{2.111}$$

where the binary signal $x(m\Delta t)$ can be 0 or 1. In Eq. (2.111) it is assumed that the impulses happen at the beginning of each interval. This assumption becomes more valid as Δt becomes smaller. The expectation of $y(t)$ is obtained as

$$\mathcal{E}[y(t)] = \sum_{k=-\infty}^{\infty} \mathcal{E}[x(m\Delta t)]h(t - m\Delta t)$$

$$= \sum_{k=-\infty}^{\infty} \lambda \Delta t\, h(t - m\Delta t) \tag{2.112}$$

and

$$r_{yy}(t_1, t_2) = \mathcal{E}[y(t_1)y(t_2)]$$

$$= \sum_{m=-\infty}^{\infty} \sum_{n=-\infty}^{\infty} \mathcal{E}[x(m\Delta t)x(n\Delta t)]h(t_1 - n\Delta t)h(t_2 - m\Delta t) \tag{2.113}$$

Using Eq. (2.110) the autocorrelation of $y(t)$ can be obtained as

$$r_{yy}(t_1, t_2) = \sum_{n=-\infty}^{\infty} \lambda\,\Delta t\, h(t_1 - m\Delta t)h(t_2 - m\Delta t) + \sum_{m=-\infty}^{\infty} \sum_{\substack{n=-\infty \\ n \neq m}}^{\infty} (\lambda\,\Delta t)^2\, h(t_1 - m\Delta t)h(t_2 - n\Delta t) \tag{2.114}$$

2.5.7   Poisson-Gaussian Model for Clutters and Impulsive Noise

An impulsive noise process consists of short duration pulses of random amplitude and time of occurrence whose shape and duration depends on the characteristics of the channel through which the impulse propagates. A Poisson process can be used to model the random time of occurrence of impulsive noise. The random amplitude of impulses can be modelled by a Gaussian process. Finally, the finite duration character of real impulsive noise may be modelled by the impulse response of linear filter. The Poisson-Gaussian impulsive noise model is given by

$$x(m) = \sum_{k=-\infty}^{\infty} A_k h(m - \tau_k) \tag{2.115}$$

where $h(m)$ is the response of a linear filter that models the shape of impulsive noise, A_k is a zero mean Gaussian process of variance σ^2 and τ_k is a Poisson process. The

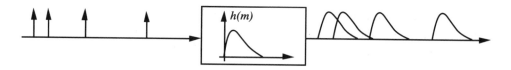

Figure 2.11 Shot noise is modelled as the output of a filter excited with a process.

output of a filter excited by a Poisson distributed sequence of Gaussian amplitude impulses can also be used to model radar clutters. Clutters are due to reflection of radar pulses from a multitude of background surfaces and objects other than the radar target.

2.5.8 Markov Processes

A first order discrete-time Markov process is defined as one for which the state of the process at time m depends only on its state at the time $m-1$ and is independent of the process history before $m-1$. In probabilistic terms, a first order Markov process can be defined as

$$f_X\big(x(m) = x_m \,\big|\, x(m-1) = x_{m-1}, \ldots, x(m-N) = x_{m-N}\big)$$
$$= f_X\big(x(m) = x_m \,\big|\, x(m-1) = x_{m-1}\big) \tag{2.116}$$

The marginal density of a Markov process at time k can be obtained by integrating the conditional density over all values of $x(k-1)$ as

$$f_X\big(x(m) = x_m\big) = \int_{-\infty}^{\infty} f_X\big(x(m) = x_m \,\big|\, x(m-1) = x_{m-1}\big)\big) f_X\big(x(m-1) = x_{m-1}\big) dx_{m-1}$$

$$\tag{2.117}$$

A process in which the present state of the system depends on the past n states may be described in terms of n first order Markov processes and is known as an nth order Markov process. The term Markov process usually refers to a first order process.

Example 2.10 A simple example of a Markov process is a first order auto-regressive process defined as

$$x(m) = a\,x(m-1) + e(m) \tag{2.118}$$

In Eq. (2.118) $x(m)$ depends on the previous value $x(m-1)$ and the input $e(m)$. The conditional pdf of $x(m)$ given the previous sample value can be expressed as

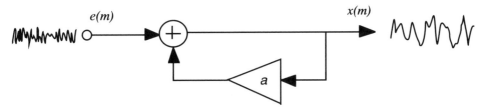

Figure 2.12 A first order autoregressive (Markov) process.

$x(m)$ given previous sample values can be expressed as

$$f_X(x(m)|x(m-1)...,x(m-N)) = f_X(x(m)|x(m-1))$$
$$= f_E(e(m) = x(m) - ax(m-1)) \qquad (2.119)$$

where $f_E(e(m))$ is the pdf of the input signal $e(m)$. Assuming that input $e(m)$ is a zero mean Gaussian process with variance σ_e^2 we have

$$f_X(x(m)|x(m-1)...,x(m-N)) = f_X(x(m)|x(m-1))$$
$$= f_E(x(m) - ax(m-1)) \qquad (2.120)$$
$$= \frac{1}{\sqrt{2\pi}\,\sigma_e} \exp\left(-\frac{1}{2\sigma_e^2}(x(m) - ax(m-1))^2\right)$$

when the input to a Markov process is Gaussian the output is known as a Gauss-Markov process.

2.5.9   Markov Chain Processes

A discrete-time Markov process $x(m)$, with N allowable states may be modelled by a Markov chain of N states. Each state can be associated with one of the N values that $x(m)$ may assume. In a Markov chain, the Markovian property is modelled by a set of state transition probabilities defined as

$$a_{ij}(m-1,m) = Prob(x(m) = j|x(m-1) = i) \qquad (2.121)$$

where $a_{ij}(m,m-1)$ is the probability that at time $m-1$ the process is in the state i and then at time m it moves to state j.

In Eq. (2.121) the transition probability is expressed in a general time-dependent form. The marginal probability that a Markov process is in the state j at time m, $P_j(m)$

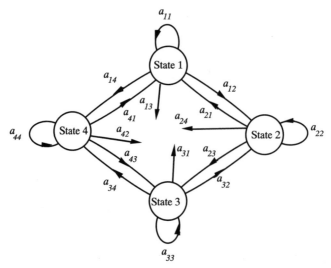

Figure 2.13 A Markov chain model of a four-state discrete time Markov process.

can be expressed as

$$P_j(m) = \sum_{i=1}^{N} P_i(m-1) a_{ij}(m-1, m) \tag{2.122}$$

A Markov chain is defined by the following set of parameters :

Number of states N
State probability vector

$$p^T(m) = [p_1(m), p_2(m), \ldots, p_N(m)]$$

and state transition matrix

$$A(m-1, m) = \begin{pmatrix} a_{11}(m-1,m) & a_{12}(m-1,m) & \cdots & a_{1N}(m-1,m) \\ a_{21}(m-1,m) & a_{22}(m-1,m) & \cdots & a_{2N}(m-1,m) \\ \vdots & \vdots & \ddots & \vdots \\ a_{N1}(m-1,m) & a_{N2}(m-1,m) & \cdots & a_{NN}(m-1,m) \end{pmatrix}$$

Homogenous and Inhomogeneous Markov Chains

A Markov chain with time-invariant state transition probabilities is known as a homogenous Markov chain. For a homogenous Markov process the probability of a transition, from a state i to a state j of the process, is independent of the time of the transition m as expressed in the following equation

$$Prob(x(m) = j|x(m-1) = i) = a_{ij}(m-1,m) = a_{ij} \qquad (2.123)$$

Inhomgeneous Markov chains have time-dependent transition probabilities. In most applications of Markov chains homogenous models are used because they are usually an adequate model of the signal process, and because homogenous Markov models are easier to train and use. Markov models are considered in Chapter 4.

2.6 Transformation of a Random Process

In this section we consider the effect of filtering or transformation of a random process on its probability density function. Figure 2.13 shows a generalised mapping operator $h()$ that transforms a random input process X into an output process Y. Signals $x(m)$ and $y(m)$ are realisations of the random processes X and Y respectively. If $x(m)$ and $y(m)$ are both discrete such that $x(m) \in \{x_1,...,x_N\}$ and $y(m) \in \{y_1,...,y_M\}$ then we have

$$P_Y\big(y(m) = y_j\big) = \sum_{x_i \to y_j} P_X\big(x(m) = x_i\big) \qquad (2.124)$$

where in Eq. (2.124) the summation is taken over all values of $x(m)$ that map to $y(m)=y_j$. Now consider the transformation of a discrete-time, *continuous-valued*, process. The probability that the output process Y has a value in the range $y(m)<Y<y(m)+\Delta y$ is

$$Prob[y(m)< Y < y(m) + \Delta y] = \int_{x(m)|y(m)<Y<y(m)+\Delta y} f_X(x(m))\ dx(m) \qquad (2.125)$$

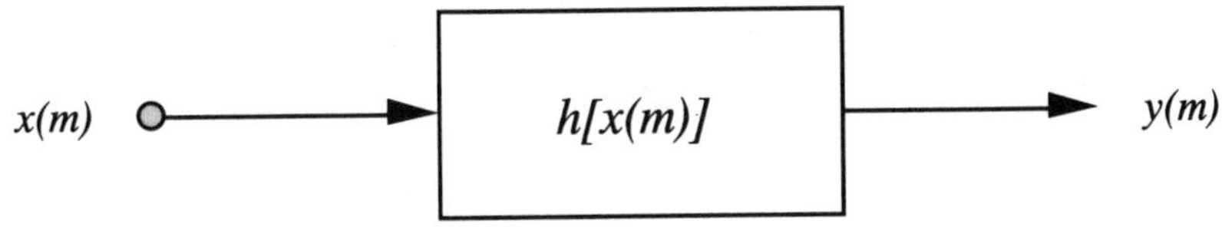

$$x(m) \circ \longrightarrow \boxed{h[x(m)]} \longrightarrow y(m)$$

Figure 2.14 Transformation of a random process $x(m)$ to an output process $y(m)$.

Where the integration is taken over all the values of *x(m)* that yield an output in the range *y(m)* to *y(m)+Δy* .

2.6.1 Monotonic Transformation of Random Signals

Now for a monotonic, one-to-one transformation, such as the one in Figure 2.15, Eq. (2.125) becomes

$$Prob(y(m)< Y < y(m) + \Delta y)= Prob(x(m)< X < x(m) + \Delta x) \qquad (2.126)$$

or in terms of the cumulative distribution functions

$$F_Y(y(m) + \Delta y)- F_Y(y(m)) = F_X(x(m) + \Delta x)- F_X(x(m)) \qquad (2.127)$$

Multiplication the left hand side of Eq. (2.127) by *Δy/Δy* and the right hand side by *Δx/Δx* and re-arrangement of the terms yields

$$\frac{F_Y(y(m) + \Delta y)- F_Y(y(m))}{\Delta y} = \frac{\Delta x}{\Delta y} \frac{F_X(x(m) + \Delta x)- F_X(x(m))}{\Delta x} \qquad (2.128)$$

Now as the intervals *Δx* and *Δy* tend to zero Eq. (2.128) becomes

$$f_Y(y(m)) = \left|\frac{\partial x(m)}{\partial y(m)}\right| \times f_X(x(m)) \qquad (2.129)$$

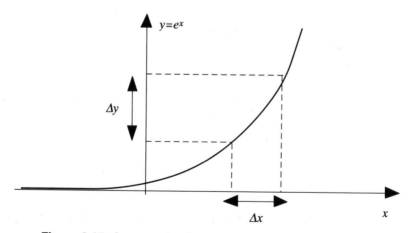

Figure 2.15 An example of a monotonic one-to-one mapping.

Now substitution of $x(m)=h^{-1}(y(m))$ in Eq. (129) yields

$$f_Y(y(m)) = \left| \frac{\partial h^{-1}(y(m))}{dy(m)} \right| f_X(h^{-1}(y(m))) \qquad (2.130)$$

Eq. (2.130) gives the pdf of the output signal in terms of the pdf of the input signal.

Example 2.11 Transformation of a Gaussian Process to a Log-normal Process, Log-normal pdfs are used for modelling positive-valued processes such as power spectrum. If a random variable x has a Gaussian pdf as in Eq. (2.80), then the non-negative valued variable $y(m)=exp(x(m))$ has a log-normal distribution obtained using Eq. (2.130) as

$$f_Y(y) = \frac{1}{\sqrt{2\pi}\ \sigma_x\ y(m)} \exp\left(-\frac{(\ln(y(m)) - \mu_x)^2}{2\sigma_x^2} \right) \qquad (2.131)$$

Conversely, if the input y to a logarithmic function has a log-normal distribution then the output $x=\ln(y)$ is Gaussian. The mapping functions for translating the mean and variance of a log normal distribution to a normal distribution can be derived as

$$\mu_x = \ln(\mu_y) - \frac{1}{2} \ln\left(1 + \sigma_y^2 / \mu_y^2 \right) \qquad (2.132)$$

$$\sigma_x^2 = \ln\left(1 + \sigma_y^2 / \mu_y^2 \right) \qquad (2.133)$$

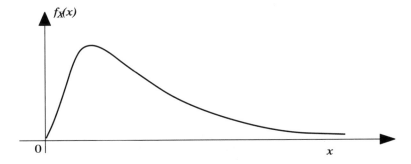

Figure 2.16 A log-normal distribution.

(μ_x, σ_x^2), and (μ_y, σ_y^2) are the mean and the variance of x and y respectively. The inverse mapping relations for the translation of mean and variances of normal to log-normal variables are

$$\mu_y = \exp(\mu_x + \sigma_x^2 / 2) \tag{2.134}$$

$$\sigma_y^2 = \mu_x^2 \, [\exp(\sigma_y^2) - 1] \tag{2.135}$$

2.6.2 Many-to-one Mapping of Random Signals

Now consider the case when the transformation $h()$ is a non-monotonic function such as that shown in Figure 2.17. Assuming that equation $y(m)=h[x(m)]$ has K roots, there are K different values of $x(m)$ that map to the same $y(m)$. The probability that a realisation of the output process Y has a value in the range $y(m)$ to $y(m)+\Delta y$ is given by

$$Prob(y(m) < Y < y(m) + \Delta y) = \sum_{k=1}^{K} Prob(x_k(m) < X < x_k(m) + \Delta x_k) \tag{2.136}$$

where x_k is the kth root of $y(m)=h(x(m))$. Similar to the development of Section 2.6.1 Equation (2.136) can be written as

$$\frac{F_Y(y(m) + \Delta y) - F_Y(y(m))}{\Delta y} \Delta y = \sum_{k=1}^{K} \frac{F_X(x_k(m) + \Delta x_k) - F_X(x_k(m))}{\Delta x_k} \Delta x_k \tag{2.137}$$

Eq. (2.137) can be rearranged as

$$\frac{F_Y(y(m) + \Delta y) - F_Y(y(m))}{\Delta y} = \sum_{k=1}^{K} \frac{\Delta x_k}{\Delta y} \frac{F_X(x_k(m) + \Delta x_k) - F_X(x_k(m))}{\Delta x_k} \tag{2.138}$$

Now as the intervals Δx and Δy tend to zero Eq. (2.138) becomes

$$f_Y(y(m)) = \sum_{k=1}^{K} \left| \frac{\partial x_k(m)}{\partial y(m)} \right| \times f_X(x_k(m))$$

$$= \sum_{k=1}^{K} \frac{1}{|h'(x_k(m))|} \times f_X(x_k(m)) \tag{2.139}$$

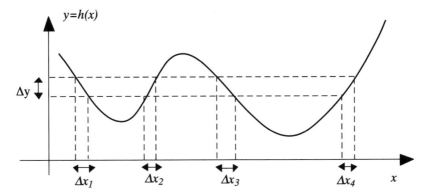

Figure 2.17 Illustration of a many to one transformation.

where $h'(x_k(m)) = \partial h(x_k(m)) / \partial x_k(m)$. Note that for a monotonic function $K=1$ and Eq. (2.139) becomes the same as Eq. (2.130). Eq. (2.139) can be expressed as

$$f_Y(y(m)) = \sum_{k=1}^{K} |J(x_k(m))|^{-1} f_X(x_k(m)) \qquad (2.140)$$

where $J(x_k(m)) = h'(x_k(m))$ is called the Jacobian of the transformation. For a multi-variate transformation of a vector-valued process such as

$$y(m) = H(x(m)) \qquad (2.141)$$

the pdf of the output $y(m)$ is given by

$$f_Y(y(m)) = \sum_{k=1}^{K} |J(x_k(m))|^{-1} f_X(x_k(m)) \qquad (2.142)$$

where $J(x)$, the Jacobian of the transformation $H()$, is the determinant of a matrix of derivatives defined as

$$J(x) = \begin{vmatrix} \dfrac{\partial y_1}{\partial x_1} & \dfrac{\partial y_1}{\partial x_2} & \cdots & \dfrac{\partial y_1}{\partial x_P} \\ \vdots & \vdots & \ddots & \vdots \\ \dfrac{\partial y_P}{\partial x_1} & \dfrac{\partial y_P}{\partial x_2} & \cdots & \dfrac{\partial y_P}{\partial x_P} \end{vmatrix} \qquad (2.143)$$

For a monotonic linear vector transformation such as

$$y = Hx \tag{2.144}$$

the pdf of y becomes

$$f_Y(y) = |J|^{-1} f_X(H^{-1}y) \tag{2.145}$$

where J is the Jacobian of the transformation.

Example 2.12 The input-output relation of a $P \times P$ linear transformation matrix H is given by

$$y = H\ x \tag{2.146}$$

The Jacobian of the linear transformation H is $J = |H|$. Assume that the input x is a zero mean Gaussian P-variate process with a covariance matrix of Σ_{xx} and a probability density function

$$f_X(x) = \frac{1}{(2\pi)^{P/2}|\Sigma_{xx}|^{1/2}} \exp\left(-\frac{1}{2} x^T \Sigma_{xx}^{-1} x\right) \tag{2.147}$$

From Eqs. (2.145), (2.146) and (2.147) the pdf of the output y is given by

$$
\begin{aligned}
f_Y(y) &= \frac{1}{(2\pi)^{P/2}|\Sigma_{xx}|^{1/2}} \exp\left(-\frac{1}{2} y^T H^{-1T} \Sigma_{xx}^{-1} H^{-1} y\right) \|H\|^{-1} \\
&= \frac{1}{(2\pi)^{P/2}|\Sigma_{xx}|^{1/2}\|H\|} \exp\left(-\frac{1}{2} y^T \Sigma_{yy}^{-1} y\right)
\end{aligned}
\tag{2.148}
$$

where $\Sigma_{yy} = H\Sigma_{xx}H^T$. Note that a linear transformation of a Gaussian process yields another Gaussian process.

Summary

The theory of stochastic signal processes is central to development of signal processing algorithms. We began this chapter with basic definitions of deterministic signals, random signals and random processes. A random process generates random signals, and the collection of all signals that can be generated by a random process is the space of the process. Probabilistic models and statistical measures, originally developed for random variables, were extended to model random signals. Although random signals are completely described in terms of probabilistic models, for many applications it is sufficient to characterise a process in terms of a set of relatively simple statistics such as the mean, the autocorrelation function, the covariance and the power spectrum. Much of the theory and application of signal processing is concerned with the identification, extraction, and utilisation of structures and patterns in a signal process. The correlation and its Fourier transform the power spectrum are particularly important because they can be used to identify the patterns in a stochastic process.

We considered the concepts of stationary, ergodic stationary, and nonstationary processes. The concept of a stationarity process is central to the theory of linear time-invariant systems, and furthermore even nonstationary signals can be modelled with a chain of stationary sub-processes as described in Chapter 4 on hidden Markov models. For signal processing applications, a number of useful pdfs including the Gaussian, the mixture Gaussian, the Markov and the Poisson process were considered. These pdf models are extensively employed in the remainder of this book. Signal processing normally involves the filtering or transformation of an input signal to an output signal. We derived general expressions for the pdf of the output of a system in terms of the pdf of the input. We also considered some applications of stochastic processes for modelling random noise such as white noise, clutters, shot noise and impulsive noise.

Bibliography

ANDERSON O.D. (1976), Time Series Analysis and Forecasting, The Box-Jenkins Approach, Butterworth, London.

AYRE A.J. (1972), Probability and Evidence, Columbia University Press.

BARTLETT M.S. (1960), Stochastic Processes, Cambridge University Press, Cambridge.

BOX G.E.P, JENKINS G.M. (1976), Time Series Analysis: Forecasting and Control, Holden-Day, San Francisco.

BREIPHOL A.M. (1970), Probabilistic System Analysis, Wiley, New York.

CARTER G. (1987),Coherence and Time Delay Estimation, Proc. IEEE, Vol. 75, No. 2 Pages 236-55.

CHUNG K. L. (1974), Elementary Probability Theory, Springer-Verlag.

CLARK A. B., DISNEY R. L. (1985), Probability and Random Processes, 2nd Ed. Wiley, New York.

COOPER G. R., McGILLEM C.D. (1986), Probabilistic Methods of Signal and System Analysis Holt, Rinehart and Winston, New York.

DAVENPORT W.B., ROOT W. L. (1958), Introduction to Random Signals and Noise, McGraw-Hill, New York.

DAVENPORT W.B., WILBUR B., (1970), Probability and Random Processes: An Introduction for Applied Scientists and Engineers., McGraw-Hill, New York.

EINESTEIN A. (1956) Investigation on the Theory of Brownian Motion, Dover, New York.

GAUSS K. G. (1963), Theory of Motion of Heavenly Bodies, Dover, New York.

JEFFREY H. (1961), Scientific Inference, 3rd ed. Cambridge University Press, Cambridge.

JEFFREY H. (1973), Theory of Probability, 3rd ed. Clarendon Press, Oxford.

GARDENER W.A.(1986), Introduction to Random Processes: With Application to Signals and Systems, Macmillan, New York.

HELSTROM C.W. (1991), Probability and Stochastic Processes for Engineers, Macmillan, New York.

ISAACSON D., MASDEN R. (1976), Markov Chains Theory and Applications Wiley, New York.

KAY S. M. (1993), Fundamentals of Statistical Signal Processing, Estimation Theory Prentice-Hall, Englewood Cliffs, N. J.

KOLMOGOROV A.N. (1956), Foundations of the Theory of Probability, Chelsea Publishing Company, New York.

KENDALL M., STUART A. (1977), The Advanced Theory of Statistics Macmillan.

LEON-GARCIA A. (1994), Probability and Random Processes for Electrical Engineering Addison Wesley, Reading, Mass.

MARKOV A. A. (1913), An Example of Statistical Investigation in the text of *Eugen Onyegin* Illustrating Coupling of Tests in Chains, Proc. Acad. Sci. St Petersburg VI Ser., Vol. 7, Pages 153-162.

MEYER P. L. (1970), Introductory Probability and Statistical Applications, Addison-Wesley, Reading, Mass.

PEEBLES P.Z. (1987), Probability, Random Variables and Random Signal Principles McGraw-Hill, New York.

PARZEN E. (1962), Stochastic Processes, Holden-Day, San Francisco.

POPULIS A. (1984), Probability, Random Variables and Stochastic Processes, McGraw-Hill, New York.

POPULIS A. (1977), Signal Analysis, McGraw-Hill, New York.

RAO C. R. (1973), Linear Statistical Inference and Its Applications, Wiley, New York.

ROZANOV Y. A. (1969), Probability Theory : A Concise Course, Dover Publications, New York.

SHANMUGAN K. S., BREIPOHL A. M. (1988), Random Signals : Detection, Estimation and Data Analysis, Wiley, New York.

THOMAS J.B. (1988), An introduction to Applied probability and Random Processes, Huntington, Krieger Publishing, New York.

WOZENCRAFT J. M., JACOBS I. M. (1965), Principles of Communication Engineering, Wiley, New York.

3

Bayesian Estimation and Classification

3.1 Estimation Theory : Basic Definitions
3.2 Bayesian Estimation
3.3 Estimate-Maximise Method
3.4 Cramer-Rao Bound on the Minimum Estimator Variance
3.5 Bayesian Classification
3.6 Modelling the Space of a Random Signal

B ayesian estimation is a framework for formulation of statistical inference problems, and includes the classical estimators such as the maximum a posterior, maximum likelihood, minimum mean squared error, and minimum mean absolute value of error as its special cases. The hidden Markov model, widely used in statistical signal processing, is also an example of a Bayesian model. Bayesian inference is based on the minimisation of a so called Bayes risk function which includes; a posterior model of the unknown parameters given the observation, and a cost of error function. This chapter begins with an introduction to the basic concepts of estimation theory, and considers the statistical measures that are used to quantify the performance of an estimator. We study the Bayesian estimation methods and consider the effects of using a prior model on the mean and the variance of an estimate. Estimation of discrete-valued parameters, and parameters from a finite-state process, are studied within the frame work of Bayesian classification. The chapter concludes with a study of the methods for the modelling of a random signal space.

3.1 Estimation Theory: Basic Definitions

Estimation theory is concerned with the determination of the best estimate of an unknown parameter vector, or the recovery of a number of distorted samples. An estimator takes as the input a set of noisy or incomplete observations and, using a predictive or a statistical model of the process, estimates the unknown parameters. The estimation accuracy depends on the available information and on the efficiency of the estimator. In this chapter the Bayesian estimation of continuous-valued parameters is studied and then extended to the classification of discrete, and finite-state-valued, parameters. The Bayesian theory is a general estimation/classification framework in which, for a given problem, we can include specific prior knowledge of the observation signal and the unknown parameters. First, in this section some basic concepts of estimation theory are introduced.

3.1.1 Predictive and Statistical Models in Estimation

Optimal estimation algorithms utilise predictive and statistical models of the signals. A predictive model captures the correlation structure of a signal, and models the dependency of the present and future values of the signal on its past trajectory. A statistical model characterises the random fluctuations of a signal in terms of the average value such as, the mean, the covariance, and most completely in terms of a probabilistic model. Conditional probabilistic models in addition to modelling the random fluctuations of a signal can also model the dependency of the signal on its past values or on some other process. As an illustration consider the estimation of a P-dimensional parameter vector $\theta=[\theta_0, \theta_1, ..., \theta_{P-1}]$ from a noisy observation vector $y=[y(0), y(1), ..., y(N-1)]$ modelled as

$$y = h(\theta, x, e) + n \tag{3.1}$$

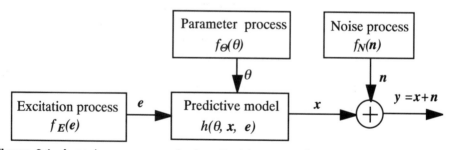

Figure 3.1 A random process **y** is described in terms of a predictive model $h(\ )$, and statistical models $f_E(\ )$, $f_\Theta(\ )$ and $f_N(\ )$.

where, as illustrated in Figure 3.1 the function $h()$ with random input e, output x, and parameter vector θ, is a predictive model of the signal process, and n is an additive random noise. In Figure 3.1 the random noise n, the random input signal e, and the distribution of the parameter vectors θ are modelled in terms of the probability density functions, $f_N(n)$, $f_E(e)$, and $f_\Theta(\theta)$.

Predictive and statistical models of a process, *guide* the estimator towards the set of values of the unknown parameters that are most consistent with both the models and the noisy observation. In general, the more modelling information used in an estimation process the better the results, provided that the models are an accurate characterisation of the observation and the parameter process.

3.1.2   Parameter Space and Signal Space

The parameter space of a process Θ, is the collection of all the values that the process parameter θ can assume. The parameters of a random process determine the "character" (i.e. the mean, the variance, the power spectrum etc.) of the signals generated by the process. As the process parameters change, so do the characteristics of the signals generated by the process. Each value of the parameter vector θ of a process has an associated signal space Y which is the collection of all signal realisations of the process with the parameter value θ.

For example, consider a 3-dimensional vector-valued Gaussian process with parameter vector $\theta=[\mu,\Sigma]$,   where μ is the mean vector and Σ is the covariance

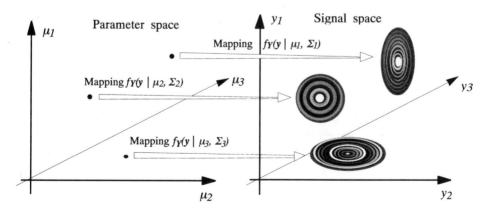

Figure 3.2 Illustration of three points in the parameter space of a Gaussian process and the associated signal spaces, for simplicity the variances are not shown in parameter space.

matrix of the Gaussian process. Figure 3.2 illustrates three mean-vectors in a three-dimensional parameter space, also shown is the signal space associated with each parameter. As shown, the signal space of each parameter vector of a Gaussian process contains an infinite number of points, centred on the mean vector μ, and with a spatial volume and orientation which is determined by the covariance matrix Σ.

3.1.3   Parameter Estimation and Signal Restoration

Parameter estimation and signal restoration are closely related. The main difference is that normally more averaging can be afforded in parameter estimation than in signal restoration. This is due to the rapid fluctuations of most signals in comparison to the relatively slow variations of most parameters. As an example, consider the problem of interpolation of a number of lost samples of a signal given an N sample record of a stationary process, as illustrated in Figure 3.3. Assume that the signal can be modelled by an autoregressive (AR) process expressed as

$$y = X\theta + e + n \qquad\qquad (3.2)$$

where y is the observation, X is the signal matrix, θ is the AR parameter, e is the random input to the AR model and n is the noise. Using Eq. (3.2), signal restoration involves the estimation of both the model parameters, and the random input e during the instances of lost samples. Assuming the parameter vector θ is time-invariant, the estimate of θ can be averaged over the entire N observation samples, and as N becomes infinitely large a consistent estimate should approach the true parameter value. The difficulty in signal interpolation is that the underlying excitation e of the signal x is purely random and, unlike θ, it can not be estimated through an averaging operation. In this chapter we are concerned with the parameter estimation problem, although the same ideas also apply to signal interpolation considered in Chapter 11.

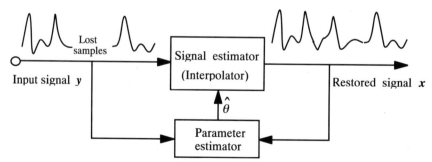

Figure 3.3 Signal restoration using estimates of a parametric model of the process.

3.1.4 Performance Measures and Desirable Properties of Estimators

In estimation of a parameter vector θ, from N observation samples y, a set of performance measures is used to quantify and compare the characteristics of different estimators. In general an estimate of a parameter vector is a function of the observation vector y, the length of the observation N, and the process model M. This dependency may be expressed as

$$\hat{\theta} = f(y, N, \mathcal{M}) \tag{3.3}$$

Different parameter estimators produce different results depending on the estimation method and utilisation of the observation and the prior information. Due to randomness of observations, even the same estimator would produce different results with different observations from the same process. Therefore an estimate is itself a random variable, and may be described by a probability density function. However for most cases it is sufficient to characterise an estimator in terms of the mean and the variance of the estimation error. The most commonly used performance measures for an estimator are the following :

(a) *Expected value* of estimate : $\mathcal{E}[\hat{\theta}]$

(b) *Bias* of estimate : $\mathcal{E}[\hat{\theta} - \theta] = \mathcal{E}[\hat{\theta}] - \theta$

(c) *Covariance* of estimate : $Cov[\hat{\theta}] = \mathcal{E}[(\hat{\theta} - \mathcal{E}[\hat{\theta}])(\hat{\theta} - \mathcal{E}[\hat{\theta}])^T]$

Optimal estimators aim for zero bias and minimum estimation error covariance. The desirable properties of an estimator are as follows :

(a) Unbiased estimator : An estimator of θ is unbiased if the expectation of the estimate is equal to the true parameter value :

$$\mathcal{E}[\hat{\theta}] = \theta \tag{3.4}$$

An estimator is *asymptotically unbiased* if for increasing length of observations N we have

$$\lim_{N \to \infty} \mathcal{E}[\hat{\theta}] = \theta \tag{3.5}$$

(b) Efficient estimator : An unbiased estimator of θ is an efficient estimator if it has the smallest covariance matrix compared with all other unbiased estimates of θ :

$$\mathrm{cov}\,[\hat{\theta}_{Efficient}] \leq \mathrm{cov}\,[\hat{\theta}] \tag{3.6}$$

where $\hat{\theta}$ is any other estimate of θ.

(c) Consistent estimator : An estimator is consistent if the estimate improves with increasing length of the observation N, such that the estimate $\hat{\theta}$ converges probabilistically to the true value θ as N becomes infinitely large :

$$\lim_{N \to \infty} P[|\hat{\theta} - \theta| > \varepsilon] = 0 \tag{3.7}$$

where ε is arbitrary small.

Example 3.1 Consider the bias in the time-averaged estimates of the mean μ_y and the variance σ_y^2 of N observation samples $[y(0), ..., y(N-1)]$, of an ergodic random process, given as

$$\hat{\mu}_y = \frac{1}{N} \sum_{m=0}^{N-1} y(m) \tag{3.8}$$

$$\hat{\sigma}_y^2 = \frac{1}{N} \sum_{m=0}^{N-1} \left(y(m) - \hat{\mu}_y\right)^2 \tag{3.9}$$

It is easy to show that $\hat{\mu}_y$ is an unbiased estimate

$$\mathcal{E}\left[\hat{\mu}_y\right] = \frac{1}{N} \sum_{m=0}^{N-1} \mathcal{E}[y(m)] = \mu_y \tag{3.10}$$

The expectation of the estimate of the variance can be expressed as

$$\mathcal{E}\left[\hat{\sigma}_y^2\right] = \mathcal{E}\left[\frac{1}{N} \sum_{m=0}^{N-1} \left(y(m) - \frac{1}{N} \sum_{k=0}^{N-1} y(k) \right)^2 \right]$$

$$= \sigma_y^2 - \frac{2}{N} \sigma_y^2 + \frac{1}{N} \sigma_y^2 \tag{3.11}$$

$$= \sigma_y^2 - \frac{1}{N} \sigma_y^2$$

From Eq. (3.11) the bias in the estimate of the variance is inversely proportional to

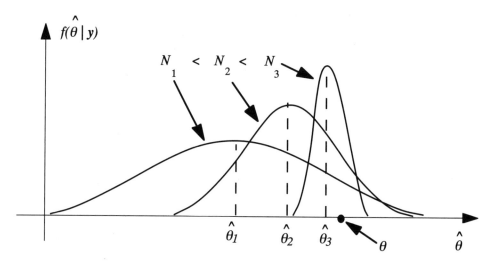

Figure 3.4 Illustration of the decrease in the bias and variance of an asymptotically unbiased estimate of the parameter θ with the increasing length of observation.

the signal length N, and vanishes as N tends to infinity, hence the estimate is asymptotically unbiased. In general the bias and the variance of an estimate decrease with the increasing number of observation samples N, and with improved modelling. Figure 3.4 illustrates the general dependency of the bias and the variance of an asymptotically unbiased estimator on the number of observation samples N.

3.1.5 Prior, and Posterior Spaces and Distributions

The *prior space* of a signal or a parameter is the collection of all possible values that the signal can assume. The *posterior signal space* is the collection of the likely values of a signal that are consistent with an *observation* and the prior. Consider a random process with a parameter space Θ, an observation space Y, and a joint pdf $f_{Y,\Theta}(y, \theta)$. From Bayes rule the posterior pdf of the parameter vector θ, given an observation vector y, $f_{\Theta|Y}(\theta \mid y)$, can be expressed as

$$
\begin{aligned}
f_{\Theta|Y}(\theta \mid y) &= \frac{f_{Y|\Theta}(y|\theta) \, f_{\Theta}(\theta)}{f_{Y}(y)} \\
&= \frac{f_{Y|\Theta}(y|\theta) \, f_{\Theta}(\theta)}{\int_{\Theta} f_{Y|\Theta}(y|\theta) \, f_{\Theta}(\theta) \, d\theta}
\end{aligned}
\tag{3.12}
$$

where for a given observation y, the pdf $f_Y(y)$ is a constant and has only a normalising effect. From Eq. (3.12), the posterior pdf is proportional to the product of the likelihood $f_{Y|\Theta}(y|\theta)$, that the observation y was generated by the parameter vector θ, and the prior pdf $f_\Theta(\theta)$. The prior pdf gives the unconditional parameter distribution *averaged* over the entire observation space as

$$f_\Theta(\theta) = \int_Y f_{\Theta,Y}(\theta,y) \; dy \tag{3.13}$$

For most applications it is relatively convenient to obtain the likelihood $f_{Y|\Theta}(y|\theta)$. The *prior* pdf *moderates* the inference drawn from the likelihood function by weighting it with $f_\Theta(\theta)$. The influence of the prior is particularly important for short length observations, where the confidence on the estimate is limited by the short length of the observation. The influence of the prior on the bias and the variance of an estimate is considered in Section 3.3.1.

A prior knowledge of the signal distribution can be used to confine the estimate to the prior signal space. The observation then guides the estimator to focus on the posterior space which is the subspace consistent with both the prior and the observation. Figure 3.5 illustrates the joint pdf of a signal $y(m)$ and a parameter θ. The prior of θ can be obtained by integrating $f_{Y,\Theta}(y(m),\theta)$ with respect to $y(m)$. As shown an observation $y(m)$ cuts a posterior pdf $f_{\Theta|Y}(\theta|y(m))$ through the joint distribution.

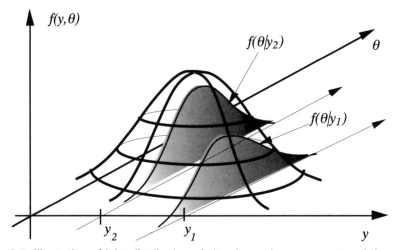

Figure 3.5 Illustration of joint distribution of signal y and parameter θ and the posterior distribution of θ given y.

Example 3.2 A noisy observation signal vector of length N is modelled as

$$y(m) = x(m) + n(m) \tag{3.14}$$

Assume that the signal $x(m)$ is Gaussian distributed with a mean vector μ_x and a covariance matrix Σ_{xx}, and that the noise $n(m)$ is also Gaussian with a mean vector of μ_n and a covariance matrix Σ_{nn}. The signal and noise pdfs model the prior spaces of the signal and the noise respectively. Given an observation $y(m)$, the underlying signal $x(m)$ would have a likelihood distribution with a mean vector of $y(m)-\mu_n$ and covariance matrix Σ_{nn}. The likelihood function is given by

$$f_{Y|X}(y(m)|x(m)) = f_N(y(m) - x(m))$$

$$= \frac{1}{(2\pi)^{N/2}|\Sigma_{nn}|^{1/2}} \exp\left(-\frac{1}{2}(x(m) - (y(m) - \mu_n))^T \Sigma_{nn}^{-1}(x(m) - (y(m) - \mu_n))\right) \tag{3.15}$$

Hence the posterior pdf can be expressed as

$$f_{X|Y}(x(m)|y(m)) = \frac{f_{Y|X}(y(m)|x(m))f_X(x(m))}{f_Y(y(m))} = \frac{1}{f_Y(y(m))} \frac{1}{(2\pi)^N |\Sigma_{nn}|^{1/2}|\Sigma_{xx}|^{1/2}} \times$$

$$\exp\left(-\frac{1}{2}\left((x(m) - (y(m) - \mu_n))^T \Sigma_{nn}^{-1}(x(m) - (y(m) - \mu_n)) + (x(m) - \mu_x)^T \Sigma_{xx}^{-1}(x(m) - \mu_x)\right)\right) \tag{3.16}$$

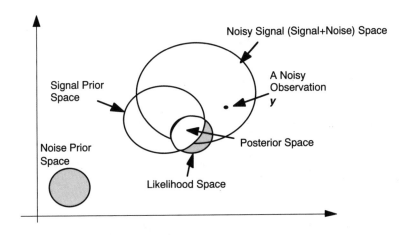

Figure 3.6 A sketch of a two-dimensional signal and noise spaces, and the likelihood and posterior spaces of a noisy observation y.

For a two-dimensional signal and noise process, the prior spaces of the signal, the noise, and the noisy signal, are illustrated in Figure 3.6. Also illustrated are the likelihood and the posterior spaces for a noisy observation vector y.

3.2 Bayesian Estimation

Bayesian estimation of a parameter vector θ is based on the minimisation of a Bayesian risk function, defined as an average cost of error function as

$$
\begin{aligned}
\mathcal{R}(\hat{\theta}) &= \mathcal{E}[C(\hat{\theta},\theta)] \\
&= \int_\Theta \int_Y C(\hat{\theta},\theta)\, f_{Y,\Theta}(y,\theta)\, dy\, d\theta \\
&= \int_\Theta \int_Y C(\hat{\theta},\theta)\, f_{\Theta|Y}(\theta|y)\, f_Y(y)\, dy\, d\theta
\end{aligned}
\tag{3.17}
$$

Where the cost of error function $C(\hat{\theta},\theta)$ allows the appropriate weighting of various outcomes to achieve desirable objective or subjective properties. The cost function can be chosen to associate a high cost with outcomes that are undesirable. For a given observation vector y, $f_Y(y)$ is a constant and has no effect on the risk minimisation process. Hence Eq. (3.17) may be written as a conditional risk function as

$$
\mathcal{R}(\hat{\theta}|y) = \int_\Theta C(\hat{\theta},\theta)\, f_{\Theta|Y}(\theta|y)\, d\theta
\tag{3.18}
$$

The Bayesian estimate obtained as the minimum risk parameter vector is given by

$$
\hat{\theta}_{Bayesian} = \arg\min_{\hat{\theta}} \mathcal{R}(\hat{\theta}|y) = \arg\min_{\hat{\theta}} \left(\int_\Theta C(\hat{\theta},\theta)\, f_{\Theta|Y}(\theta|y)\, d\theta \right)
\tag{3.19}
$$

Using the Bayes rule Eq. (3.19) can be written as

$$
\hat{\theta}_{Bayesian} = \arg\min_{\hat{\theta}} \left(\int_\Theta C(\hat{\theta},\theta)\, f_{Y|\Theta}(y|\theta)\, f_\Theta(\theta)\, d\theta \right)
\tag{3.20}
$$

Assuming that the risk function is differentiable, and has a well defined minimum, the Bayesian estimate can be obtained as

$$
\hat{\theta}_{Bayesian} = \arg\mathrm{zero}_{\hat{\theta}} \frac{\partial \mathcal{R}(\hat{\theta}|y)}{\partial \hat{\theta}} = \arg\mathrm{zero}_{\hat{\theta}} \left(\frac{\partial}{\partial \hat{\theta}} \int_\Theta C(\hat{\theta},\theta)\, f_{Y|\Theta}(y|\theta)\, f_\Theta(\theta)\, d\theta \right)
\tag{3.21}
$$

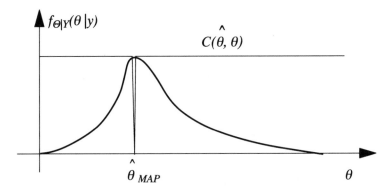

Figure 3.7 Illustration of the Bayesian cost function for the MAP estimate.

3.2.1 Maximum a Posterior Estimation

The maximum a posterior (MAP) estimate $\hat{\theta}_{MAP}$ is obtained as the parameter vector that maximises the posterior pdf $f_{\Theta|Y}(\theta|y)$. The MAP estimate corresponds to a Bayesian estimate with a so called uniform cost function defined as

$$C(\hat{\theta},\ \theta) = 1 - \delta(\hat{\theta},\ \theta) \tag{3.22}$$

where $\delta(\hat{\theta},\ \theta)$ is the Kronecker delta function. Substitution of the cost function in the Bayesian risk equation yields

$$\mathcal{R}_{MAP}(\hat{\theta}|y) = \int_{\Theta}[1 - \delta(\hat{\theta},\ \theta)]\, f_{\Theta|Y}(\theta|y)\ d\theta$$
$$= 1 - f_{\Theta|Y}(\hat{\theta}|y) \tag{3.23}$$

From Eq. (3.23) the minimum Bayesian risk estimate corresponds to the maximum of the posterior function. Hence the MAP estimate of the parameter vector θ is obtained from a minimisation of the risk (Eq. (3.23)) as

$$\hat{\theta}_{MAP} = \arg\max_{\theta}\ f_{\Theta|Y}(\theta\,|\,y)$$
$$= \arg\max_{\theta}\ [f_{Y|\Theta}(y|\theta)\, f_{\Theta}(\theta)] \tag{3.24}$$

Figure 3.7 illustrates the uniform cost function which is actually a notch shaped function.

3.2.2 Maximum Likelihood Estimation

The maximum likelihood (ML) estimate $\hat{\theta}_{ML}$ is obtained as the parameter vector that maximises the likelihood function $f_{Y|\Theta}(y|\theta)$.. The ML estimator corresponds to a Bayesian estimator with a uniform cost function and a uniform parameter prior pdf as

$$
\begin{aligned}
\mathcal{R}_{ML}(\hat{\theta}|y) &= \int_{\Theta} [1 - \delta(\hat{\theta}, \theta)] \, f_{Y|\Theta}(y|\theta) \, f_{\Theta}(\theta) \, d\theta \\
&= const.[1 - f_{Y|\Theta}(y|\hat{\theta})]
\end{aligned}
\tag{3.25}
$$

where $f_{\Theta}(\theta)=const$. Within a Bayesian framework, the main difference between the ML and the MAP estimators is that the ML assumes that the prior pdf of θ is uniform. Note that a uniform prior, in addition to modelling genuinely uniform pdfs, is also used when the parameter prior pdf is unknown, or when the parameter is an unknown constant.

From Eq. (3.25) it is evident that minimisation of the risk function is achieved by maximisation of the likelihood function as

$$
\hat{\theta}_{ML} = \underset{\theta}{\operatorname{argmax}} \; f_{Y|\Theta}(y|\theta)
\tag{3.26}
$$

In practice it is convenient to maximise the log-likelihood function instead of the likelihood. The log-likelihood is usually chosen because : (a) logarithm is a monotonic function and hence log-likelihood has the same turning points as the likelihood function, (b) the joint log-likelihood of independent variables is the sum of the log-likelihood of individual elements, and (c) unlike the likelihood function, the log-likelihood has a dynamic range which does not cause computational under flow.

Example 3.3 ML Estimation of the Mean and Variance of a Gaussian Process Consider the problem of maximum likelihood estimation of the mean vector μ_y and the covariance matrix Σ_{yy} of a P-dimensional Gaussian vector process from N observation vectors $[y(0), y(1), \cdots, y(N-1)]$. Assuming the observation vectors are uncorrelated, the pdf of the observation sequence is given by

$$
f_Y(y(0), \cdots, y(N-1)) = \prod_{m=0}^{N-1} \frac{1}{(2\pi)^{P/2}|\Sigma_{yy}|^{1/2}} \exp\left(-\frac{1}{2}(y(m) - \mu_y)^T \Sigma_{yy}^{-1}(y(m) - \mu_y)\right)
\tag{3.27}
$$

and the log-likelihood equation is given by

$$\ln f_Y(y(0), \cdots, y(N-1)) = \sum_{m=0}^{N-1} \left(-\frac{P}{2}\ln(2\pi) - \frac{1}{2}\ln|\Sigma_{yy}| - \frac{1}{2}(y(m) - \mu_y)^T \Sigma_{yy}^{-1}(y(m) - \mu_y) \right)$$

$$(3.28)$$

Taking the derivative of the log-likelihood equation with respect to the mean vector μ_y yields

$$\frac{\partial \ln f_Y(y(0), \cdots, y(N-1))}{\partial \mu_y} = \sum_{m=0}^{N-1} \left(2\Sigma_{yy}^{-1}\mu_y - 2\Sigma_{yy}^{-1}y(m) \right) = 0 \qquad (3.29)$$

From Eq. (3.29) we have

$$\hat{\mu}_y = \frac{1}{N} \sum_{m=0}^{N-1} y(m) \qquad (3.30)$$

To obtain the ML estimate of the covariance matrix we take the derivative of the log-likelihood equation with respect to Σ_{yy}^{-1} as

$$\frac{\partial \ln f_Y(y(0), \cdots, y(N-1))}{\partial \Sigma_{yy}^{-1}} = \sum_{m=0}^{N-1} \left(\frac{1}{2}\Sigma_{yy} - \frac{1}{2}(y(m) - \mu_y)(y(m) - \mu_y)^T \right) = 0$$

$$(3.31)$$

From Eq. (3.31) we have an estimate of the covariance matrix as

$$\hat{\Sigma}_{yy} = \frac{1}{N} \sum_{m=0}^{N-1} (y(m) - \hat{\mu}_y)(y(m) - \hat{\mu}_y)^T \qquad (3.32)$$

Example 3.4 ML and MAP Estimation of a Gaussian Random Parameter

Consider the estimation of a P-dimensional random parameter vector θ from an N-dimensional observation vector y. Assume that the relation between the signal vector and the parameter vector is described by a linear model as

$$y = G\theta + e \qquad (3.33)$$

where e is a random excitation signal. Assuming the matrix G is known, the pdf of the parameter vector θ given an observation vector y can be described, using the Bayes rule, as

$$f_{\Theta|Y}(\theta|y) = \frac{1}{f_Y(y)} f_{Y|\Theta}(y|\theta) f_\Theta(\theta) \qquad (3.34)$$

The likelihood of y given θ is the pdf of the random vector e

$$f_{Y|\Theta}(y|\theta) = f_E(e = y - G\theta) \tag{3.35}$$

Now assume the input e is a zero mean, Gaussian distributed, random process with a diagonal covariance matrix, and the parameter vector θ is also a Gaussian process with a mean of μ_θ and a covariance matrix $\Sigma_{\theta\theta}$. Therefore we have

$$f_{Y|\Theta}(y|\theta) = f_E(e) = \frac{1}{(2\pi\sigma_e^2)^{N/2}} \exp\left(-\frac{1}{2\sigma_e^2}(y - G\theta)^T (y - G\theta)\right) \tag{3.36}$$

and

$$f_\Theta(\theta) = \frac{1}{(2\pi)^{P/2}\left|\Sigma_{\theta\theta}\right|^{1/2}} \exp\left(-\frac{1}{2}(\theta - \mu_\theta)^T \Sigma_{\theta\theta}^{-1}(\theta - \mu_\theta)\right) \tag{3.37}$$

The ML estimate obtained from maximisation of the log-likelihood function $\ln[f_{Y|\Theta}(y|\theta)]$ with respect to θ is given by

$$\hat{\theta}_{ML}(y) = (G^T G)^{-1} G^T y \tag{3.38}$$

To obtain the MAP estimate we first form the posterior distribution by substitution of Eqs. (3.37) and (3.36) in Eq. (3.34) as

$$f_{\Theta|Y}(\theta|y) = \frac{1}{f_Y(y)} \frac{1}{(2\pi\sigma_e^2)^{N/2}} \frac{1}{(2\pi)^{P/2}\left|\Sigma_{\theta\theta}\right|^{1/2}} \times$$

$$\exp\left(-\frac{1}{2\sigma_e^2}(y - G\theta)^T (y - G\theta) - \frac{1}{2}(\theta - \mu_\theta)^T \Sigma_{\theta\theta}^{-1}(\theta - \mu_\theta)\right) \tag{3.39}$$

The MAP parameter estimate is obtained by differentiating the log-likelihood function $\ln[f_{\Theta|Y}(\theta|y)]$ and setting the derivative to zero as

$$\hat{\theta}_{MAP}(y) = \left(G^T G + \sigma_e^2 \Sigma_{\theta\theta}^{-1}\right)^{-1} \left(G^T y + \sigma_e^2 \Sigma_{\theta\theta}^{-1}\mu_\theta\right) \tag{3.40}$$

Note that as the covariance of the Gaussian distributed parameter increases, or equivalently as $\Sigma_{\theta\theta}^{-1} \to 0$, the Gaussian prior tends to a uniform prior and the MAP solution Eq. (3.40) tends to the ML solution given by Eq. (3.38).

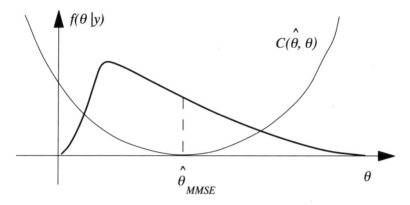

Figure 3.8 Illustration of mean squared error cost function and estimate.

3.2.3 Minimum Mean Squared Error Estimation

The Bayesian minimum mean squared error (MMSE) estimate is obtained as the parameter vector that minimises a mean squared error cost function defined as

$$
\begin{aligned}
\mathcal{R}_{MSE}(\hat{\theta}|y) &= \mathcal{E}\left[(\hat{\theta}-\theta)^2|y\right] \\
&= \int_{\theta}(\hat{\theta}-\theta)^2 \, f_{\Theta|Y}(\theta|y) \, d\theta
\end{aligned}
\tag{3.41}
$$

In the following it is shown that *the Bayesian MMSE estimate is the conditional mean of the posterior pdf*. Assuming that the mean squared error risk function is differentiable and has a well defined minimum, the MMSE solution can be obtained by setting the gradient of the mean squared error risk function to zero :

$$
\frac{\partial \mathcal{R}_{MSE}(\hat{\theta}|y)}{\partial \hat{\theta}} = 2\,\hat{\theta}\int_{\theta} f_{\Theta|Y}(\theta|y) \, d\theta - 2\int_{\theta}\theta \, f_{\Theta|Y}(\theta|y) \, d\theta
\tag{3.42}
$$

Since the first integral in the right hand side of Eq. (3.42) is equal to 1.0, we have

$$
\frac{\partial \mathcal{R}_{MSE}(\hat{\theta}|y)}{\partial \hat{\theta}} = 2\,\hat{\theta} - 2\int_{\theta}\theta \, f_{\Theta|Y}(\theta|y) \, d\theta
\tag{3.43}
$$

The MMSE solution is obtained by setting Eq. (3.43) to zero as

$$\hat{\theta}_{MMSE}(y) = \int_\theta \theta \, f_{\Theta|Y}(\theta|y) \, d\theta \qquad (3.44)$$

For cases where we do not have a pdf model of the parameter process, the minimum mean squared error estimate is obtained through minimisation of a mean squared error function $\mathcal{E}[e^2(\theta|y)]$ as

$$\hat{\theta}_{MMSE} = \underset{\hat{\theta}}{\operatorname{argmin}} \, \mathcal{E}[e^2(\hat{\theta}|y)] \qquad (3.45)$$

The MMSE estimation of Eq. (3.45) does not use any prior knowledge of the distribution of the signals and the parameters. This can be considered a strength in situations where the prior pdfs are unknown, but it can also be considered a weakness in cases where fairly accurate models of the priors are available but not utilised by the MMSE.

Example 3.5 Consider the MMSE estimation of a parameter vector θ assuming a linear model of the observation y as

$$y = G\theta + e \qquad (3.46)$$

The MMSE estimate is obtained as the parameter vector at which the gradient of the mean squared error with respect to θ is zero :

$$\frac{\partial e^T e}{\partial \theta} = (y^T y - 2\, \theta^T G^T y + \theta^T G^T G\theta)\Big|_{\theta_{MMSE}} = 0 \qquad (3.47)$$

From Eq. (3.47) the MMSE parameter estimate is given by

$$\theta_{MMSE} = [G^T G]^{-1} G^T y \qquad (3.48)$$

Note that for a Gaussian likelihood function, the MMSE solution is the same as the ML solution of Eq. (3.38).

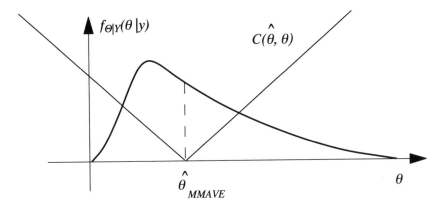

Figure 3.9 Illustration of mean absolute value of error cost function. Note that the MAVE estimate coincides with the conditional median of the posterior function.

3.2.4 Minimum Mean Absolute Value of Error Estimation

The minimum mean absolute value of error (MAVE) estimate is obtained through minimisation of a Bayesian risk function defined as

$$\mathcal{R}_{\text{MAVE}}(\hat{\theta}|y) = \mathcal{E}\big[|\hat{\theta}-\theta|\big]$$
$$= \int_{\Theta} |\hat{\theta}-\theta|\, f_{\Theta|Y}(\theta|y)\, d\theta \qquad (3.49)$$

In the following it is shown that the minimum mean absolute value estimate is the median of the parameter process. Eq. (3.49) can be re-expressed as

$$\mathcal{R}_{\text{MAVE}}(\hat{\theta}|y) = \int_{-\infty}^{\hat{\theta}(y)} [\hat{\theta}-\theta]\, f_{\Theta|Y}(\theta|y)\, d\theta + \int_{\hat{\theta}(y)}^{\infty} [\theta-\hat{\theta}]\, f_{\Theta|Y}(\theta|y)\, d\theta$$
$$(3.50)$$

Taking the derivative of the risk function with respect to $\hat{\theta}$ we have

$$\frac{\partial \mathcal{R}_{\text{MAVE}}(\hat{\theta}|y)}{\partial \hat{\theta}} = \int_{-\infty}^{\hat{\theta}(y)} f_{\Theta|Y}(\theta|y)\, d\theta - \int_{\hat{\theta}(y)}^{\infty} f_{\Theta|Y}(\theta|y)\, d\theta \qquad (3.51)$$

The minimum absolute value of error is obtained by setting Eq. (3.51) to zero as

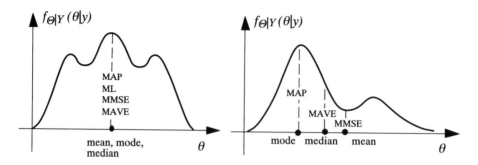

Figure 3.10 Illustration of a symmetric and an asymmetric pdf and their respective mode, mean and median and the relations to MAP, MMAVE and MMSE estimates.

$$\int_{-\infty}^{\hat{\theta}_{MAVE}(y)} f_{\Theta|Y}(\theta|y) \, d\theta = \int_{\hat{\theta}_{MAVE}(y)}^{\infty} f_{\Theta|Y}(\theta|y) \, d\theta \tag{3.52}$$

From Eq. (3.52) we note the MAVE estimate is the median of the posterior density.

3.2.5 Equivalence of MAP, ML, MMSE and MAVE for Gaussian Processes with Uniform Parameter Priors

Example 3.4 shows that for a Gaussian distributed process the MMSE estimate and the ML estimate are identical. Furthermore Eq. (3.40), for the MAP estimate of a Gaussian distributed parameter, shows that as the parameter variance increases, or equivalently as the parameter prior pdf tends to a uniform distribution, the MAP estimate tends to the ML and the MMSE estimates. In general, for any symmetric distribution, centred around the maximum, the mode, the mean and the median are identical. Hence, for a process with a symmetric pdf, if the prior distribution of the parameter is uniform, then the MAP, the ML, the MMSE and the MMAVE parameter estimates are identical. Figure 3.10 illustrates a symmetric pdf, an asymmetric pdf, and the relative position of various estimates.

3.2.6 Influence of the Prior on Estimation Bias and Variance

The use of a prior pdf, introduces a bias in the estimate towards the range of parameter values with a relatively high prior pdf, and reduces the variance of the estimate. To illustrate the effects of the prior pdf on the bias and the variance of an estimate, we consider the following examples in which the bias and the variance of the ML and the MAP estimates of the mean of a process are compared.

Example 3.6 Consider the ML estimation of a random scalar parameter θ, observed in a zero mean additive white Gaussian noise (AWGN) $n(m)$, and expressed as

$$y(m) = \theta + n(m) \qquad\qquad m = 0,\ldots,\ N{-}1 \qquad (3.53)$$

It is assumed that, for each realisation of the parameter θ, N observation samples are available. Note that, as the noise is assumed to be a zero mean process, this problem is equivalent to estimation of the mean of the process $y(m)$. The likelihood of an observation vector $y = [y(0), y(1), \ldots, y(N-1)]$ and a parameter value of θ is given by

$$
\begin{aligned}
f_{Y|\Theta}(y|\theta) &= \prod_{m=0}^{N-1} f_N(y(m) - \theta) \\
&= \frac{1}{(2\pi\sigma_n^2)^{N/2}} \exp\left(-\frac{1}{2\sigma_n^2} \sum_{m=0}^{N-1} (y(m) - \theta)^2\right)
\end{aligned} \qquad (3.54)
$$

From Eq. (3.54) the log-likelihood function is given by

$$\ln f_{Y|\Theta}(y|\theta) = -\frac{N}{2}\ln(2\pi\sigma_n^2) - \frac{1}{2\sigma_n^2}\sum_{m=0}^{N-1}[y(m)-\theta]^2 \qquad (3.55)$$

The ML estimate of θ, obtained by setting the derivative of $\ln f_{Y|\Theta}(y|\theta)$ to zero, is

$$\hat{\theta}_{ML} = \frac{1}{N}\sum_{m=0}^{N-1} y(m) = \bar{y} \qquad (3.56)$$

where $\bar{y}$ denotes the time-average of $y(m)$. From Eq. (3.56) we note that the ML solution is an unbiased estimate

$$\mathcal{E}[\hat{\theta}_{ML}] = \mathcal{E}\left(\frac{1}{N}\sum_{m=0}^{N-1}[\theta + n(m)]\right) = \theta \qquad (3.57)$$

and the variance of the ML estimate is given by

$$\text{Var}[\hat{\theta}_{ML}] = \mathcal{E}[(\hat{\theta}_{ML} - \theta)^2] = \mathcal{E}\left[\left(\frac{1}{N}\sum_{m=0}^{N-1} y(m) - \theta\right)^2\right] = \frac{\sigma_n^2}{N} \qquad (3.58)$$

Note that the variance of the ML estimate decreases with the increasing length of the observation.

Example 3.7 Estimation of a Uniform-Distributed Parameter Observed in AWGN

Consider the effects of using a uniform parameter prior on the mean and the variance of the estimate in Example 3.6. Assume that the prior for the parameter θ is given by

$$f_\Theta(\theta) = \begin{cases} 1/(\theta_{max} - \theta_{min}) & \theta_{min} \le \theta \le \theta_{max} \\ 0 & otherwise \end{cases} \tag{3.59}$$

as illustrated in Figure 3.11. From Bayes rule the posterior pdf is given by

$$f_{\Theta|Y}(\theta|y) = \frac{1}{f_Y(y)} f_{Y|\Theta}(y|\theta) f_\Theta(\theta)$$

$$= \begin{cases} \dfrac{1}{f_Y(y)} \dfrac{1}{(2\pi\sigma_n^2)^{N/2}} \exp\left(-\dfrac{1}{2\sigma_n^2} \sum_{m=0}^{N-1}(y(m) - \theta)^2\right) & \theta_{min} \le \theta \le \theta_{max} \\ 0 & otherwise \end{cases}$$

$$\tag{3.60}$$

The MAP estimate is obtained by maximising the posterior pdf as

$$\hat{\theta}_{MAP}(y) = \begin{cases} \theta_{min} & if \quad \hat{\theta}_{ML}(y) < \theta_{min} \\ \hat{\theta}_{ML}(y) & if \quad \theta_{min} \ge \hat{\theta}_{ML}(y) \ge \theta_{max} \\ \theta_{max} & if \quad \hat{\theta}_{ML}(y) > \theta_{max} \end{cases} \tag{3.61}$$

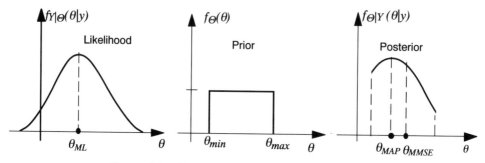

Figure 3.11 Illustration of the effects of a uniform prior.

Note that the MAP estimate is constrained to the range θ_{min} to θ_{max}. This constraint is desirable and moderates those estimates, which, due to say low signal to noise ratio, fall outside the range of possible values of θ. It is easy to see that the variance of an estimate constrained to a range θ_{min} to θ_{max} is less than the variance of the ML estimate in which there is no constraint on the range of the parameter estimate

$$\text{Var}[\hat{\theta}_{MAP}] = \int_{\theta_{min}}^{\theta_{max}} (\hat{\theta}_{MAP} - \theta)^2 f_{Y|\Theta}(y|\theta)\,dy \;\le\; \text{Var}[\hat{\theta}_{ML}] = \int_{-\infty}^{\infty} (\hat{\theta}_{ML} - \theta)^2 f_{Y|\Theta}(y|\theta)\,dy$$

$$(3.62)$$

Example 3.8   Estimation of a Gaussian-Distributed Parameter Observed in AWGN In this example we consider the effect of a Gaussian prior on the mean and the variance of the MAP estimate. Assume that the parameter θ is Gaussian distributed with a mean μ_θ and a variance σ_θ^2 as

$$f_\Theta(\theta) = \frac{1}{(2\pi\sigma_\theta^2)^{1/2}} \exp\left(-\frac{(\theta - \mu_\theta)^2}{2\sigma_\theta^2}\right)$$

$$(3.63)$$

From Bayes rule the posterior pdf is given by

$$f_{\Theta|Y}(\theta|y) = \frac{1}{f_Y(y)} f_{Y|\Theta}(y|\theta) f_\Theta(\theta)$$

$$= \frac{1}{f_Y(y)} \frac{1}{(2\pi\sigma_n^2)^{N/2}(2\pi\sigma_\theta^2)^{1/2}} \exp\left(-\frac{1}{2\sigma_n^2}\sum_{m=0}^{N-1}(y(m)-\theta)^2 - \frac{1}{2\sigma_\theta^2}(\theta-\mu_\theta)^2\right)$$

$$(3.64)$$

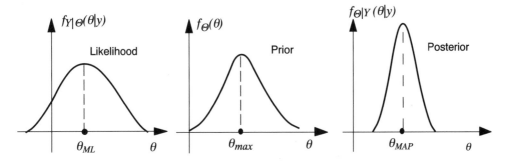

Figure 3.12 Illustration of the effect of a Gaussian prior.

The maximum posterior solution is obtained by setting the derivative of the log-posterior function, $\ln f_{\Theta|Y}(\theta|y)$, with respect to θ to zero as

$$\hat{\theta}_{MAP}(y) = \frac{\sigma_\theta^2}{\sigma_\theta^2 + \sigma_n^2/N} \bar{y} + \frac{\sigma_n^2/N}{\sigma_\theta^2 + \sigma_n^2/N} \mu_\theta \qquad (3.65)$$

where $\bar{y} = \sum_{m=0}^{N-1} y(m)/N$. The expectation of the MAP estimate is obtained by noting that the only random variable on the right hand side of Eq. (3.65) is the term $\bar{y}$, and that $E[\bar{y}] = \theta$

$$\mathcal{E}[\hat{\theta}_{MAP}(y)] = \frac{\sigma_\theta^2}{\sigma_\theta^2 + \sigma_n^2/N} \theta + \frac{\sigma_n^2/N}{\sigma_\theta^2 + \sigma_n^2/N} \mu_\theta \qquad (3.66)$$

and the variance of the MAP estimate is given as

$$Var[\hat{\theta}_{MAP}(y)] = \frac{\sigma_\theta^2}{\sigma_\theta^2 + \sigma_n^2/N} Var[\bar{y}] = \frac{\sigma_n^2/N}{1 + \sigma_n^2/N\sigma_\theta^2} \qquad (3.67)$$

Substitution of Eq. (3.58) in Eq. (3.67) yields

$$Var[\hat{\theta}_{MAP}(y)] = \frac{Var[\hat{\theta}_{ML}(y)]}{1 + Var[\hat{\theta}_{ML}(y)]/\sigma_\theta^2} \qquad (3.68)$$

Note that as σ_θ^2, the variance of the parameter θ, increases the influence of the prior decreases, and the variance of the MAP estimate tends towards the variance of the ML estimate.

3.2.7 The Relative Importance of the Prior and the Observation

The influence of the prior pdf on Bayesian estimation, depends on the confidence on the observation, which in turn depends on the length of the observation, and on the signal to noise ratio (SNR). In general, as the number of observation samples, and the SNR, increase the variance of the estimate, and the influence of the prior, decreases. From Eq. (3.65), in estimation of a Gaussian distributed parameter observed in AWGN, as the length of the observation N increases, the importance of the prior decreases, and the MAP estimate tends to the ML estimate as

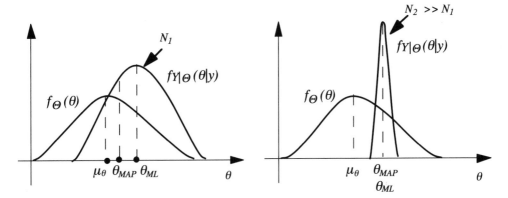

Figure 3.13 Illustration of the effect of increasing length of observation on the variance an estimator.

$$\lim_{N \to \infty} \hat{\theta}_{MAP}(y) = \lim_{N \to \infty} \left(\frac{\sigma_\theta^2}{\sigma_\theta^2 + \sigma_n^2/N} \bar{y} + \frac{\sigma_n^2/N}{\sigma_\theta^2 + \sigma_n^2/N} \mu_\theta \right) = \bar{y} = \hat{\theta}_{ML} \quad (3.69)$$

Example   3.9 MAP Estimation of a Signal in Additive Noise
Consider the estimation of a scalar-valued Gaussian signal $x(m)$, observed in an additive Gaussian white noise $n(m)$, and modelled as

$$y(m) = x(m) + n(m) \quad (3.70)$$

The posterior pdf of the signal $x(m)$ is given by

$$f_{X|Y}(x(m)|y(m)) = \frac{1}{f_Y(y(m))} f_{Y|X}(y(m)|x(m)) f_X(x(m))$$

$$= \frac{1}{f_Y(y(m))} f_N(y(m) - x(m)) f_X(x(m)) \quad (3.71)$$

where $f_X(x(m)) = \mathcal{N}\left(x(m), \mu_x, \sigma_x^2 \right)$ and $f_N(n(m)) = \mathcal{N}\left(n(m), \mu_n, \sigma_n^2 \right)$ are the Gaussian pdfs of the signal and noise respectively. Substitution of the signal and noise pdfs in Eq. (3.71) yields

$$f_{X|Y}(x(m)|y(m)) = \frac{1}{f_Y(y(m))} \; \frac{1}{\sqrt{2\pi}\sigma_n} \exp\left(-\frac{(y(m)-x(m)-\mu_n)^2}{2\sigma_n^2}\right) \times$$

$$\frac{1}{\sqrt{2\pi}\sigma_x} \exp\left(-\frac{(x(m)-\mu_x)^2}{2\sigma_x^2}\right)$$

(3.72)

This equation can be rewritten as

$$f_{X|Y}(x(m)|y(m)) = \frac{1}{f_Y(y(m))}\frac{1}{2\pi\sigma_n\sigma_x} \exp\left(-\frac{\sigma_x^2(y(m)-x(m)-\mu_n)^2 + \sigma_n^2(x(m)-\mu_x)^2}{2\sigma_x^2\sigma_n^2}\right)$$

(3.73)

To obtain the MAP estimate we set the derivative of the log likelihood function $\ln f_{X|Y}(x(m)|y(m))$ with respect to $x(m)$ to zero as

$$\frac{\partial[\ln f_{X|Y}(x(m)|y(m))]}{\partial\hat{x}(m)} = -\frac{2\sigma_n^2(x(m)-\mu_x)-2\sigma_x^2(y(m)-x(m)-\mu_n)}{2\sigma_x^2\sigma_n^2} = 0 \quad (3.74)$$

From Eq. (3.74) the MAP signal estimate is given by

$$\hat{x}(m) \; = \; \frac{\sigma_x^2}{\sigma_x^2 + \sigma_n^2} \; (y(m)-\mu_n) \; + \; \frac{\sigma_n^2}{\sigma_x^2 + \sigma_n^2} \; \mu_x \qquad (3.75)$$

Note that the estimate $\hat{x}(m)$ is a weighted linear interpolation between the unconditional mean of $x(m)$, μ_x, and the observed value $(y(m)-\mu_n)$. As expected, at a poor SNR, when $\sigma_x^2 \gg \sigma_n^2$, $\hat{x}(m) \approx \mu_x$, and on the other hand for noise-free signal $\hat{x}(m) = y(m)$. It is also easy to show that $E[\hat{x}(m)]=\mu_x$ and $\text{Var}[\hat{x}(m)]=\sigma_x^2$

Example 3.10   MAP estimate of a Gaussian-AR process observed in AWGN Consider a vector of N samples x from an autoregressive (AR) process observed in an additive Gaussian noise, and modelled as

$$y \; = \; x + n \qquad (3.76)$$

From Chapter 7 a vector x from an AR process may be expressed as

$$e \; = \; Ax \qquad (3.77)$$

where A is a matrix composed of the AR model coefficients, and e is a vector composed of the input samples of the AR model. Assuming that the signal x is Gaussian, and that the P initial samples x_0 are known, the pdf of x is given by

$$f_X(x|x_0) = f_E(e) = \frac{1}{(2\pi\sigma_e^2)^{N/2}} \exp\left(-\frac{1}{2\sigma_e^2} x^T A^T A x\right) \tag{3.78}$$

where it is assumed that the input signal e, of the AR model, is a zero mean uncorrelated process with variance σ_e^2. The pdf of a zero-mean Gaussian noise vector n, with covariance matrix Σ_{nn}, is given by

$$f_N(n) = \frac{1}{(2\pi)^{N/2}|\Sigma_{nn}|^{1/2}} \exp\left(-\frac{1}{2} n^T \Sigma_{nn}^{-1} n\right) \tag{3.79}$$

From the Bayes rule the pdf of the signal given the noisy observation is

$$f_{X|Y}(x|y) = \frac{f_{Y|X}(y|x)\, f_X(x)}{f_Y(y)}$$
$$= \frac{1}{f_Y(y)} f_N(y-x)\, f_X(x) \tag{3.80}$$

Substitution of the pdfs of the signal and noise in Eq. (3.80) yields

$$f_{X|Y}(x|y) = \frac{1}{f_Y(y)(2\pi)^N \sigma_e^{N/2}|\Sigma_{nn}|^{1/2}} \exp\left(-\frac{1}{2}\left((y-x)^T \Sigma_{nn}^{-1}(y-x) + \frac{x^T A^T A x}{\sigma_e^2}\right)\right) \tag{3.81}$$

The MAP estimate corresponds to the minimum of the argument of the exponential function in Eq. (3.81). Assuming that the argument of the exponential function is differentiable, and has a well defined minimum, we can obtain the MAP estimate from

$$\hat{x}_{MAP}(y) = \underset{x}{\text{argzero}} \left(\frac{\partial}{\partial x}\left((y-x)^T \Sigma_{nn}^{-1}(y-x) + \frac{x^T A^T A x}{\sigma_e^2}\right)\right) \tag{3.82}$$

The MAP estimate is

$$\hat{x}_{MAP}(y) = \left(I + \frac{1}{\sigma_e^2}\Sigma_{nn} A^T A\right)^{-1} y \tag{3.83}$$

3.3 Estimate-Maximise (EM) Method

The EM algorithm is an iterative likelihood maximisation method with applications in blind deconvolution, model-based signal interpolation, spectral estimation from noisy observations, estimation of the parameters of a data set etc. The EM is a framework for solving problems where it is difficult to obtain a direct ML estimate either because the data is incomplete or because the problem is difficult.

To define the term *incomplete data,* consider a signal x from a random process X with an unknown parameter vector θ and a pdf $f_{X;\Theta}(x;\theta)$. The notation $f_{X;\Theta}(x;\theta)$ expresses the dependency of the pdf of X on the value of the unknown parameter θ. The signal x is the so called *complete data* and the ML estimate of the parameter vector θ may be obtained from $f_{X;\Theta}(x;\theta)$. Now assume that the signal x goes through a many-to-one non-invertible transformation (for example when a number of samples of the vector x are lost) and is observed as y. The observation y is the so called incomplete data.

Maximisation of the likelihood of the incomplete data $f_{Y;\Theta}(y;\theta)$ with respect to θ is often a difficult task, whereas maximisation of the likelihood of the complete data $f_{X;\Theta}(x;\theta)$ is relatively easy. Since the complete data is unavailable, the parameter estimate is obtained through maximisation of the conditional *expectation* of the log-likelihood of the complete data defined as

$$\mathcal{E}\left[\ln f_{X;\Theta}(x;\theta)|\ y\right] = \int_X f_{X|Y;\Theta}(x|y;\theta) \ln f_{X;\Theta}(x;\theta)\ dx \qquad (3.84)$$

In Eq. (3.78), the computation of the term $f_{X|Y;\Theta}(x|y;\theta)$ requires an estimate of the unknown parameter vector θ. For this reason, the expectation of the likelihood function is maximised iteratively starting with an initial estimate of θ, and updating the estimate as described in the following steps :

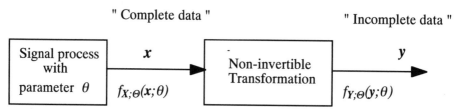

Figure 3.14 Illustration of transformation of a complete data to an incomplete data.

EM Algorithm :

> *Step 1 Initialisation :* select an initial parameter estimate $\hat{\theta}_0$, and
> For $i = 0, 1, ...$ until convergence :

Step 2 Expectation : compute

$$
\begin{aligned}
U(\theta, \hat{\theta}_i) &= \mathcal{E}[\ln f_{X,\Theta}(x;\theta)|y;\hat{\theta}_i] \\
&= \int_X \ln f_{X,\Theta}(x;\theta) \, f_{X|Y;\Theta}(x|y;\hat{\theta}_i) \, dx
\end{aligned}
\tag{3.85}
$$

Step 3 Maximisation : select

$$
\hat{\theta}_{i+1} = \arg\max_\theta \; U(\theta, \hat{\theta}_i)
\tag{3.86}
$$

Step 4 Convergence Test : if not converged then go to step 2.

3.3.1   Convergence of the EM algorithm

In this section it is shown that the EM algorithm converges to a maximum of the likelihood of the incomplete data $f_{Y;\Theta}(y;\theta)$. The likelihood of the complete data can be written as

$$
f_{X,Y;\Theta}(x,y;\theta) = f_{X|Y;\Theta}(x|y;\theta) \, f_{Y;\Theta}(y;\theta)
\tag{3.87}
$$

where $f_{X,Y;\Theta}(x,y;\theta)$ is the likelihood of x and y with θ as a parameter. From Eq. (3.87) the log-likelihood of the incomplete data is obtained as

$$
\ln f_{Y;\Theta}(y;\theta) = \ln f_{X,Y;\Theta}(x,y;\theta) - \ln f_{X|Y;\Theta}(x|y;\theta)
\tag{3.88}
$$

Using an estimate $\hat{\theta}_i$ of the parameter vector θ, and taking the expectation of Eq. (3.88) over the space of the complete signal x we obtain

$$
\ln f_{Y;\Theta}(y;\theta) = U(\theta;\hat{\theta}_i) - V(\theta;\hat{\theta}_i)
\tag{3.89}
$$

where for a given y, the expectation of $\ln f_{Y;\Theta}(y;\theta)$ is itself, and the function $U(\theta;\hat{\theta})$ is the conditional expectation of $\ln f_{X,Y;\Theta}(x,y;\theta)$ given as

$$U(\theta, \hat{\theta}_i) = \mathcal{E}[\ln f_{X,\Theta}(x, y; \theta)|y; \hat{\theta}_i]$$

$$= \int_X f_{X,Y;\Theta}(x|y; \hat{\theta}_i) \ln f_{X,\Theta}(x; \theta) \ dx \tag{3.90}$$

and the function $V(\theta, \hat{\theta})$ is the conditional expectation of $\ln f_{X|Y;\Theta}(x|y; \theta)$ given as

$$V(\theta, \hat{\theta}_i) = \mathcal{E}\left[\ln f_{X|Y;\Theta}(x|y; \theta)\big|y; \hat{\theta}_i\right]$$

$$= \int_X f_{X|Y;\Theta}(x|y; \hat{\theta}_i) \ln f_{X|Y;\Theta}(x|y; \theta) \ dx \tag{3.91}$$

Now from Eq. (3.89) the log likelihood of the incomplete data y with parameter estimate $\hat{\theta}_i$ at iteration i is

$$\ln f_{Y;\Theta}(y; \hat{\theta}_i) = U(\hat{\theta}_i; \hat{\theta}_i) - V(\hat{\theta}_i; \hat{\theta}_i) \tag{3.92}$$

It can be shown, [DEMPSTER et al. 1977], that the function V satisfies the following inequality

$$V(\hat{\theta}_{i+1}, \hat{\theta}_i) \leq V(\hat{\theta}_i, \hat{\theta}_i) \tag{3.93}$$

and in the maximisation step of EM we chose $\hat{\theta}_{i+1}$ such that

$$U(\hat{\theta}_{i+1}, \hat{\theta}_i) \geq U(\hat{\theta}_i, \hat{\theta}_i) \tag{3.94}$$

From Eq. (3.92) and the inequalities (3.93) and (3.94) it follows that

$$\ln f_{Y;\Theta}(y; \hat{\theta}_{i+1}) \geq \ln f_{Y;\Theta}(y; \hat{\theta}_i) \tag{3.95}$$

Therefore at every iteration of the EM algorithm, the conditional likelihood of the estimate increases until the estimate converges to a local maximum of the log likelihood function $\ln f_{Y;\Theta}(y; \theta)$.

The EM algorithm is applied to solution of a number of problems in this book. In Section 3.6.4 of this chapter the estimation of the parameters of a mixture Gaussian model for a data base is formulated in an EM framework. In Chapter 4 the EM is used for estimation of the parameters of a hidden Markov model.

3.4   Cramer-Rao Bound on the Minimum Estimator Variance

An important performance measure for an estimator, is the variance of the estimate with the observation vector y and the parameter vector θ. The minimum estimation variance depends on the distributions of the parameter and the observations. In this section, we first consider the lower bound on the variance of the estimates of a constant parameter, and then extend the results to random parameters. The Cramer-Rao lower bound on the variance of estimate of the ith coefficient, θ_i, of a parameter vector θ is given as

$$\text{Var}[\hat{\theta}_i(y)] \geq \frac{\left(1 + \dfrac{\partial \theta_{Bias}}{\partial \theta_i}\right)^2}{\mathcal{E}\left[\left(\dfrac{\partial \ln f_{Y|\Theta}(y|\theta)}{\partial \theta_i}\right)^2\right]} \tag{3.96}$$

An estimator that achieves the lower bound on the variance is called the minimum variance, or the most efficient, estimator.

Proof : The bias in the estimate $\hat{\theta}_i(y)$ of the ith coefficient of the parameter vector θ is defined as

$$\mathcal{E}[\hat{\theta}_i(y) - \theta_i] = \int_{-\infty}^{\infty} [\hat{\theta}_i(y) - \theta_i] f_{Y|\Theta}(y|\theta)\ dy = \theta_{Bias} \tag{3.97}$$

Differentiation of Eq. (3.97) with respect to θ_i gives

$$\int_{-\infty}^{\infty} \left([\hat{\theta}_i(y) - \theta_i]\frac{\partial f_{Y|\Theta}(y|\theta)}{\partial \theta_i} - f_{Y|\Theta}(y|\theta)\right)\ dy = \frac{\partial \theta_{Bias}}{\partial \theta_i} \tag{3.98}$$

For a pdf we have

$$\int_{-\infty}^{\infty} f_{Y|\Theta}(y|\theta)\ dy = 1 \tag{3.99}$$

Therefore Eq. (3.98) can be written as

$$\int_{-\infty}^{\infty} [\hat{\theta}_i(y) - \theta_i] \frac{\partial f_{Y|\Theta}(y|\theta)}{\partial \theta_i} \, dy \; = \; 1 + \frac{\partial \theta_{Bias}}{\partial \theta_i} \tag{3.100}$$

Now, since the derivative of the integral of a pdf is zero, multiplying the derivative of Eq. (3.99) by θ_{Bias} yields

$$\theta_{Bias} \int_{-\infty}^{\infty} \frac{\partial f_{Y|\Theta}(y|\theta)}{\partial \theta_i} \, dy = 0 \tag{3.101}$$

Substituting $\partial f_{Y|\Theta}(y|\theta)/\partial \theta_i = f_{Y|\Theta}(y|\theta) \partial \ln f_{Y|\Theta}(y|\theta)/\partial \theta_i$ in Eq. (3.100), and using Eq. (3.101) we obtain

$$\int_{-\infty}^{\infty} [\hat{\theta}_i(y) - \theta_{Bias} - \theta_i] \frac{\partial \ln f_{Y|\Theta}(y|\theta)}{\partial \theta_i} f_{Y|\Theta}(y|\theta) \, dy \; = \; 1 + \frac{\partial \theta_{Bias}}{\partial \theta_i} \tag{3.102}$$

Now square both sides of Eq. (3.102) to obtain

$$\left(\int_{-\infty}^{\infty} [\hat{\theta}_i(y) - \theta_{Bias} - \theta_i] \frac{\partial \ln f_{Y|\Theta}(y|\theta)}{\partial \theta_i} f_{Y|\Theta}(y|\theta) \, dy \right)^2 = \left(1 + \frac{\partial \theta_{Bias}}{\partial \theta_i} \right)^2 \tag{3.103}$$

For the left hand side of Eq. (3.103) the Schwartz inequality can be written as

$$\left(\int_{-\infty}^{\infty} \left([\hat{\theta}_i(y) - \theta_{Bias} - \theta_i] f_{Y|\Theta}^{1/2}(y|\theta) \right) \left(\frac{\partial \ln f_{Y|\Theta}(y|\theta)}{\partial \theta_i} f_{Y|\Theta}^{1/2}(y|\theta) \right) \, dy \right)^2 \; \leq \;$$

$$\left(\int_{-\infty}^{\infty} \left([\hat{\theta}_i(y) - \theta_{Bias} - \theta_i]^2 f_{Y|\Theta}(y|\theta) \right) \, dy \right) \left(\int_{-\infty}^{\infty} \left(\frac{\partial \ln f_{Y|\Theta}(y|\theta)}{\partial \theta_i} \right)^2 f_{Y|\Theta}(y|\theta) \, dy \right) \tag{3.104}$$

From Eqs. (3.103) and (3.104) we have

$$Var[\hat{\theta}_i(y)] \times \mathcal{E} \left(\left(\frac{\partial \ln f_{Y|\Theta}(y|\theta)}{\partial \theta_i} \right)^2 \right) \; \geq \; \left(1 + \frac{\partial \theta_{Bias}}{\partial \theta_i} \right)^2 \tag{3.105}$$

The Cramer-Rao inequality (3.96) results directly from inequality (3.105).

3.4.1 Cramer-Rao Bound for Random Parameters

For random parameters the Cramer-Rao bound may be obtained using the same procedure as above, with the difference that in Eq. (3.96) instead of the likelihood $f_{Y|\Theta}(y|\theta)$ we use the joint pdf $f_{Y,\Theta}(y,\theta)$, and also use the logarithmic relation

$$\frac{\partial \ln f_{Y,\Theta}(y,\theta)}{\partial \theta_i} = \frac{\partial \ln f_{Y|\Theta}(y|\theta)}{\partial \theta_i} + \frac{\partial \ln f_{\Theta}(\theta)}{\partial \theta_i} \tag{3.106}$$

to obtain the Cramer-Rao bound for random parameters as

$$Var[\hat{\theta}_i(y)] \geq \frac{\left(1 + \dfrac{\partial \theta_{Bias}}{\partial \theta_i}\right)^2}{\mathcal{E}\left[\left(\dfrac{\partial \ln f_{Y|\Theta}(y|\theta)}{\partial \theta_i}\right)^2 + \left(\dfrac{\partial \ln f_{\Theta}(\theta)}{\partial \theta_i}\right)^2\right]} \tag{3.107}$$

Where the second term in the denominator of Eq. (3.107) describes the effect of the prior pdf of θ. As expected the use of the prior, $f_{\Theta}(\theta)$, can result in a decrease in the variance of the estimate. An alternative form of the minimum bound on estimation variance can be obtained by using the likelihood relation

$$\mathcal{E}\left(\left(\frac{\partial \ln f_{Y,\Theta}(y,\theta)}{\partial \theta_i}\right)^2\right) = -\mathcal{E}\left(\frac{\partial^2 \ln f_{Y,\Theta}(y,\theta)}{\partial \theta_i^2}\right) \tag{3.108}$$

as

$$Var[\hat{\theta}_i(y)] \geq -\frac{\left(1 + \dfrac{\partial \theta_{Bias}}{\partial \theta_i}\right)^2}{\mathcal{E}\left[\dfrac{\partial^2 \ln f_{Y|\Theta}(y|\theta)}{\partial \theta_i^2} + \dfrac{\partial^2 \ln f_{\Theta}(\theta)}{\partial \theta_i^2}\right]} \tag{3.109}$$

3.4.2 Cramer-Rao Bound for a Vector Parameter

For real-valued P-dimensional vector parameters, the Cramer-Rao bound on the covariance matrix of an unbiased estimator of θ is given by

$$Cov[\hat{\theta}] \geq J^{-1}(\theta) \tag{3.110}$$

where J is the $P \times P$ Fisher's information matrix with its elements given by

$$[J(\theta)]_{ij} = -\mathcal{E}\left(\frac{\partial^2 \ln f_{Y,\Theta}(y,\theta)}{\partial\theta_i\partial\theta_j}\right) \tag{3.111}$$

The lower bound on the variance of the ith element of the vector θ is given by

$$Var(\hat{\theta}_i) \geq [J^{-1}(\theta)]_{ii} = \frac{1}{\mathcal{E}\left(\dfrac{\partial^2 \ln f_{Y,\Theta}(y,\theta)}{\partial\theta_i^2}\right)} \tag{3.112}$$

where $(J^{-1}(\theta)_{ii})$ is the ith diagonal element of the inverse of Fishers matrix.

3.5 Bayesian Classification

Classification is the *labelling* of an observation with one of N classes of signals. Classifiers are used in applications such as decoding of discrete-valued symbols in digital receivers, speech recognition, character recognition, impulsive noise detection, signal/noise classification *etc*. In this section we study the Bayesian classification of discrete-valued and composite-valued processes. A composite-valued process is formed from a finite number of continuous-valued processes.

Figures 3.15.a and 3.15.b illustrate two examples of a simple binary classification problem in a 2-dimensional signal space. In each case the observation is the result of a random mapping (e.g. signal plus a random noise) from the binary source to the continuous observation space. In Figure 3.15.a the binary sources and the observation space associated with each source are well separated, and it is possible to make an error-free classification of each observation. In Figure 3.15.b the sources are spaced closer to each other and the observation signals have a greater spread (or variance). This results in some overlap of the signal spaces and classification error can occur.

3.5.1 Classification of Discrete-valued Parameters

Let the set $\Theta=\{\theta_i, i =1, ..., M\}$ denote the values that a discrete P-dimensional parameter vector θ. can assume. In general, the observation space Y associated with a parameter space Θ may be a discrete or a continuous space. Assuming that the observation space is continuous, the pdf of the parameter vector θ_i, conditioned on a

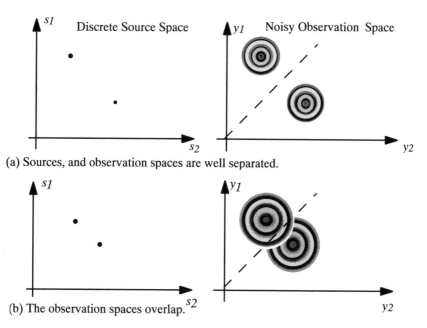

(a) Sources, and observation spaces are well separated.

(b) The observation spaces overlap.

Figure 3.15 Illustration of the binary classification problem : (a) The two classes are well separated and error-free classification is possible, (b) The two classes overlap and some error is inevitable.

given observation vector y, may be expressed, using the Bayes rule, as

$$P_{\Theta|Y}(\theta_i \mid y) = \frac{f_{Y|\theta_i}(y \mid \theta_i)P_{\Theta}(\theta_i)}{f_Y(y)}$$

$$= \frac{f_{Y|\theta_i}(y \mid \theta_i)P_{\Theta}(\theta_i)}{\sum_{i=1}^{M} f_{Y|\theta_i}(y \mid \theta_i)P_{\Theta}(\theta_i)} \tag{3.113}$$

For the case when the observation space Y is discrete-valued, the probability density functions are replaced by the appropriate probability mass functions. The Bayesian risk in selecting the parameter vector θ_i given the observation y is defined as

$$\mathcal{R}(\theta_i|y) = \sum_{j=1}^{M} C(\theta_i|\theta_j)P_{\Theta|Y}(\theta_j|y) \tag{3.114}$$

where $C(\theta_i|\theta_j)$ is the cost of selecting the parameter θ_i when the true parameter is θ_j. The maximum a posterior, the maximum likelihood and the minimum mean squared error classifiers can be obtained from the Bayesian classification (Eq. (3.114)) in much the same way as for continuous-valued parameter estimation discussed in previous sections.

3.5.2 Maximum a Posterior Classification

MAP classification corresponds to Bayesian classification with a uniform cost function defined as

$$C(\theta_i|\theta_j) = 1 - \delta(\theta_i, \theta_j) \tag{3.115}$$

Substitution of this cost function in the Bayesian risk function yields

$$\mathcal{R}_{MAP}(\theta_i|y) = \sum_{j=1}^{M} [1 - \delta(\theta_i, \theta_j)] P_{\Theta|y}(\theta_j|y)$$
$$= 1 - P_{\Theta|y}(\theta_i|y) \tag{3.116}$$

Note that the MAP risk in selecting θ_i is the classification error probability; that is the sum of the probabilities of all other candidates. From Eq. (3.116) minimisation of the MAP risk function is achieved by maximisation of the posterior pmf as

$$\hat{\theta}_{MAP}(y) = \underset{\theta_i}{\mathrm{argmax}} \; P_{\Theta|Y}(\theta_i|y)$$
$$= \underset{\theta_i}{\mathrm{argmax}} \; P_{\Theta}(\theta_i) f_{Y|\Theta}(y|\theta_i) \tag{3.117}$$

3.5.3 Maximum Likelihood Classification

ML classification corresponds to Bayesian classification when the parameter θ has a uniform prior pmf and the cost function is also uniform as

$$\mathcal{R}_{ML}(\theta_i|y) = \sum_{j=1}^{M} [1 - \delta(\theta_i, \theta_j)] \frac{1}{f_Y(y)} f_{Y|\Theta}(y|\theta_j) P_{\Theta}(\theta_j)$$
$$= 1 - \frac{1}{f_Y(y)} f_{Y|\Theta}(y|\theta_i) P_{\Theta} \tag{3.118}$$

where P_Θ is the uniform pmf of θ. Minimisation of the ML risk function (3.118) is equivalent to maximisation of the likelihood $f_{Y|\Theta}(y|\theta_i)$

$$\hat{\theta}_{ML}(y) = \underset{\theta_i}{\mathrm{argmax}} \ f_{Y|\Theta}(y|\theta_i) \qquad (3.119)$$

3.5.4   Minimum Mean Squared Error Classification

The Bayesian minimum mean squared error classification results from minimisation of the following risk function :

$$\mathcal{R}_{MSE}(\theta_i|y) = \sum_{j=1}^{M} |\theta_i - \theta_j|^2 \ P_{\Theta|Y}(\theta_j|y) \qquad (3.120)$$

For the case when $P_{\Theta|Y}(\theta_j|y)$ is not available, the MMSE classifier is given by

$$\hat{\theta}_{MMSE}(y) = \underset{\theta_i}{\mathrm{argmin}} \ |\theta_i - \theta(y)|^2 \qquad (3.121)$$

where $\theta(y)$ is an estimate based on the observation y.

3.5.5   Bayesian Classification of Finite State Processes

In this section the classification problem is formulated within the framework of a finite state random process. A finite state process is composed of a probabilistic chain of a number of different random processes. Finite state processes are used for modelling nonstationary signals such as speech, image, and impulsive noise.
Consider a process with a set of M states denoted as $S=\{s_1, s_2, \ldots, s_M\}$, where each state has some distinct statistical property. In its simplest form, a state is just a single vector, and the finite state process is equivalent to a discrete-valued random process with M outcomes. In this case the Bayesian state estimation is identical to the Bayesian classification of a signal into one of M discrete-valued vectors. More generally a state generates continuous-valued, or discrete-valued, vectors from a pdf, or a pmf, associated with the state. Figure 13.16 illustrates an M-state process, where the output of the ith state is expressed as

$$x(m) = h_i(\theta_i, e(m)) \qquad\qquad i= 1, \ldots, M \quad (3.122)$$

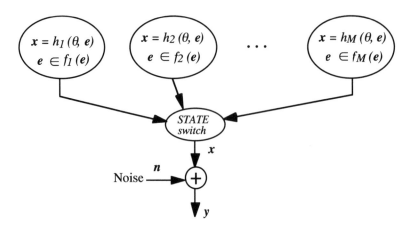

Figure 3.16 Illustration of a random process generated by a finite state system.

where in each state the signal $x(m)$ is modelled as the output of a state-dependent function $h_i()$ with parameter θ_i, input $e(m)$ and an input pdf $f_{Ei}(e(m))$. The prior probability of each state is given by

$$P_S(s_i) = \mathcal{E}[N(s_i)] \Big/ \mathcal{E}\left[\sum_{j=1}^{M} N(s_j)\right] \tag{3.123}$$

where $E[N(s_i)]$ is the expected number of observation from state s_i. The pdf of the output of a finite state process is a weighted combination of the pdf of each state and is given by

$$f_X(x(m)) = \sum_{i=1}^{M} P_S(s_i) f_{X|S}(x|s_i) \tag{3.124}$$

In Figure 3.16, the noisy observation $y(m)$ is the sum of the process output $x(m)$ and an additive noise $n(m)$. From Bayes rule the posterior probability of the state s_i given the observation $y(m)$ can be expressed as

$$
P_{S|Y}(s_i|y(m)) = \frac{f_{Y|S}(y(m)|s_i)P_S(s_i)}{f_Y(y(m))}
$$
$$
= \frac{f_{Y|S}(y(m)|s_i)P_S(s_i)}{\sum_{j=1}^{M} f_{Y|S}(y(m)|s_j)P_S(s_j)} \tag{3.125}
$$

In MAP classification, the state with the maximum posterior probability is selected as

$$s_{MAP}(y(m)) = \underset{s_i}{\operatorname{argmax}} \; P_{S|Y}(s_i| y(m)) \qquad (3.126)$$

The general Bayesian state classifier, in addition to using the posterior probability, also assigns a misclassification cost function $C(s_i|s_j)$ to the action of selecting the state s_i when the true state is s_j. The risk function for the Bayesian classification is given by

$$\mathcal{R}(s_i|y(m)) = \sum_{j=1}^{M} C(s_i|s_j)P_{S|Y}(s_j|y(m)) \qquad (3.127)$$

3.5.6 Bayesian Estimation of the Most Likely State Sequence

Consider the estimation of the most likely state sequence $s = [s_{i_0}, s_{i_1}, \ldots, s_{i_{T-1}}]$ of a finite state process, given a sequence of T observation vectors $Y = [y_0, y_1, \ldots, y_{T-1}]$. A state sequence s, of length T, is itself a random integer-valued vector process with N^T possible values. From the Bayes rule, the posterior pmf of a state sequence s, given an observation sequence Y, can be expressed as

$$P_{S|Y}(s_{i_0}, \ldots, s_{i_{T-1}}|y_0, \ldots, y_{T-1}) = \frac{f_{Y|S}(y_0, \ldots, y_{T-1}|s_{i_0}, \ldots, s_{i_{T-1}}) \, P_S(s_{i_0}, \ldots, s_{i_{T-1}})}{f_Y(y_0, \ldots, y_{T-1})}$$

$$(3.128)$$

where $P_S(s)$ is the pmf of the state sequence s, and for a given observation sequence, the denominator $f_Y(y_0, \ldots, y_{T-1})$ is a constant. The Bayesian risk in selecting a state sequence s_i is expressed as

$$\mathcal{R}(s_i|y) = \sum_{j=1}^{N^T} C(s_i|s_j)P_{S|Y}(s_j|y) \qquad (1.129)$$

For a statistically independent process the state of the process at any time is independent of the previous states, and hence the conditional probability of a state sequence can be written as

$$P_{S|Y}(s_{i_0}, \ldots, s_{i_{T-1}}|y_0, \ldots, y_{T-1}) = \prod_{k=0}^{T-1} f_{Y|S}(y_k|s_{i_k})P_S(s_{i_k}) \qquad (3.130)$$

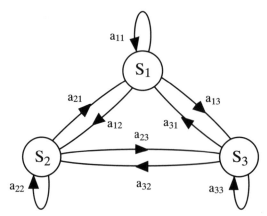

Figure 3.17 A three state Markov Process.

where s_{ik} denotes state s_i at time instant k. A particular case of a finite state process is the Markov chain where the state transition is governed by a Markovian process such that the probability of the state i at time m depends on the state of the process at time m-1. The conditional pmf of a Markov state sequence can be expressed as

$$P_{S|Y}(s_{i_0},\ldots,s_{i_{T-1}}|y_0,\ldots,y_{T-1}) = \prod_{k=0}^{T-1} a_{i_{k-1}i_k} f_{S|Y}(s_{i_k}|y_k) \qquad (3.131)$$

where $a_{i_{k-1}i_k}$ is the probability that the process moves from state $s_{i_{k-1}}$ to state s_{i_k} Finite state random processes and computationally efficient methods of state sequence estimation are described in detail in Chapter 4.

3.6 Modelling the Space of a Random Signal

In this section we consider the training of statistical models for a data base of P-dimensional vectors from a random process. The vectors in the data base can be visualised as forming a number of clusters or regions in a P-dimensional space. The statistical modelling method consists of two steps : (a) the partitioning of the data base into a number of regions, or clusters, and (b) the estimation of the parameters of a statistical model for each cluster. A simple method for modelling the space of a random signal is to use a set of prototype vectors that represent the centroids of the signal space. This method effectively quantises the space of a random process into a relatively small number of typical vectors, and is known as *vector quantisation* (VQ).

In the following we first consider a VQ model of a random process, and then extend this model to a pdf model, based on a mixture of Gaussian densities.

3.6.1 Vector Quantisation of a Random Process

In vector quantisation, the space of a random vector process X is partitioned into K clusters or regions $[X_1, X_2, ..., X_K]$, and each cluster X_i is represented by a cluster centroid c_i. The set of centroid vectors $[c_1, c_2, ..., c_K]$ form a VQ codebook model of the process X. The VQ codebook can then be used to classify an unlabelled vector x with the nearest centroid. The codebook is searched to find the centroid vector with the minimum distance from x, then x is labelled with the index of the minimum distance centroid as

$$Label(x) = \arg\min_i \, d(x, c_i) \qquad (3.132)$$

Where $d(x, c_i)$ is a measure of distance between the vectors x and c_i. The most commonly used distance measure is the mean squared distance.

3.6.2 Design of a Vector Quantiser : K-Means Algorithm

The K-means algorithm, illustrated in Figure 3.18, is an iterative method for the design of a VQ codebook. Each iteration consists of two basic steps : (a) Partition the training signal space into K regions or clusters and (b) compute the centroid of each region. The steps in K-Means method are as follows :

Step-1 : Initialisation : Use a suitable method to choose a set of K initial centroids $[c_i]$. For $m = 1, 2, \ldots$

Step-2 : Classification : Classify the training vectors $\{x\}$ into K clusters $\{[x_1], [x_2], \ldots [x_K]\}$ using the so called nearest-neighbour rule that is Eq. (3.132).

Step-3 : Centroid computation : Use the vectors $[x_i]$ associated with the ith cluster to compute an updated cluster centroid c_i, and calculate the cluster distortion defined as

$$D_i(m) = \frac{1}{N_i} \sum_{j=1}^{N_i} d(x_i(j), c_i(m)) \qquad (3.133)$$

where it is assumed that a set of N_i vectors $[x_i(j) \, j=0, ..., N_i]$ are associated with cluster i. The total distortion is given by

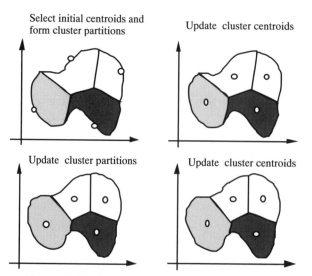

Figure 3.18 Illustration of K-means clustering method.

$$D(m) = \sum_{i=1}^{K} D_i(m) \qquad (3.134)$$

Step 4 : Convergence test : if

$$D(m-1) - D(m) \geq \textit{Threshold} \qquad (3.135)$$

stop, else goto step-2.

A vector quantiser models the regions, or the clusters, of the signal space with a set of cluster centroids. A more complete description of the signal space can be achieved by modelling each cluster with a Gaussian density as described in the next section.

3.6.3 Design of a Mixture Gaussian Model

A practical method for modelling the probability density function of an arbitrary data space is to use a mixture of a number of Gaussian densities. Figure 3.19 illustrates the modelling of a two-dimensional signal space with a number of circular and elliptically shaped Gaussian processes. Note that the Gaussian densities can be overlapping, with the result that in an area of overlap a data point can be associated with various probabilities to different components of the Gaussian mixture.

A main advantage of the use of a mixture Gaussian model is that it results in mathematically tractable signal processing solutions. A mixture Gaussian pdf model for a process X is defined as

$$f_X(x) = \sum_{k=1}^{K} P_k \mathcal{N}_k(x;\mu_k, \Sigma_k) \tag{3.136}$$

where $\mathcal{N}_k(x;\mu_k, \Sigma_k)$ denotes the kth component of the mixture Gaussian pdf, with the mean vector μ_k and the covariance matrix Σ_k. The parameter P_k is the prior probability of the kth mixture, and it can be interpreted as the expected fraction of the number of vectors from the process X that is associated with the kth mixture.

In general there are an infinite number of different K-mixture Gaussian densities that can be used to "tile up" a signal space. Hence the modelling of a signal space with a K-mixture pdf space can be regarded as a many-to-one, non-invertible, mapping, and the expectation-maximisation (EM) method can be applied to estimation of the parameters of the Gaussian pdf models.

3.6.4 The EM Algorithm for Estimation of Mixture Gaussian Densities

The EM algorithm, discussed in Section 3.4, is an iterative maximum-likelihood estimation method, and can be employed to calculate the parameters of a K-mixture Gaussian pdf model for a given data set. To apply the EM we need to define the so called complete and incomplete data sets. As usual the observation vectors $[y(m)$ $m=0, ..., N-1]$ form the incomplete data. The complete data may be viewed as the observation vectors with a *label* attached to each vector $y(m)$ to indicate the component of the mixture Gaussian that generated the vector. Note that if each vector $y(m)$ had a mixture component label attached, then the computation of the mean vector and the covariance matrix of each component of the mixture would be a relatively simple exercise. Therefore the complete and incomplete data can be defined as

The incomplete data $\quad y(m)\ \ m = 0, ..., N-1$
The complete data $\quad x(m)=[y(m),k]=y_k(m)\quad m = 0, ..., N-1,\ k \in (1, ..., K)$

The probability of the complete data is the probability that an observation vector $y(m)$ has a label k. The main step in the application of the EM method is to define the expectation of the complete data, given the observations and a current estimate of the parameter vector, as

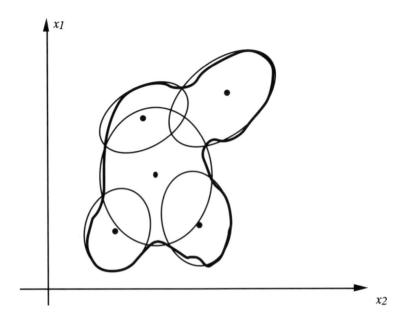

Figure 3.19 Illustration of probabilistic modelling of a two dimensional signal space with a mixture of five bi-variate Gaussian densities.

$$U(\Theta, \hat{\Theta}_i) = \mathcal{E}[\ln f_{X,\Theta}(y(m), k; \Theta)|y(m); \hat{\Theta}_i]$$

$$= \sum_{m=0}^{N-1} \sum_{k=1}^{K} \frac{f_{Y,K|\Theta}\big(y(m), k|\hat{\Theta}_i\big)}{f_{Y|\Theta}\big(y(m)|\hat{\Theta}_i\big)} \ln\big(f_{Y,K;\Theta}(y(m), k; \Theta)\big) \tag{3.137}$$

Where $\Theta=\{\theta_k=[P_k,\ \mu_k,\ \Sigma_k]\ k=1,...,\ K\}$, are the parameters of the Gaussian mixture as in Eq. (3.136). Now the joint pdf of $y(m)$ and the kth Gaussian component of the mixture density can be written as

$$f_{Y,K|\Theta}\big(y(m),\ k\ |\hat{\Theta}_i\big) = P_{k_i} f_k\big(y(m)|\hat{\theta}_{k_i}\big)$$

$$= P_{k_i} \mathcal{N}_k\big(y(m); \hat{\mu}_{k_i},\ \hat{\Sigma}_{k_i}\big) \tag{3.138}$$

where $\mathcal{N}_k(y(m); \mu_k,\ \Sigma_k)$ is a Gaussian density of mean vector μ_k and covariance matrix Σ_k given as

$$\mathcal{N}_k(y(m);\mu_k, \Sigma_k) = \frac{1}{(2\pi)^{P/2}|\Sigma_k|^{1/2}} \exp\left(-\frac{1}{2}(y(m)-\mu_k)^T \Sigma_k^{-1}(y(m)-\mu_k)\right) \quad (3.139)$$

The pdf of $y(m)$ and the mixture of K Gaussian densities is given by

$$f_{Y|\Theta}\left(y(m)|\hat{\Theta}_i\right) = \mathcal{N}\left(y(m)|\hat{\Theta}_i\right)$$
$$= \sum_{k=1}^{K} \hat{P}_{k_i} \mathcal{N}_k\left(y(m);\hat{\mu}_{k_i}, \hat{\Sigma}_{k_i}\right) \quad (3.140)$$

Substitution of the Gaussian densities of Eq. (3.138) and Eq. (3.140) in Eq. (3.137) yields

$$U[(\mu, \Sigma, P),(\hat{\mu}_i, \hat{\Sigma}_i, \hat{P}_i)] = \sum_{m=0}^{N-1} \sum_{k=1}^{K} \frac{\hat{P}_{k_i} \mathcal{N}_k(y(m);\hat{\mu}_{k_i}, \hat{\Sigma}_{k_i})}{\mathcal{N}\left(y(m)|\hat{\Theta}_i\right)} \ln[P_k \mathcal{N}_k(y(m);\mu_k, \Sigma_k)]$$

$$= \sum_{m=0}^{N-1} \sum_{k=1}^{K} \left(\frac{\hat{P}_{k_i} \mathcal{N}_k(y(m);\hat{\mu}_{k_i}, \hat{\Sigma}_{k_i})}{\mathcal{N}\left(y(m)|\hat{\Theta}_i\right)} \ln P_k + \frac{\hat{P}_{k_i} \mathcal{N}_k(y(m);\hat{\mu}_{k_i}, \hat{\Sigma}_{k_i})}{\mathcal{N}\left(y(m)|\hat{\Theta}_i\right)} \ln \mathcal{N}_k(y_k;\mu_k, \Sigma_k)\right)$$
$$(3.141)$$

Eq. (3.141) is maximised with respect to the parameter P_k using the constrained optimisation method. This involves subtracting the constant term $\Sigma P_k=1$ form the right hand side of Eq. (3.141) and then setting the derivative of this equation with respect to P_k to zero, this process yields

$$\hat{P}_{k_{i+1}} = \underset{P_k}{\arg\max}\ U[(\mu, \Sigma, P),(\hat{\mu}_i, \hat{\Sigma}_i, \hat{P}_i)]$$
$$= \frac{1}{N} \sum_{m=0}^{N-1} \frac{\hat{P}_{k_i} \mathcal{N}_k(y(m);\hat{\mu}_{k_i}, \hat{\Sigma}_{k_i})}{\mathcal{N}\left(y(m)|\hat{\Theta}_i\right)} \quad (3.142)$$

The parameters μ_k, Σ_k that maximise the function $U()$ are obtained, by setting the derivative of the function with respect to these parameters to zero, as

$$\hat{\mu}_{k_{i+1}} = \arg\max_{\mu_k} \; U[(\mu, \Sigma, P), (\hat{\mu}_i, \hat{\Sigma}_i, \hat{P}_i)]$$

$$= \frac{\displaystyle\sum_{m=0}^{N-1} \frac{\hat{P}_{k_i} \mathcal{N}_k(y(m); \hat{\mu}_{k_i}, \hat{\Sigma}_{k_i})}{\mathcal{N}\left(y(m)|\hat{\Theta}_i\right)} \, y(m)}{\displaystyle\sum_{m=0}^{N-1} \frac{\hat{P}_{k_i} \mathcal{N}_k(y(m); \hat{\mu}_{k_i}, \hat{\Sigma}_{k_i})}{\mathcal{N}\left(y(m)|\hat{\Theta}_i\right)}} \qquad (3.143)$$

and

$$\hat{\Sigma}_{k_{i+1}} = \arg\max_{\Sigma_k} \; U[(\mu, \Sigma, P), (\hat{\mu}_i, \hat{\Sigma}_i, \hat{P}_i)]$$

$$= \frac{\displaystyle\sum_{m=0}^{N-1} \frac{\hat{P}_{k_i} \mathcal{N}_k(y(m); \hat{\mu}_{k_i}, \hat{\Sigma}_{k_i})}{\mathcal{N}\left(y(m)|\hat{\Theta}_i\right)} (y(m) - \hat{\mu}_{k_i})(y(m) - \hat{\mu}_{k_i})^T}{\displaystyle\sum_{m=0}^{N-1} \frac{\hat{P}_{k_i} \mathcal{N}_k(y(m); \hat{\mu}_{k_i}, \hat{\Sigma}_{k_i})}{\mathcal{N}\left(y(m)|\hat{\Theta}_i\right)}} \qquad (3.144)$$

Equations (3.142-144) are estimates of the parameters of a mixture Gaussian pdf model. These equations can be used in further iterations of the EM until the parameter estimates converge.

Summary

This chapter began with an introduction to the basic concepts in estimation theory; such as the signal space and the parameter space, the prior and posterior spaces, and the statistical measures that are used to quantify the performance of an estimator. The Bayesian inference method, with its ability to include as much information as is available, provides a general framework for statistical signal processing problems. The minimum mean squared error, the maximum likelihood, the maximum a posterior, and the minimum absolute value of error methods were derived from the Bayesian formulation. Further examples of the applications of Bayesian type models in this book include the hidden Markov models for nonstationary processes studied in Chapter 4, and blind equalisation of distorted signals studied in Chapter 14.

We considered a number of examples for estimation of a signal observed in noise, and derived the expressions for the effects of using prior pdfs on the mean and the variance of the estimates. The choice of the prior pdf is an important consideration in Bayesian estimation. Many processes, for example speech or the response of a telecommunication channel, are not uniformly distributed in space, but are constrained to a particular region of the signal or the parameter space. The use of a prior pdf can guide the estimator to focus on the posterior space which is the subspace consistent with both the likelihood and the prior pdfs. The choice of the prior, depending on how well it fits the process, can have a significant influence on the solutions.

The iterative estimate-maximise method, studied in Section 3.3, provides a practical framework for solving many statistical signal processing problems, such as the modelling of a signal space with a mixture Gaussian densities, and the training of hidden Markov models in Chapter 4. In Section 3.4 the Cramer-Rao lower bound on the variance of an estimator was derived, and it was shown that the use of a prior pdf can reduce the minimum estimator variance.

In Section 3.5 the Bayesian statistical classification was considered within the general frame work of a finite state process. The finite state structure is developed further in the next chapter on hidden Markov models. In Section 3.6 we considered the vector quantisation method for modelling a random signal space. Finally we considered the modelling of a data space with a mixture Gaussian process, and used the EM method to derive a solution for the parameters of the mixture Gaussian model.

Bibliography

ANDERGERG M.R. (1973), Cluster Analysis for Applications, Academic Press, New York.

ABRAMSON, N. (1963), Information Theory and Coding, McGraw Hill, New York.

BAUM L.E., PETRIE T., SOULES G. WEISS N. (1970), A Maximisation Technique occurring in the Statistical Analysis of Probabilistic Functions of Markov Chains, Ann. Math. Stat. Vol. 41, pages 164-171.

BAYES T. (1763) An Essay Towards Solving a Problem in the Doctrine of Changes, Philosophical Transactions of the Royal Society of London, Vol. 53, pages 370-418. and reprinted in 1958 in Biometrika Vol.45, pages 293-315.

CHOU P. LOOKABAUGH T. and GRAY R. (1989), Entropy-Constrained Vector Quantisation, IEEE Trans. Acoustics, Speech and Signal Processing, Vol. ASSP-37, pages 31-42.

BEZDEK J.C. (1981), Pattern Recognition with Fuzzy Objective Function Algorithms, Plenum Press, New York.

CRAMER H. (1974), Mathematical Methods of Statistics. Princeton University Press, New Jersey.

DEUTSCH R. (1965), Estimation Theory, Prentice-Hall, Englewood Cliffs, N. J.

DEMPSTER A.P., LAIRD N.M., RUBIN D.B. (1977), Maximum Likelihood from Incomplete Data via the EM Algorithm, J. Royal Stat. Soc. ser. B Vol. 39, Pages 1-38.

DUDA R.O. HART R.E. (1973), Pattern Classification, Wiley, New York.

FEDER M., WEINSTEIN E. (1988), Parameter Estimation of Superimposed Signals using the EM algorithm, IEEE Trans. Acoustics, Speech and Signal Processing, Vol. ASSP-36(4), pages 477.489.

FISHER R. A. (1922), On the Mathematical Foundations of the Theoretical Statistics, Phil Trans. Royal. Soc. London, Vol. 222, Pages 309-368.

GERSHO A. (1982), On the Structure of Vector Quantisers, IEEE Trans. Information Theory, Vol. IT-28, pages 157-166.

GRAY R.M. (1984), Vector Quantisation, IEEE ASSP Magazine pages 4-29.

GRAY R.M., KARNIN E.D, Multiple local Optima in Vector Quantisers, IEEE Trans. Information Theory, Vol. IT-28, pages 256-261.

JEFFREY H. (1961), Scientific Inference, 3rd ed. Cambridge University Press.

LARSON H.J and BRUNO O.S. (1979), Probabilistic Models in Engineering Sciences, Vol. I and II. Wiley, New York.

LINDE Y., BUZO A., GRAY R.M. (1980), An Algorithm for Vector Quantiser Design, IEEE Trans. Comm. Vol. COM-28, pages 84-95.

MAKHOUL J., ROUCOS S., GISH H. (1985), Vector Quantisation in Speech Coding, Proc. IEEE, Vol. 73, pages 1551-1588.

MOHANTY N. (1986), Random Signals, Estimation and Identification, Van Nostrand, New York.

RAO C.R. (1945), Information and Accuracy Attainable in the Estimation of Statistical Parameters, Bull Calcutta Math. Soc., Vol. 37, Pages 81-91.

RENDER R.A., WALKER H.F.(1984), Mixture Densities, Maximum Likelihood and the EM algorithm, SIAM review, Vol. 26, Pages 195-239.

SCHARF L.L. (1991), Statistical Signal Processing : Detection, Estimation, and Time Series Analysis, Addison Wesley, Reading, Mass.

4

Hidden Markov Models

4.1 Statistical Models for Nonstationary Processes
4.2 Hidden Markov Models
4.3 Training Hidden Markov Models
4.4 Decoding of Signalsusing Hidden Markov Models
4.5 HMM-based Estimation of Signals in Noise

Hidden Markov models (HMMs), are used for statistical modelling of nonstationary stochastic processes such as speech and time-varying noise. An HMM models the time-variations of the statistics of a random process, with a Markovian chain of state-dependent stationary sub-processes. An HMM is essentially a Bayesian finite state process, with a Markovian prior for modelling the transitions between the states, and a set of state pdfs for the modelling of the random variations of the stochastic process within each state.

This chapter begins with a brief introduction to continuous and finite-state nonstationary models, before concentrating on the theory and applications of hidden Markov models. We study the Baum-Welch method for the maximum likelihood training of the parameters of an HMM, and then consider the use of HMMs and the Viterbi decoding algorithm for the classification and decoding of an unlabelled observation sequence. Finally the application of HMMs in signal enhancement is considered.

4.1 Statistical Models for Nonstationary Processes

A nonstationary process can be defined as one whose statistical parameters vary over time. Most "naturally generated" stochastic signals, such as audio signals, biomedical signals, seismic signals *etc.,* are nonstationary, in that the parameters of the systems that generate the signals, and the environments in which the signals propagate, change with time. A nonstationary process can be modelled as a double-layered stochastic process, with a hidden process that controls the time-variations of the statistics of an observable process, as illustrated in Figure 4.1. In general, a nonstationary process can be classified into two broad categories of : *continuously variable state* and *finite-state* processes. A continuously variable state process is defined as one whose underlying statistics vary in a continuous manner with time. Examples of this class of random processes are audio signals such as speech and music whose loudness and spectral composition vary continuously with time. A finite state process is one whose statistical characteristics can *switch* between a finite number of, usually stationary, states. For example, impulsive noise is a binary-state nonstationary process.

Figure 4.2(a) illustrates a nonstationary first order autoregressive (AR) process. This process is modelled as the combination of a *hidden*, stationary AR model of the signal parameters, and an observable time-varying AR model of the signal. The hidden model controls the time variations of the parameters of the nonstationary AR model. For this model, the observation signal equation, and the parameter state equation, can be expressed as

$$x(m) = a(m)x(m-1) + e(m) \tag{4.1}$$

$$a(m) = \beta\, a(m-1) + \varepsilon(m) \tag{4.2}$$

where $a(m)$ is the time-varying coefficient of the observable AR process, and β is the coefficient of the hidden state-control process.

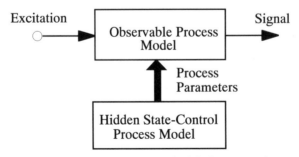

Figure 4.1 Illustration of a two-layered model of a nonstationary process.

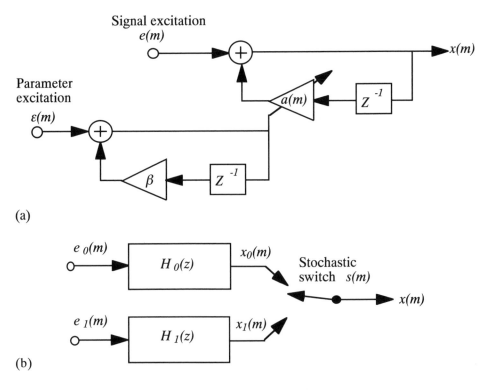

Figure 4.2 (a) A continuously variable state AR process, (b) a binary-state AR process.

A simple example of a finite state nonstationary model, is the binary state autoregressive process illustrated in Figure 4.2.b, where at each time instant a random switch selects one of the two models for connection to the output terminal. For this model the output signal $x(m)$ can be expressed as

$$x(m) = \bar{s}(m)x_0(m) + s(m)x_1(m) \tag{4.3}$$

where the binary switch $s(m)$ selects the state of the process at time m, and $\bar{s}(m)$ denotes the Boolean complement of $s(m)$.

In practical modelling applications, continuously variable state processes, such as speech and noise, are approximated by finite state models. This is mainly due to the existence of a well developed and powerful method for finite state models, namely the hidden Markov model which is the main subject of this chapter.

In the remainder of this chapter, for compatibility with the general literature on HMMs, the letter t is used to denote discrete time index.

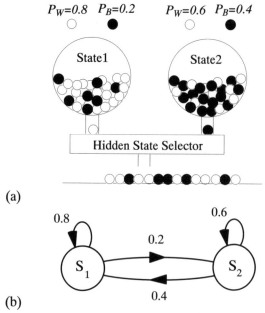

Figure 4.3 (a) Illustration of a two-layered random process, (b) an HMM model of the process in (a).

4.2 Hidden Markov Models

A hidden Markov model (HMM) is a double-layered finite state stochastic process, with a hidden Markovian process that controls the selection of the states of an observable process. As a simple illustration of a binary-state Markovian process, consider Figure 4.3, which shows two containers of different mixtures of black and white balls. The probability of the black and the white balls in each container, denoted as P_B and P_W respectively, is as shown in the Figure 4.3. Assume that at successive time intervals a hidden selection process selects one of the two containers to release a ball. The balls released are replaced so that the mixture density of the black and the white balls in each container remains unaffected. Each container can be considered as an underlying state of the output process. Now assume that the hidden, container-selection, process is governed by the following rule : at any time if the output from the currently selected container is a white ball then the same container is selected to output the next ball, otherwise the other container is selected. This is an example of a Markovian process because the next state of the process depends on the current state as shown in the binary state model of Figure 4.4(b). Note that in this example the outcome does not unambiguously signify the state, because both states are capable of releasing black and white balls.

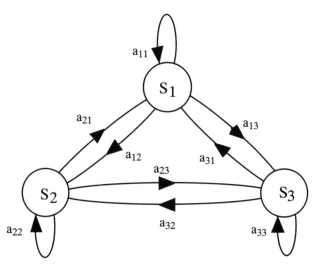

Figure 4.4 A three-state ergodic HMM structure

In general a hidden Markov model has N sates, with each state trained to model a distinct segment of a stochastic random process. A hidden Markov model can be used to model a time-varying random process as a probabilistic Markovian chain of N stationary, or quasi-stationary, elementary sub-processes. A general form of a three-state HMM is shown in Figure 4.4. This structure is known as an *ergodic* HMM. In the context of an HMM the term ergodic implies that there is no structural constraints for connecting any state to any other state.

A more constrained form of an HMM is the left-right model of Figure 4.5, so called because the allowed state transitions are from a left state to a right state. The left-right constraint is useful for characterisation of temporal or sequential structures of stochastic signals such as speech signals, because time may be visualised as having a direction from left to right.

4.2.1   A Physical Interpretation of Hidden Markov Models

For a physical interpretation of the use of HMMs in stochastic signal processing consider the illustration of Figure 4.5 which shows a left-right HMM of a spoken letter "C". In general, there are two main types of variation in speech and other stochastic signals : these are, variations in the spectral composition, and variations in articulation rate. In a hidden Markov model these variations are modelled by the state observation and the state transition probabilities. A useful way of interpreting and using HMMs is to consider each state of an HMM as a model of a segment of a stochastic process. For example in Figure 4.5, the state S_1 models the first segment

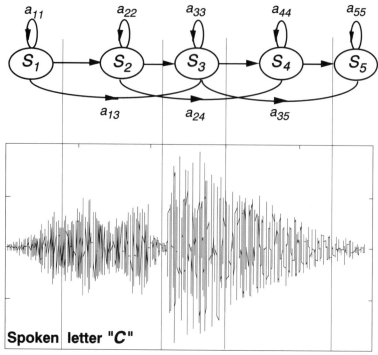

Figure 4.5 A 5-state left-right HMM speech model.

of the spoken letter "C", state S_2 models the second segment and so on. Each state must have a mechanism to accommodate the variations in different realisations of the segments that it models. The state transition probabilities provide a mechanism for connection of various states, and for modelling the variations in the duration and time-scales of the signals in each state. If a segment of an utterance is elongated, due to say slow articulation, then this can be accommodated by more self-loop transitions into the state that models the segment. Conversely, if a segment of a word is omitted, due to say fast speaking, then the skip-next-state connection accommodates that situation. The state observation pdfs model the distributions of the spectral composition of the speech segments associated with the state.

4.2.2 Hidden Markov Model As a Bayesian Method

A hidden Markov model M is a Bayesian structure, with a Markovian state transition probability, and a state observation likelihood which can be either a discrete pmf or a continuous pdf. The *posterior* pmf of a state sequence s of a model M, given an observation sequence X, can be expressed using the Bayes rule as the product of a state *prior* and an observation *likelihood* as

$$P_{S|X,\mathcal{M}}(s|X,\mathcal{M}) = \frac{1}{f_X(X)} P_{S|\mathcal{M}}(s|\mathcal{M}) f_{X|S,\mathcal{M}}(X|s,\mathcal{M}) \qquad (4.4)$$

and the posterior probability that an observation sequence X was generated by the model M is summed over all likely state sequences, and may also be weighted by the model prior $P_{\mathcal{M}}(\mathcal{M})$, as

$$P_{\mathcal{M}|X}(\mathcal{M}|X) = \frac{1}{f_X(X)} \underbrace{P_{\mathcal{M}}(\mathcal{M})}_{Model\ Prior} \sum_s \underbrace{P_{S|\mathcal{M}}(s|\mathcal{M})}_{State\ Prior} \underbrace{f_{X|S,\mathcal{M}}(X|s,\mathcal{M})}_{Observation\ Likelihood} \qquad (4.5)$$

The Markovian state transition prior can be used to model, often with considerable success, the time variations and the sequential dependency of most nonstationary processes. However, for many applications, such as speech recognition, the state observation likelihood has more influence on the posterior probability than the state transition prior.

4.2.3   Parameters of a Hidden Markov Model

A hidden Markov model has the following parameters :
Number of states N. This is usually set to the number of distinct, or elementary, stochastic events in a signal process. For example, in modelling a binary-state process such as impulsive noise N is set to two, and in isolated-word speech modelling N is set between 5 to 10.

State transition-probability matrix $A=\{a_{ij},\ i,j=1,\ ...\ N\}$ provides a Markovian connection network between the states, and models the variations in the duration of the signals associated with each state. For a left-right HMM, see Figure 4.5, $a_{ij}=0$ for $i>j$ and hence the transition matrix A is upper-triangular.

State observation vectors $\{\mu_{i1},\ \mu_{i2},\ ...,\ \mu_{iM},\ i=1,\ ...,\ N\}..$ For each state a set of M prototype vectors model the centroids of the signal space associated with each state.

State observation vector probability models. This can be either a discrete model composed of the M prototype vectors and their associated pmf $P=\{P_{ij}()\ i=1,\ ...\ N,\ j=1,\ ...\ M\}$, or it may be a continuous pdf model $F=\{f_{ij}(),\ i=1,\ ...\ N,\ j=1,\ ...\ M\}$.

Initial state probability vector $\pi=[\pi_1,\ \pi_2,\ ...,\ \pi_N]$.

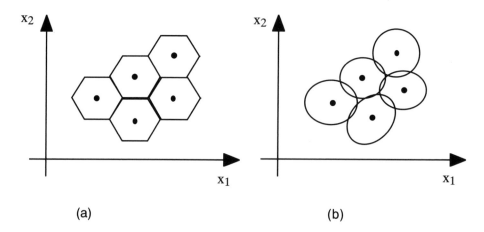

Figure 4.6 Modelling a random signal space using : (a) a discrete-valued pmf, (b) a continuous-valued mixture Gaussian density.

4.2.4   State Observation Models

Depending on whether a signal process is discrete-valued or continuous-valued, the state observation model for the process can be a discrete-valued probability mass function (pmf), or a continuous-valued probability density function (pdf). The discrete models can also be used for the modelling of the space of a continuous-valued process quantised into a number of discrete points.

First, we consider a discrete state observation density model. Assume that associated with the ith state of an HMM there are M discrete vectors $[\mu_{i1}, ..., \mu_{iM}]$ with a pmf of $[P_{i1}, ..., P_{iM}]$. These centroid vectors and their probabilities are normally obtained through clustering of a set of training signals associated with each state.

For the modelling of a continuous-valued process, the signal space associated with each state is partitioned into a number of clusters as in Figure 4.6. If the signals within each cluster are modelled by a uniform distribution, then each cluster is described by the centroid vector and the cluster probability, and the state observation model consists of M cluster centroids and the associated pmf $\{\mu_{ik}, P_{ik}, i=1, ..., N, k=1, ..., M\}$. In effect, this results in a discrete state observation HMM for a continuous-valued process. Figure 4.6.a shows a partitioning, and quantisation, of a signal space into a number of centroids.

Now if each cluster of the state observation space is modelled by a continuous pdf, such as a Gaussian pdf, then a continuous density HMM results. The most widely used state observation pdf for an HMM is the mixture Gaussian density defined as

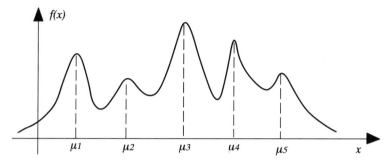

Figure 4.7 A mixture Gaussian probability density function.

$$f_{X|S}(x|s = i) = \sum_{k=1}^{M} P_{ik}\, \mathcal{N}(x, \mu_{ik}, \Sigma_{ik})\tag{4.6}$$

where $\mathcal{N}(x, \mu_{ik}, \Sigma_{ik})$ is a Gaussian density of mean vector μ_{ik} and variance matrix Σ_{ik}, and P_{ik} is a weighting factor for the kth Gaussian pdf of the state i. Note that P_{ik} is the prior probability of the kth mode of the mixture pdf for the state i. Figure 4.6.b shows the space of a mixture Gaussian model of an observation signal space. A 5-mode mixture Gaussian pdf is shown in Figure 4.7.

4.2.5   State Transition Probabilities

The first order Markovian property of an HMM entails that the transition probability to any state $s(t)$ at time t depends only on $s(t-1)$ the state of the process at time $t-1$, and is independent of the previous states of the HMM. This can be expressed as

$$Prob(s(t) = j\,|\,s(t-1) = i, s(t-2) = k, \ldots, s(t-N) = l) = Prob(s(t) = j\,|\,s(t-1) = i)$$
$$= a_{ij}$$
$$\tag{4.7}$$

where $s(t)$ denotes the state of HMM at time t. The transition probabilities provide a probabilistic mechanism for connecting the states of an HMM, and for modelling the variations in the duration of the signals associated with each state. The probability of occupancy of a state i for d consecutive time units, $P_i(d)$, can be expressed in terms of the state self-loop transition probability a_{ii} as

$$P_i(d) = a_{ii}^{d-1}(1 - a_{ii}) \tag{4.8}$$

Note from Eq. (4.8) that the probability of state occupancy for a duration of d time units decays exponentially with d. From Eq. (4.8) the mean occupancy for each state of an HMM can be derived as

$$Mean\ occupancy\ of\ state\ i = \sum_{d=0}^{\infty} d\,P_i(d) = \frac{1}{1 - a_{ii}} \tag{4.9}$$

4.2.6   State-Time Trellis Diagram

A state-time diagram shows the HMM states together with all the different paths that can be taken through various states as time unfolds. Figures 4.8.a and 4.8.b illustrate a 4-state HMM and its state-time diagram. Since the number of states and the state parameters of an HMM are time-invariant, a state-time diagram is a repetitive and regular trellis structure. Note that in Figure 4.8 for a left-right HMM the state-time trellis has to diverge from the first state and converge into the last state. In general, there are many different state sequences that start from the initial state and end in the

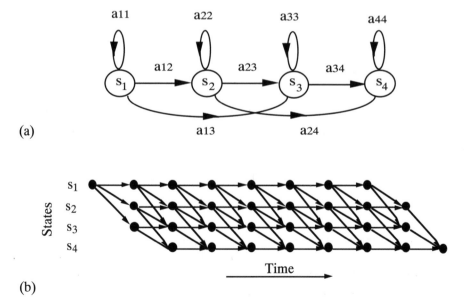

Figure 4.8 (a) A 4-state left-right HMM, and (b) its state-time trellis diagram.

final state. Each state sequence has a prior probability that can be obtained by multiplication of the state transition probabilities of the sequence For example the probability of the state sequence $s = [S_1, S_1, S_2, S_2, S_3, S_3, S_4,]$ is $P(s) = a_{11} \cdot a_{12} \cdot a_{22} \cdot a_{23} \cdot a_{33} \cdot a_{34}$. Since each state has a different set of prototype observation vectors, different state sequences model different observation sequences. In general an N-state HMM can reproduce N^T different realisations of the random process that it is trained to model.

4.3 Training Hidden Markov Models

The first step in training the parameters of an HMM, is to collect a data-base of a sufficiently large number of different examples of the process to be modelled. Assume that the examples in a training data-base consist of L vector-valued sequences $[X]=[X_k, k=0, ..., L–1]$ with each sequence $X_k=[x(t), t=0, ..., T_k–1]$ having a variable number of vectors. The objective is to train the parameters of an HMM to model the signals in the training data set. In a probabilistic sense, the fitness of a model is measured by the posterior probability $P_{\mathcal{M}|X}(\mathcal{M}|X)$ of the model M given the training data X. The training process aims to maximise the posterior probability of the model M and the training data $[X]$ which can be expressed, using the Bayes rule, as

$$P_{\mathcal{M}|X}(\mathcal{M}|X) = \frac{1}{f_X(X)} f_{X|\mathcal{M}}(X|\mathcal{M}) P_{\mathcal{M}}(\mathcal{M}) \qquad (4.10)$$

where the denominator $f_X(X)$ on the right hand side of Eq. (4.10) has only a normalising effect and $P_M(M)$ is the prior probability of the model M. For a given training data set$[X]$ and a given model M, maximisation of Eq. (4.10) is equivalent to maximising the likelihood function $f_{X|\mathcal{M}}(X|\mathcal{M})$. The likelihood of an observation vector sequence X given a model M can be expressed as

$$f_{X|\mathcal{M}}(X|\mathcal{M}) = \sum_s f_{X|S,\mathcal{M}}(X|s,\mathcal{M}) P_{s|\mathcal{M}}(s|\mathcal{M}) \qquad (4.11)$$

where $f_{X|S,\mathcal{M}}(X|s,\mathcal{M})$, the pdf of the signal sequence X along the state sequence $s = [s(0), s(1), ..., s(T-1)]$ of the model M, is given by

$$f_{X|S,\mathcal{M}}(X|s,\mathcal{M}) = f_{X|S}(x(0)|s(0)) f_{X|S}(x(1)|s(1)) ... f_{X|S}(x(T-1)|s(T-1)) \qquad (4.12)$$

where $s(t)$, the state at time t, can be one of N states, and $f_{X|S}(x(t)|s(t))$ (a short hand for $f_{X|S,\mathcal{M}}(x(t)|s(t),\mathcal{M}))$ is the pdf of $x(t)$ given the state $s(t)$ of the model $\mathbf{M}$. The Markovian probability of the state sequence s is given by

$$P_{S|\mathcal{M}}(s|\mathcal{M}) = \pi_{s(0)} a_{s(0)s(1)} a_{s(1)s(2)} \cdots a_{s(T-2)s(T-1)} \tag{4.13}$$

Substituting Eqs. (4.12) and (4.13) in Eq. (4.11) yields

$$f_{X|\mathcal{M}}(X|\mathcal{M}) = \sum_s f_{X|S,\mathcal{M}}(X|s,\mathcal{M}) P_{s|\mathcal{M}}(s|\mathcal{M})$$

$$= \sum_s \pi_{s(0)} f_{X|S}(x(0)|s(0)) \, a_{s(0)s(1)} f_{X|S}(x(1)|s(1)) \cdots a_{s(T-2)s(T-1)} f_{X|S}(x(T-1)|s(T-1))$$

$$\tag{4.14}$$

where the summation is taken over all state sequences s. In the training process, the transition probabilities and the parameters of the observation pdfs must be estimated to maximise the model likelihood of Eq. (4.14). Direct maximisation of Eq. (4.14) with respect to the model parameters is a nontrivial task. Furthermore, for an observation sequence of length T vectors, the computational load of Eq. (4.14) is in the order of $O(N^T)$. This is an impractically large load, even for such modest values as $N=6$ and $T=30$. However, the repetitive structure of the trellis state time diagram of an HMM implies that there is a large amount of repeated computation in Eq. (4.14) that can be avoided in an efficient implementation. In the next section we consider the forward-backward method of model likelihood calculation, and then proceed to describe an iterative maximum likelihood model optimisation method.

4.3.1 Forward-Backward Probability Computation

An efficient recursive algorithm for the computation of the likelihood function $f_{X|\mathcal{M}}(X|\mathcal{M})$ is the so called forward-backward algorithm. The forward-backward computation method exploits the highly regular structure of the state-time trellis diagram of Figure 4.8.

In this method, a forward probability variable, $\alpha_t(i)$ is defined as the joint probability of the partial observation sequence $X=[x(0), x(1), ..., x(t)]$ and the state i at time t, of the model $\mathbf{M}$, as

$$\alpha_t(i) = f_{X,S|\mathcal{M}}(x(0),x(1), \cdots, x(t), \; s(t) = i \,|\mathcal{M}) \tag{4.15}$$

The forward probability variable $\alpha_t(i)$ of Eq. (4.15) can be expressed in a recursive equation in terms the forward probabilities at time $t-1$, $\alpha_{t-1}(i)$, as

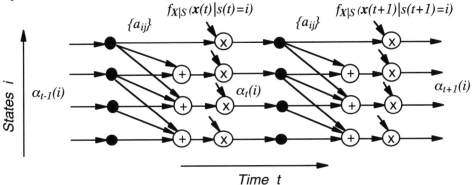

Figure 4.9 A network for computation of forward probabilities for a left-right HMM.

$$\alpha_t(i) = f_{X,S|\mathcal{M}}(x(0),x(1),\ldots,x(t),\ s(t)=i|\mathcal{M})$$

$$= \left(\sum_{j=1}^{N} f_{X,S|\mathcal{M}}(x(0),x(1),\ldots,x(t-1),\ s(t-1)=j|\mathcal{M})\ a_{ji}\right) f_{X|S,\mathcal{M}}(x(t)|s(t)=i,\mathcal{M})$$

$$= \sum_{j=1}^{N}\left(\alpha_{t-1}(j)\ a_{ji}\right) f_{X|S,\mathcal{M}}(x(t)|s(t)=i,\mathcal{M})$$

$$(4.16)$$

Figure 4.9 illustrates, a network for computation of the forward probabilities for the 4-state left-right HMM of Figure 4.8. The likelihood of an observation sequence $X=[x(0), x(1), \ldots, x(T-1)]$ given a model M can be expressed in terms of the forward probabilities as

$$f_{X|\mathcal{M}}(x(0),x(1),\ldots,x(T-1)|\mathcal{M}) = \sum_{i=1}^{N} f_{X,S|\mathcal{M}}(x(0),x(1),\ldots,x(T-1),\ s(T-1)=i|\mathcal{M})$$

$$= \sum_{i=1}^{N}\alpha_{T-1}(i)$$

$$(4.17)$$

Similar to the definition of the forward probability concept, a backward probability is defined as the probability of the state i at time t followed by the partial observation sequence $[x(t+1), x(t+2), \ldots, x(T-1)]$ as

$$\beta_t(i) = f_{X,S|\mathcal{M}}(s(t) = i, x(t+1), x(t+2), \ldots, x(T-1)|\mathcal{M})$$

$$= \sum_{j=1}^{N} a_{ij} f_{X,S|\mathcal{M}}(s(t+1) = j, x(t+2), x(t+3), \ldots, x(T-1)) f_{X|S}(x(t+1)|s(t+1) = j, \mathcal{M})$$

$$= \sum_{j=1}^{N} a_{ij} \beta_{t+1}(j) f_{X|S,\mathcal{M}}(x(t+1)|s(t+1) = j, \mathcal{M})$$

$$(4.18)$$

In the next section, forward and backward probabilities are used for the training of HMM parameters.

4.3.2   Baum-Welch Model Re-Estimation

The HMM training problem is the estimation of the model parameters $M=(\pi, A, F)$ for a given data set. These parameters are the initial state probabilities π, the state transition probability matrix A, and the continuous or the discrete density state observation pdfs. The HMM parameters are estimated from a set of training examples $\{X=[x(0), \ldots, x(T-1)]\}$, with the objective of maximising $f_{X|\mathcal{M}}(X|\mathcal{M})$ the likelihood of the model and the training data. The Baum-Welch method of training HMMs is an iterative likelihood maximisation method based on the forward-backward probabilities defined in the preceding Section. The Baum-Welch method is an instance of the EM algorithm described in Chapter 3. For an HMM M, the posterior probability of a transition at time t from state i to state j of the model M, given an observation sequence X, can be expressed as

$$\gamma_t(i, j) = P_{S|X,\mathcal{M}}(s(t) = i, s(t+1) = j|X, \mathcal{M})$$

$$= \frac{f_{S,X|\mathcal{M}}(s(t) = i, s(t+1) = j, X|\mathcal{M})}{f_{X|\mathcal{M}}(X|\mathcal{M})}$$

$$= \frac{\alpha_t(i) a_{ij} f_{X|S}(x(t+1)|s(t+1) = j, \mathcal{M}) \beta_{t+1}(j)}{\sum_{i=1}^{N} \alpha_{T-1}(i)}$$

$$(4.19)$$

where $f_{S,X|\mathcal{M}}(s(t) = i, s(t+1) = j, X|\mathcal{M})$ is the joint pdf of the states $s(t)$ and $s(t+1)$ and the observation sequence X, and $f_{X|S}(x(t+1)|s(t+1) = i)$ is the state observation pdf for the state i. Note that for a discrete observation density HMM the state observation pdf in Eq. (4.19) is replaced with the discrete state observation pmf

$P_{X|S}(x(t+1)|s(t+1) = i)$. The posterior probability of state i at time t given the model M and the observation X is

$$
\begin{aligned}
\gamma_t(i) &= P_{S|X,\mathcal{M}}(s(t) = i | X, \mathcal{M}) \\
&= \frac{f_{S,X|\mathcal{M}}(s(t) = i, X|\mathcal{M})}{f_{X|\mathcal{M}}(X|\mathcal{M})} \\
&= \frac{\alpha_t(i)\beta_t(i)}{\sum\limits_{j=1}^{N} \alpha_{T-1}(j)}
\end{aligned}
\tag{4.20}
$$

Now the state transition probability a_{ij} can be interpreted as

$$
a_{ij} = \frac{Expected\ number\ of\ transitions\ from\ state\ i\ to\ state\ j}{Expected\ number\ of\ transitions\ from\ state\ i}
\tag{4.21}
$$

From Equations (4.19), (4.20) and (4.21) the state transition probability can be re-estimated as the ratio

$$
\bar{a}_{ij} = \frac{\sum\limits_{t=0}^{T-2} \gamma_t(i,j)}{\sum\limits_{t=0}^{T-2} \gamma_t(i)}
\tag{4.22}
$$

Note that for an observation sequence $[x(0), ..., x(T-1)]$ of length T the last transition occurs at time $T-2$ as indicated in the upper limits of the summations in Eq. (4.22). The initial state probabilities are estimated as

$$
\bar{\pi}_i = \gamma_0(i)
\tag{4.23}
$$

4.3.3   Training Discrete Observation Density HMMs

In a discrete density HMM the observation signal space of each state is modelled with a set of discrete symbols or vectors. Assume that a set of M vectors $[\mu_{i1}, \mu_{i2}, ..., \mu_{iM}]$ model the space of the signal associated with the ith state. These vectors may be obtained from a clustering process; as the centorids of the clusters of the training signals associated with each state. The objective in training discrete density HMMs is to compute the state transition probabilities and the state observation probabilities.

The forward-backward equation for discrete density HMMs are the same as those for continuous HMMs, derived in the previous sections, with the difference that the probability density functions such as $f_{X|S}(x(t)|s(t) = i)$ are substituted with probability mass functions $P_{X|S}(x(t)|s(t) = i)$ defined as

$$P_{X|S}(x(t)|s(t) = i) = P_{X|S}(Q[x(t)]|s(t) = i) \tag{4.24}$$

where the function $Q[x(t)]$ quantises the observation vector $x(t)$ to the nearest discrete vector in the set $[\mu_{i1}, \mu_{i2}, ..., \mu_{iM}]$. For discrete density HMMs, the probability of a state vector μ_{ik} can be defined as the ratio of the number of occurrences of μ_{ik} (or vectors quantised to μ_{ik}) in the state i, divided by the total number of occurrences of all other vectors in state i as

$$\bar{P}_{ik}(\mu_{ik}) = \frac{Expected\ number\ of\ times\ in\ state\ i\ and\ observing\ \mu_{ik}}{Expected\ number\ of\ times\ in\ state\ i}$$

$$= \frac{\displaystyle\sum_{\substack{t \in x(t) \to \mu_{ik}}}^{T-1} \gamma_t(i)}{\displaystyle\sum_{t=0}^{T-1} \gamma_t(i)} \tag{4.25}$$

In Eq. (4.25) the summation in the numerator is taken over those time instances t where the kth symbol μ_{ik} is observed in the state i.

For statistically reliable results an HMM must be trained on a large data set X consisting of a sufficient number of independent realisations of the process to be modelled. Assume that the training data set consists of L realisation $X=[X(0), X(1), ..., X(L-1)]$ where $X(k)=[x(0), x(1), ..., x(T_k-1)]$. The re-estimation formula can be averaged over the entire data set as

$$\hat{\pi}_i = \frac{1}{L} \sum_{l=0}^{L-1} \gamma_0^l(i) \tag{4.26}$$

$$\hat{a}_{ij} = \frac{\displaystyle\sum_{l=0}^{L-1} \sum_{t=0}^{T_l-2} \gamma_t^l(i,j)}{\displaystyle\sum_{l=0}^{L-1} \sum_{t=0}^{T_l-2} \gamma_t^l(i)} \tag{4.27}$$

and

$$\hat{P}_i(\mu_{ik}) = \frac{\sum_{l=0}^{L-1} \sum_{\substack{t=0 \\ t \in x(t) \to \mu_{ik}}}^{T_l-1} \gamma_t^l(i)}{\sum_{l=0}^{L-1} \sum_{t=0}^{T_l-1} \gamma_t^l(i)} \tag{4.28}$$

The parameter estimates of Eqs. (4.26) to (4.28) can be used in further iteration of the estimation process until the model converges.

4.3.4 HMMs with Continuous Observation PDFs

In continuous density HMMs, continuous probability density functions (pdfs) are used to model the observation signals associated with each state. Baum *et al.* generalised the parameter re-estimation method to HMMs with concave continuous pdfs such a Gaussian pdf. A continuous P-variate Gaussian pdf for the state i of an HMM can be defined as

$$f_{X|S}(x(t)|s(t) = i) = \frac{1}{(2\pi)^{P/2}|\Sigma_i|^{1/2}} \exp\left((x(t) - \mu_i)^T \Sigma_i^{-1}(x(t) - \mu_i)\right) \tag{4.29}$$

where μ_i and Σ_i are the mean vector and the covariance matrix associated with the state i. The re-estimation formula for the mean vector of the state Gaussian pdf, can be derived as

$$\overline{\mu}_i = \frac{\sum_{t=0}^{T-1} \gamma_t(i)x(t)}{\sum_{t=0}^{T-1} \gamma_t(i)} \tag{4.30}$$

Similarly the covariance matrix is estimated as

$$\overline{\Sigma}_i = \frac{\sum_{t=0}^{T-1} \gamma_t(i)(x(t) - \overline{\mu}_i)(x(t) - \overline{\mu}_i)^T}{\sum_{t=0}^{T-1} \gamma_t(i)} \tag{4.31}$$

The proof that the Baum-Welch re-estimation algorithm leads to maximisation of the likelihood function $f(X|M)$ can be found in Baum.

4.3.5 HMMs with Mixture Gaussian pdfs

The modelling of the space of a signal process with a mixture of Gaussian pdfs is considered in Section 3.6.3. In HMMs with mixture-Gaussian pdfs for state observation, the signal space associated with the ith state is modelled with a mixtures of M Gaussian densities as

$$f_{X|S}(x(t)|s(t) = i) = \sum_{k=1}^{M} P_{ik} \, \mathcal{N}\left(x(t), \mu_{ik}, \Sigma_{ik}\right) \qquad (4.32)$$

where P_{ik} is the prior probability of the kth component of the mixture and $\sum_{k=1}^{M} P_{ik} = 1$.
The posterior probability of state i at time t and state j at time $t+1$ of the model M, given an observation sequence $X = [x(0), ..., x(T-1)]$, can be expressed as

$$\begin{aligned}
\gamma_t(i,j) &= P_{S|X,\mathcal{M}}\left(s(t) = i, s(t+1) = j | X, \mathcal{M}\right) \\
&= \frac{\alpha_t(i) a_{ij} \left[\sum_{k=1}^{M} P_{jk} \, \mathcal{N}\left(x(t+1), \mu_{jk}, \Sigma_{jk}\right) \right] \beta_{t+1}(j)}{\sum_{i=1}^{N} \alpha_{T-1}(i)}
\end{aligned} \qquad (4.33)$$

and the posterior probability of state i at time t given the model M and the observation X is

$$\begin{aligned}
\gamma_t(i) &= P_{S|X,\mathcal{M}}\left(s(t) = i | X, \mathcal{M}\right) \\
&= \frac{\alpha_t(i) \beta_t(i)}{\sum_{j=1}^{N} \alpha_{T-1}(j)}
\end{aligned} \qquad (4.34)$$

Now we define the posterior of the state i and the kth mixture of state i at time t as

$$\begin{aligned}
\zeta_t(i,k) &= P_{S,K|X,\mathcal{M}}\left(s(t) = i, m(t) = k | X, \mathcal{M}\right) \\
&= \frac{\sum_{j=1}^{N} \alpha_{t-1}(j) a_{ji} P_{ik} \, \mathcal{N}\left(x(t), \mu_{ik}, \Sigma_{ik}\right) \beta_t(i)}{\sum_{j=1}^{N} \alpha_{T-1}(j)}
\end{aligned} \qquad (4.35)$$

where $m(t)$ is the Gaussian mixture component at time t. Eqs (4.33) to (4.35) are used to derive the re-estimation formula for the mixture coefficients, the mean vectors, and the covariance matrices of the state mixture Gaussian pdfs as

$$\overline{P}_{ik} = \frac{Expected\ number\ of\ times\ in\ state\ i\ and\ observing\ mixture\ k}{Expected\ number\ of\ times\ in\ state\ i}$$

$$= \frac{\displaystyle\sum_{t=0}^{T-1} \xi_t(i,k)}{\displaystyle\sum_{t=0}^{T-1} \gamma_t(i)} \tag{4.36}$$

and

$$\overline{\mu}_{ik} = \frac{\displaystyle\sum_{t=0}^{T-1} \xi_t(i,k)x(t)}{\displaystyle\sum_{t=0}^{T-1} \xi_t(i,k)} \tag{4.37}$$

Note that in Eq. (4.37) the numerator is the probabilistically-weighted mean of the observation sequence X. Similarly the covariance matrix is estimated as

$$\overline{\Sigma}_{ik} = \frac{\displaystyle\sum_{t=0}^{T-1} \xi_t(i,k)\big(x(t) - \overline{\mu}_{ik}\big)\big(x(t) - \overline{\mu}_{ik}\big)^T}{\displaystyle\sum_{t=0}^{T-1} \xi_t(i,k)} \tag{4.38}$$

4.4   Decoding of Signals Using Hidden Markov Models

Hidden Markov models are used in such applications as speech recognition, signal restoration, and the decoding of the underlying states of a signal. For example, in speech recognition HMMs are trained to model the statistical variations of the acoustic realisations of the words of a vocabulary of say size V. In the recognition phase an utterance is classified and labelled an with the most likely of the V candidate HMMs as illustrated in Figure 4.10. In Chapter 11, on impulsive noise, a binary state HMM is used to model the noise process.

Consider the problem of decoding an unlabelled sequence of T signal vectors $X=[x(0), x(1), ..., X(T-1)]$, given a set of V candidate HMMs $[M_1,..., M_V]$. The probability score for the observation sequence X and the model M_k can be calculated as the likelihood :

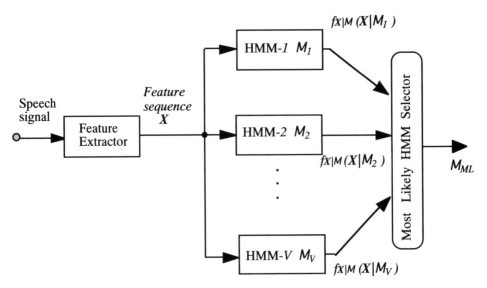

Figure 4.10 Illustration of the use of HMMs in speech recognition.

$$f_{X|\mathcal{M}}(X|\mathcal{M}_k) = \sum_s \pi_{s(0)} f_{X|S}(x(0)|s(0)) a_{s(0)s(1)} f_{X|S}(x(1)|s(1)) \cdots a_{s(T-2)s(T-1)} f_{X|S}(x(T-1)|s(T-1))$$

$$(4.39)$$

where the likelihood of the observation sequence X is summed over all possible state sequences. Eq. (4.39) can be efficiently calculated using the forward-backward method described in Section 4.3.1. The observation sequence X, is labelled with the HMM that scores the highest likelihood as

$$Label(X) = \arg\max_k \left(f_{X|\mathcal{M}}(X|\mathcal{M}_k) \right) \qquad k=1, ..., V \qquad (4.40)$$

In decoding applications often the likelihood of an observation sequence X and a model M_k, is obtained along the *single* most likely state sequence of model M_k, instead of being summed over all sequences, so Eq. (4.40) becomes

$$Label(X) = \arg\max_k \left(\max_s f_{X,S|\mathcal{M}}(X,s|\mathcal{M}_k) \right) \qquad (4.41)$$

In Section 4.5, on the use of HMMs for noise reduction, the most likely state sequence is used to obtain the maximum likelihood estimate of the underlying statistics of the signal process.

4.4.1   Viterbi Decoding Algorithm

In this section we consider the decoding of a signal to obtain the maximum a posterior (MAP) estimate of the underlying state sequence. The MAP state sequence s^{MAP} of a model M given an observation sequence $X=[x(0), ..., x(T-1)]$ is obtained as

$$
\begin{aligned}
s^{MAP} &= \arg\max_{s} f_{X,S|M}(X,s|M) \\
&= \arg\max_{s} \left(f_{X|S,M}(X|s, M) P_{S|M}(s|M) \right)
\end{aligned}
\tag{4.42}
$$

The MAP state sequence is used in such applications as : the calculation of a similarity score between an observation sequence and an HMM, segmentation of a nonstationary signal into a number of distinct quasi-stationary segments, and implementation of state-based Wiener filters for restoration of noisy signals described in the next Section.

For an N-state HMM and an observation sequence of length T, there are altogether N^T state sequences. Even for moderate values of N and T say ($N=6$ and $T=30$) an exhaustive search of the state-time trellis for the best state sequence is a computationally prohibitive exercise. The Viterbi algorithm is an efficient method for the estimation of the most likely state sequence of an HMM. In a state-time trellis diagram, such as Figure 4.8, the number of path diverging from each state of a trellis can grow exponentially by a factor of N at successive time instants. The Viterbi method prunes the trellis by selecting the most likely path to each state. At each time instant t, for each state i, the algorithm selects the most probable path to state i and prunes out the less likely branches. This procedure ensures that at any time instant only one single path *survives* into each state of the trellis. For each time instant t and

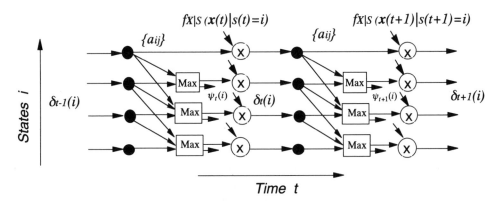

Figure 4.11 A network illustration of the Viterbi algorithm.

for each state i, the algorithm keeps a record of the state j from which the maximum likelihood path branched into i, and also records the cumulative probability of the most likely path into state i at time t.. The Viterbi algorithm follows.

Viterbi Algorithm

$\delta_t(i)$ records the cumulative probability of the best path to state i at time t.
$\psi_t(i)$ records the best state sequence to state i at time t.

Step 1 Initialisation At time $t=0$, for states $i=1$ to N
$$\delta_0(i) = \pi_i f_i(x(0))$$
$$\psi_0(i) = 0$$

Step 2 Recursive calculation of the ML state sequences and their
    probabilities
    For time $t = 1$ to $T-1$
        For states $i = 1$ to N
$$\delta_t(i) = \max_j\left(\delta_{t-1}(j)a_{ji}\right)f_i(x(t))$$
$$\psi_t(i) = \arg\max_j\left(\delta_{t-1}(j)a_{ji}\right)$$

Step 3 Termination, Retrieve the most likely final state
$$s^{MAP}(T-1) = \arg\max_i\left(\delta_{T-1}(i)\right)$$
$$Prob_{max} = \max_i\left(\delta_{T-1}(i)\right)$$

Step 4 Backtracking through the most likely state sequence
    For $t = T-2$ to 0
$$s^{MAP}(t) = \psi_{t+1}\left(s^{MAP}(t+1)\right).$$

The back tracking routine retrieves the most likely state sequence of the model M. Note that the variable $Prob_{max}$ which is the probability of the observation sequence $X=[x(0), ..., x(T-1)]$ and the most likely state sequence of the model M, can be used as the probability score for the model M and the observation X. For example in speech recognition, for each candidate word model the probability of the observation and the most likely state sequence is calculated, and then the observation is labelled with the word that achieves the highest probability score.

4.5 HMM-based Estimation of Signals in Noise

In this section we consider the use of hidden Markov models for estimation of a signal $x(t)$ observed in an additive noise $n(t)$, and modelled as

$$y(t) = x(t) + n(t) \tag{4.43}$$

From the Bayes rule, the posterior pdf of the signal $x(t)$ given the noisy observation $y(t)$ is defined as

$$
\begin{aligned}
f_{X|Y}(x(t)|y(t)) &= \frac{f_{Y|X}(y(t)|x(t)) \, f_X(x(t))}{f_Y(y(t))} \\
&= \frac{1}{f_Y(y(t))} f_N(y(t) - x(t)) \, f_X(x(t))
\end{aligned}
\tag{4.44}
$$

For a given observation $f_Y(y(t))$ is a constant, and the maximum a posterior (MAP) estimate is obtained as

$$\hat{x}^{MAP}(t) = \underset{x(t)}{\arg\max} \, f_N(y(t) - x(t)) \, f_X(x(t)) \tag{4.45}$$

From Eqs (4.44) and (4.45), the computation of the posterior pdf, or the MAP estimate, requires the pdf models of the signal and the noise processes. Stationary, continuous-valued, processes are often modelled by a Gaussian or a mixture Gaussian pdf which is equivalent to a single state HMM. For a nonstationary process an N-state HMM can model the time-varying pdf of the process as a Markovian chain of N stationary Gaussian pdfs. Now assume that we have an N_s-state HMM M for the signal, and another N_n-state HMM η for the noise. For signal estimation, we need estimates of the underlying state sequences of the signal and the noise processes. For an observation sequence of length T there are N_s^T possible signal state sequences and N_n^T possible noise state sequences that could have generated the noisy signal. Given an observation sequence $Y = [y(0), y(1), ..., y(T-1)]$, the most probable state sequences of the signal and the noise HMMs can be expressed as

$$s_{signal}^{MAP} = \underset{s_{signal}}{\arg\max} \left(\underset{s_{noise}}{\max} \, f_Y\!\left(Y, s_{signal}, s_{noise} | M, \eta\right) \right) \tag{4.46}$$

and

$$s_{noise}^{MAP} = \underset{s_{noise}}{\arg\max} \left(\underset{s_{signal}}{\max} \, f_Y\!\left(Y, s_{signal}, s_{noise} | M, \eta\right) \right) \tag{4.47}$$

Given the state sequence estimates for the signal and the noise models, the MAP estimation Eq. (4.45) becomes

$$\hat{x}^{MAP}(t) = \underset{x}{\arg\max}\left(f_{N|S,\eta}\left(y(t) - x(t)\middle| s_{noise}^{MAP}, \eta\right) f_{X|S,\mathcal{M}}\left(x(t)\middle| s_{signal}^{MAP}, \mathcal{M}\right)\right) \quad (4.48)$$

Example 4.1 Assume a signal, modelled by a binary-state HMM, is observed in an additive stationary Gaussian noise. Let the noisy observation be modelled as

$$y(t) = \bar{s}(t)x_0(t) + s(t)x_1(t) + n(t) \quad (4.49)$$

where $s(t)$ is a hidden binary state process such that : $s(t) = 0$ indicates that the signal is from the state S_0 with a Gaussian pdf of $\mathcal{N}\left(x(t), \mu_{x_0}, \Sigma_{x_0 x_0}\right)$, and $s(t) = 1$ indicates that the signal is from the state S_1 with a Gaussian pdf of $\mathcal{N}\left(x(t), \mu_{x_1}, \Sigma_{x_1 x_1}\right)$. Assume the noise can also be modelled by a stationary Gaussian process $\mathcal{N}\left(n(t), \mu_n, \Sigma_{nn}\right)$ which is equivalent to a single-state HMM. Using the Viterbi algorithm the most likely state sequence of the signal model can be estimated as

$$s_{signal}^{MAP} = \underset{s}{\arg\max}\left(f_{Y|S,\mathcal{M}}(Y|s,\mathcal{M}) P_{S|\mathcal{M}}(s|\mathcal{M})\right) \quad (4.50)$$

For a Gaussian distributed signal and additive Gaussian noise, the observation pdf of the noisy signal is also Gaussian. Hence, the state observation pdfs of the signal model can be modified to account for the additive noise as

$$f_{Y|s_0}\left(y(t)\middle| s_0\right) = \mathcal{N}\left(y(t), (\mu_{x_0} + \mu_n), (\Sigma_{x_0 x_0} + \Sigma_{nn})\right) \quad (4.51)$$

and

$$f_{Y|s_1}\left(y(t)\middle| s_1\right) = \mathcal{N}\left(y(t), (\mu_{x_1} + \mu_n), (\Sigma_{x_1 x_1} + \Sigma_{nn})\right) \quad (4.52)$$

where $\mathcal{N}(y(t), \mu, \Sigma)$ denotes a Gaussian pdf of mean vector μ and covariance matrix Σ. The MAP signal estimate, given a state sequence estimate s^{MAP}, is obtained from

$$\hat{x}^{MAP}(t) = \underset{x}{\arg\max}\left(f_{X|S,\mathcal{M}}\left(x(t)\middle| s^{MAP}, \mathcal{M}\right) f_N(y(t) - x(t))\right) \quad (4.53)$$

Substitution of the Gaussian pdf of the signal from the most likely state sequence, and the pdf of noise, in Eq. (4.53) results in the following MAP estimate

$$\hat{x}^{MAP}(t) = \left(\Sigma_{xx,s(t)} + \Sigma_{nn}\right)^{-1}\Sigma_{xx,s(t)}\left(y(t) - \mu_n\right) + \left(\Sigma_{xx,s(t)} + \Sigma_{nn}\right)^{-1}\Sigma_{nn}\,\mu_{x,s(t)} \quad (4.54)$$

where $\mu_{x,s(t)}$ and $\Sigma_{xx,s(t)}$ are the mean and variance of $x(t)$ obtained from the most likely state sequence $[s(t)]$.

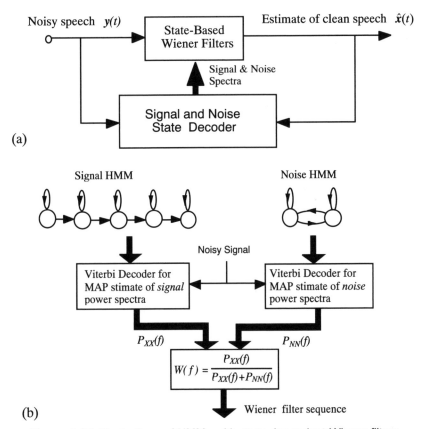

Figure 4.12 Illustrations of HMMs with state-dependent Wiener filters.

4.5.1 HMM-based Wiener Filters

The least mean squared error Wiener filter is derived in Chapter 5. For a stationary signal $x(t)$, observed in an additive noise $n(t)$, the Wiener filter equations in the time and the frequency domains are derived as :

$$w = [R_{xx} + R_{nn}]^{-1}r_{xx} \tag{4.55}$$

$$W(f) = \frac{P_{XX}(f)}{P_{XX}(f) + P_{NN}(f)} \tag{4.56}$$

Where R_{xx}, r_{xx} and $P_{XX}(f)$ denote the autocorrelation matrix, the autocorrelation vector and the power spectrum functions respectively. The implementation of the Wiener filter Eq. (4.56) requires the signal and the noise power spectra. The power spectral variables may be obtained from the states of the HMMs trained to model the power spectra of the signal and the noise. Figure 4.12.a illustrates an HMM-based state-dependent Wiener filters. To implement the state-dependent Wiener filter, we need an estimate of the state sequences for the signal and the noise. In practice, for signals such as speech there are a number of HMMs; one HMM per word, phoneme, or any other acoustic unit of the signal. In such cases it is necessary to classify the signal, so that the state-based Wiener filters are derived from the most likely HMM. Furthermore the noise process can also be modelled by an HMM. Assuming that there are V HMMs $\{M_1, ..., M_V\}$ for the signal process, and one HMM for the noise, the state-based Wiener filter can be implemented as follows :

Step 1 Using the noisy signal, obtain the maximum likelihood state sequence estimates for each signal HMM, and for the noise HMM.

Step 2 Using the signal and noise power spectra from the most likely state sequence estimates, derive a series of state-dependent Wiener filters.

Step 3 For each model use the Wiener filter sequence to filter the noisy signal.

Step 4 Obtain a probability score for the filtered signal and its respective model, and finally select the signal and the model with the highest probability score.

Figure 4.12.b is a further illustration of the use of HMMs for implementation of state dependent Wiener filters. As shown the power spectral variables required for implementation of the Wiener filter sequence are obtained from the most likely state sequences of the signal and the noise models.

4.5.2   Modelling Noise Characteristics

The implicit assumption in using a hidden Markov model for noise is that noise statistics can be modelled by a Markovian chain of a number of different stationary processes. A stationary noise process can be modelled by a single-state HMM. A single state HMM may also be adequate for a slowly varying noise process such as the helicopter noise, shown in Figure 4.13.a. For a nonstationary noise, a multi-state HMM can model the time variations of the noise process with a finite number of quasi-stationary states. In general, the number of states required to accurately model the noise depends on the nonstationary character of the noise.

An example of a non-stationary noise is the impulsive noise of Figure 4.13.b. Figure 4.14 shows a two-state HMM of the impulsive noise sequence : state S_0 models the "*off*" periods in-between the impulses, and state S_1 models an impulse. In cases where each impulse has a well defined temporal structure, it may be beneficial to use a multi-state HMM to model the pulse itself. HMMs are used in Chapter 11 for modelling impulsive noise, and in Chapter 14 for channel equalisation.

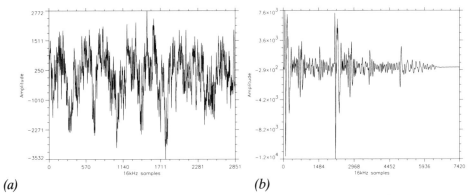

(a)   *(b)*

Figure 4.13 Examples of a helicopter noise, and impulsive noise

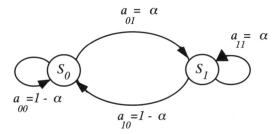

Figure 4.14 A binary-state model of an impulsive noise process.

Summary

HMMs provide a powerful method for the modelling of nonstationary processes such as speech, noise, and time-varying channels. An HMM is a Bayesian finite-state process, with a Markovian state prior, and a state likelihood function which can be either a discrete density model or a continuous Gaussian pdf model. The Markovian prior models the time evolution of a nonstationary process with a chain of stationary sub-processes. The state observation likelihood models the space of the process within each state of the HMM.

In Section 4.3 we studied the Baum-Welch method for the training of the parameters of an HMM to model a given data set, and derived the forward-backward method for efficient calculation of the likelihood of an HMM given an observation signal. In Section 4.4 we considered the use of HMMs in signal classification and in the decoding of the underlying state sequence of a signal. The Viterbi algorithm is a computationally efficient method for estimation of the most likely sequence of an HMM. Given an unlabelled observation signal, the decoding of the underlying state sequence, and the labelling of the observation with one of number of candidate HMMs, is accomplished using the Viterbi method. In Section 4.5 we considered the use of HMMs for MAP estimation of a signal observed in noise, and considered the use of HMMs in implementation of state-based Wiener filter sequence.

Bibliography

BAHL L.R., BROWN P.F., de SOUZA P.V., MERCER R.L. (1986), Maximum Mutual Information Estimation of Hidden Markov Model Parameters for Speech Recognition, IEEE Proc. Acoustics, Speech and Signal Processing, ICASSP-86 Tokyo, Pages 40-43.

BAHL L.R., BROWN P.F. de SOUZA P.V., MERCER R.L. (1989), Speech Recognition with Continuous Parameter Hidden Markov Models, IEEE Proc. Acoustics, Speech and Signal Processing, ICASSP-88 New York Pages 40-43.

BAHL L.R., JELINEK F, MERCER R.L. (1983), A Maximum Likelihood Approach to Continuous Speech Recognition, IEEE Trans. on Pattern Analysis and Machine Intelligence, Vol. 5, Pages 179-190.

BAUM L.E., EAGON J.E. (1967), An Inequality with Applications to Statistical Estimation for Probabilistic Functions of a Markov Process and to Models for Ecology, Bull. AMS, Vol. 73 Pages 360-363.

BAUM L.E., PETRIE T. (1966), Statistical Inference for Probabilistic Functions of Finite State Markov Chains. Ann. Math. Stat. 37, Pages 1554-1563.

BAUM L.E., PETRIE T., SOULES G., WEISS N.(1970), A Maximisation Technique Occurring in the Statistical Analysis of Probabilistic Functions of Markov Chains, Ann. Math. Stat. ,Vol. 41, Pages 164-171.

CONNER P. N. (1993), Hidden Markov Model with Improved Observation and Duration Modelling, Ph.D. Thesis, University of East Anglia, England.

EPHARAIM Y., MALAH D., JUANG B.H.(1989), On Application of Hidden Markov Models for Enhancing Noisy Speech., IEEE Trans. Acoustics Speech and Signal Processing, Vol. 37(12), Pages 1846-1856, Dec.

FORNEY G.D. (1973), The Viterbi Algorithm, Proc. IEEE, Vol. 61, Pages 268-278.

GALES M.J.F., YOUNG S.J. (1992), An Improved Approach to the Hidden Markov Model Decomposition of Speech and Noise, in Proc. IEEE Int. Conf. on Acoustics., Speech, Signal Processing, ICASSP-92, Pages 233-236.

GALES M.J.F., YOUNG S.J. (1993), HMM Recognition in Noise using Parallel Model Combination, EUROSPEECH-93, Pages 837-840.

HUANG X.D., ARIKI Y., JACK M.A. (1990), Hidden Markov Models for Speech Recognition, Edinburgh University Press, Edinburgh.

HUANG X.D., JACK M.A. (1989), Unified Techniques for Vector Quantisation and Hidden Markov Modelling using Semi-Continuous Models, IEEE Proc. Acoustics, Speech and Signal Processing, ICASSP-89 Glasgow, Pages 639-642.

JELINEK F, MERCER R (1980), Interpolated Estimation of Markov Source Parameters from Sparse Data, Proc. of the Workshop on Pattern Recognition in Practice. North-Holland, Amesterdam.

JELINEK F, (1976), Continuous Speech Recognition by Statistical Methods, Proc. of IEEE, Vol. 64, Pages 532-556.

JUANG B.H. (1985), Maximum-Likelihood Estimation for Mixture Multi-Variate Stochastic Observations of Markov Chain, AT&T Bell laboratories Tech J., Vol. 64, Pages 1235-1249.

JUANG B.H. (1984), On the Hidden Markov Model and Dynamic Time Warping for Speech Recognition- A unified Overview, AT&T Technical Journal, Vol. 63 Pages 1213-1243.

KULLBACK S., and LEIBLER R. A. (1951), On Information and Sufficiency, Ann. Math. Stat. Vol. 22 Pages 79-86.

LEE K.F. (1989), Automatic Speech Recognition : the Development of SPHINX System, MA: Kluwer Academic Publishers, Boston.

LEE K.F. (1989), Hidden Markov Model : Past, Present and Future, Eurospeech 89, Paris..

LIPORACE L.R. (1982), Maximum Likelihood Estimation for Multi-Variate Observations of Markov Sources, IEEE Trans. IT, Vol. IT-28 Pages 729-734.

MARKOV A. A. (1913), An Example of Statistical Investigation in the text of Eugen Onyegin Illustrating Coupling of Tests in Chains, Proc. Acad. Sci. St Petersburg VI Ser., Vol. 7, Pages 153-162.

MILNER B. P. (1995), Speech Recognition in Adverse Environments, Ph.D. Thesis, University of East Anglia, England.

PETERIE T. (1969), Probabilistic Functions of Finite State Markov Chains, Ann. Math. Stat., Vol. 40, Pages 97-115.

RABINER L. R., JUANG B. H. (1986), An Introduction to Hidden Markov Models, IEEE ASSP Magazine, Pages 4-16.

RABINER L. R., JUANG B. H., LEVINSON S. E., SONDHI M. M., (1985), Recognition of Isolated Digits using Hidden Markov Models with Continuous Mixture Densities, AT&T Technical Journal, Vol. 64, Pages 1211-1234.

RABINER L. R., JUANG B. H. (1993), Fundamentals of Speech Recognition, Prentice-Hall, Englewood Cliffs, N. J.

YOUNG S.J. , HTK : Hidden Markov Model Tool Kit, Cambridge University Engineering Department, Cambridge.

VARGA A, MOORE R.K., Hidden Markov Model Decomposition of Speech and Noise, in Proc. IEEE Int., Conf. on Acoust., Speech, Signal Processing, 1990, Pages. 845-848

VITERBI A.J. (1967), Error Bounds for Convolutional Codes and an Asymptotically Optimum Decoding Algorithm, IEEE Trans. on Information theory, Vol. IT-13 Pages 260-269.

5

Wiener Filters

5.1 Wiener Filters : Least Squared Error Estimation
5.2 Block-data Formulation of the Wiener Filter
5.3 Vector Space Interpretation of Wiener Filters
5.4 Analysis of the Least Mean Squared Error Signal
5.5 Formulation of Wiener Filters in Frequency Domain
5.6 Some Applications of Wiener Filters

Wiener theory, formulated by Norbert Wiener, forms the foundation of data-dependent linear least squared error filters. Wiener filters play a central role in a wide range of applications such as linear prediction, signal coding, echo cancellation, signal restoration, channel equalisation, system identification etc. The coefficients of a Wiener filter are calculated to minimise the average squared distance between the filter output and a desired signal. In its basic form, the Wiener theory assumes that the signals are stationary processes. However, if the filter coefficients are periodically recalculated, for every block of N samples, then the filter adapts to the average characteristics of the signals within the blocks and becomes block-adaptive. A block-adaptive filter can be used for signals that can be considered stationary over the duration of the block. In this chapter we study the theory of Wiener filters, and consider alternative methods of formulation of the Wiener filter problem. We consider the application of Wiener filters in channel equalisation, time-delay estimation, and additive noise suppression. A case study of the frequency response of a Wiener filter, for additive noise reduction, provides useful insight into the operation of the filter. We also deal with some implementation issues of Wiener filters.

5.1   Wiener Filters : Least Squared Error Estimation

Wiener studied the continuous-time, least mean squared error, estimation problem in his classic work on interpolation, extrapolation and smoothing of time series (Wiener 1949). The extension of Wiener theory to discrete time is simple, and of more practical use for implementation on digital signal processors. A Wiener filter can be either an infinite-duration impulse response (IIR) filter, or a finite-duration impulse response (FIR) filter. The formulation of an IIR Wiener filter results in a set of nonlinear equations, whereas the formulation of an FIR Wiener filter results in a set of linear equations and has a closed-form solution. In this chapter we consider FIR Wiener filters as they are relatively simple to compute, inherently stable, and more practical. Compared to IIR filters, the main drawback of FIR filters is that they may need a large number of coefficients to approximate a desired response.

As illustrated in Figure 5.1, a Wiener filter, represented by the coefficient vector w, takes as the input a signal $y(m)$, and produces an output signal $\hat{x}(m)$, where $\hat{x}(m)$ is the minimum mean squared estimate of a desired or target signal $x(m)$. The filter input-output relation is given by

$$\hat{x}(m) \; = \; \sum_{k=0}^{P-1} w_k y(m-k)$$

$$= \; w^T y$$

(5.1)

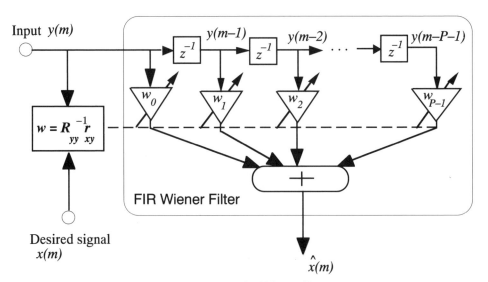

Figure 5.1 Illustration of a Wiener filter structure.

Where m is the discrete-time index, the vector $y^T=[y(m), y(m-1), ..., y(m-P-1)]$ is the filter input signal, and the parameter vector $w^T=[w_0, w_1, ..., w_{P-1}]$ is the Wiener filter coefficients vector. In Eq. (5.1), the filtering operation is expressed in two equivalent forms of : a convolutional sum, and an inner vector product. The filter error signal, $e(m)$, is defined as the difference between the desired signal $x(m)$, and the filter output $\hat{x}(m)$ as

$$
\begin{aligned}
e(m) &= x(m) - \hat{x}(m) \\
&= x(m) - w^T y
\end{aligned}
\tag{5.2}
$$

In Eq. (5.2), for a given set of input and desired signals, $y(m)$ and $x(m)$, the filter error $e(m)$ depends on the filter coefficient vector w. To explore the relation between the filter coefficients and the error signal we write Eq. (5.2) for N samples of the signals $x(m)$ and $y(m)$ as

$$
\begin{pmatrix} e(0) \\ e(1) \\ e(2) \\ \vdots \\ e(N-1) \end{pmatrix} = \begin{pmatrix} x(0) \\ x(1) \\ x(2) \\ \vdots \\ x(N-1) \end{pmatrix} - \begin{pmatrix} y(0) & y(-1) & y(-2) & \cdots & y(1-P) \\ y(1) & y(0) & y(-1) & \cdots & y(2-P) \\ y(2) & y(1) & y(0) & \cdots & y(3-P) \\ \vdots & \vdots & \vdots & \ddots & \vdots \\ y(N-1) & y(N-2) & y(N-3) & \cdots & y(N-P) \end{pmatrix} \begin{pmatrix} w_0 \\ w_1 \\ w_2 \\ \vdots \\ w_{P-1} \end{pmatrix}
\tag{5.3}
$$

In a compact vector notation this matrix equation may be written as

$$
e = x - Yw
\tag{5.4}
$$

where e is the error vector, x is the desired signal, Y is the input signal matrix and $Yw = \hat{x}$ is the filter output. It is assumed that the P initial input samples $[y(-1), . . ., y(-P-1)]$ are either known or set to zero.

In Eq. (5.3), if the number of signal samples is equal to the number of filter coefficients $N = P$, then we have a square matrix equation, and there is a unique solution w, with zero estimation error $e = 0$, such that $\hat{x} = Yw = x$. If $N < P$, then the number of samples N is insufficient to obtain a unique solution for Eq. (5.3), in this case there are an infinite number of solutions with zero estimation error, and the matrix equation is said to be *under-determined*. In practice, the number of signal samples is much larger than the filter length $N > P$, in this case the matrix equation is said to be *over-determined* and has a unique solution usually with nonzero error.

When $N > P$, the filter coefficients are calculated to minimise an average error cost function, such as the average absolute value of error $E[|e(m)|]$, or the mean squared error $E[e^2(m)]$, where $E[.]$ is the expectation operator. The choice of the error function affects the optimality and the computational complexity of the solution.

In Wiener theory, the objective criterion is the minimum mean squared error (MMSE) between the filter output and the desired signal. The minimum mean squared error criterion is optimal for Gaussian distributed signals. As shown in the following, for FIR filters the MMSE criterion leads to a linear and closed form solution. The Wiener filter coefficients are obtained by minimising the average squared error function, $\mathcal{E}[e^2(m)]$, with respect to the filter coefficients vector w. From Eq. (5.2) the mean squared estimation error is given by

$$
\begin{aligned}
\mathcal{E}[e^2(m)] &= \mathcal{E}[(x(m) - w^T y)^2] \\
&= \mathcal{E}[x^2(m)] - 2w^T \mathcal{E}[yx(m)] + w^T \mathcal{E}[yy^T]w \\
&= r_{xx}(0) - 2w^T r_{yx} + w^T R_{yy} w
\end{aligned}
\tag{5.5}
$$

Where $R_{yy} = E[y(m)y^T(m)]$ is the autocorrelation matrix of the input signal, and $r_{xy} = E[x(m)y(m)]$ is the cross correlation vector of the input and the desired signals. An expanded equivalent form of Eq. (5.5) can be obtained as

$$
\mathcal{E}[e^2(m)] = r_{xx}(0) - 2\sum_{k=0}^{P-1} w_k r_{yx}(k) + \sum_{k=0}^{P-1} w_k \sum_{j=0}^{P-1} w_j\, r_{yy}(k - j)
\tag{5.6}
$$

where $r_{yy}(k)$ and $r_{yx}(k)$ are the elements of the autocorrelation matrix R_{yy} and the cross correlation vector r_{xy} respectively. From Eq. (5.5), the mean squared error, for an FIR filter, is a quadratic function of the filter coefficient vector w and has a single minimum point. For example, for a filter with only two coefficients (w_0, w_1), the mean squared error function is a bowl shaped surface, with a single global minimum, as illustrated in Figure 5.2. The least mean squared error point corresponds to the minimum error power. At this optimal operating point the mean squared error surface has zero gradient.

From Eq. (5.5), the gradient of the mean squared error function with respect to the filter coefficient vector is given by

$$
\begin{aligned}
\frac{\partial}{\partial w}\mathcal{E}[e^2(m)] &= -2\mathcal{E}[x(m)y(m)] + 2w^T \mathcal{E}[y(m)y^T(m)] \\
&= -2\, r_{yx} + 2\, w^T R_{yy}
\end{aligned}
\tag{5.7}
$$

Where the gradient vector is defined as

$$
\frac{\partial}{\partial w} = \left[\frac{\partial}{\partial w_0},\ \frac{\partial}{\partial w_1},\ \frac{\partial}{\partial w_2},\ \cdots,\ \frac{\partial}{\partial w_{P-1}} \right]^T
\tag{5.8}
$$

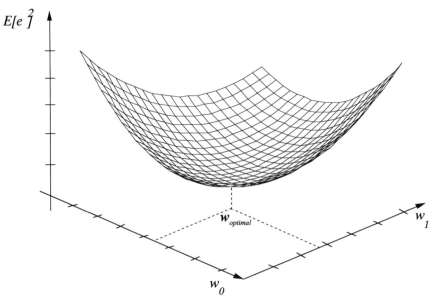

Figure 5.2 Mean squared error surface for a two tap FIR filter.

The minimum mean squared error Wiener filter is obtained by setting Eq. (5.7) to zero as

$$R_{yy}\, w\ =\ r_{yx} \tag{5.9}$$

or equivalently

$$w\ =\ R_{yy}^{-1}\, r_{yx} \tag{5.10}$$

In an expanded form, the Wiener solution (5.10) can be written as

$$
\begin{pmatrix} w_0 \\ w_1 \\ w_2 \\ \vdots \\ w_{P-1} \end{pmatrix}
=
\begin{pmatrix}
r_{yy}(0) & r_{yy}(1) & r_{yy}(2) & \cdots & r_{yy}(P-1) \\
r_{yy}(1) & r_{yy}(0) & r_{yy}(1) & \cdots & r_{yy}(P-2) \\
r_{yy}(2) & r_{yy}(1) & r_{yy}(0) & \cdots & r_{yy}(P-3) \\
\vdots & \vdots & \vdots & \ddots & \vdots \\
r_{yy}(P-1) & r_{yy}(P-2) & r_{yy}(P-3) & \cdots & r_{yy}(0)
\end{pmatrix}^{-1}
\begin{pmatrix} r_{yx}(0) \\ r_{yx}(1) \\ r_{yx}(2) \\ \vdots \\ r_{yx}(P-1) \end{pmatrix}
\tag{5.11}
$$

From Eq. (5.11), the calculation of Wiener filter coefficients requires the autocorrelation matrix of the input signal, and the cross correlation vector of the input and the desired signals.

In stochastic signal processing theory, the correlation values of a random process are obtained as the averages taken across the ensemble of different realisations of the process. However in many practical situations there is only one finite duration realisation of the signals $x(m)$ and $y(m)$. In such cases assuming the signals are correlation-ergodic we can use time-averages instead of the ensemble averages. For a signal record of length N samples, the time-averaged correlations are computed as

$$r_{yx}(k) = \frac{1}{N} \sum_{m=0}^{N-1} y(m)\, x(m+k) \qquad (5.12)$$

The autocorrelation matrix R_{yy} has a highly regular Toeplitz structure. A Toeplitz matrix has constant elements along the left-right diagonals of the matrix. Furthermore, the correlation matrix is also symmetric about the main diagonal elements. There are a number of efficient methods for solving the linear matrix Equation (5.11) including the Cholesky decomposition, the singular value decomposition and the QR decomposition methods.

5.2 Block-data Formulation of the Wiener Filter

In this section we consider the formulation of a Wiener filter for N samples of the input signal $[y(0), y(1), ..., y(N-1)]$ and the desired signal $[x(0), x(1), ..., x(N-1)]$. The set of N linear equations describing the Wiener filter input/output relation can be written as

$$
\begin{pmatrix} \hat{x}(0) \\ \hat{x}(1) \\ \hat{x}(2) \\ \vdots \\ \hat{x}(N-2) \\ \hat{x}(N-1) \end{pmatrix}
=
\begin{pmatrix}
y(0) & y(-1) & y(-2) & \cdots & y(2-P) & y(1-P) \\
y(1) & y(0) & y(-1) & \cdots & y(3-P) & y(2-P) \\
y(2) & y(1) & y(0) & \cdots & y(4-P) & y(3-P) \\
\vdots & \vdots & \vdots & \ddots & \vdots & \vdots \\
y(N-2) & y(N-3) & y(N-4) & \cdots & y(N-P) & y(N-1-P) \\
y(N-1) & y(N-2) & y(N-3) & \cdots & y(N+1-P) & y(N-P)
\end{pmatrix}
\begin{pmatrix} w_0 \\ w_1 \\ w_2 \\ \vdots \\ w_{P-2} \\ w_{P-1} \end{pmatrix}
\qquad (5.13)
$$

Eq. (5.13) can be rewritten in a compact matrix notation as

$$\hat{x} = Yw \qquad (5.14)$$

The Wiener filter error is the difference between the desired signal and the filter output defined as

$$e = x - \hat{x}$$
$$= x - Yw \tag{5.15}$$

The energy of the error vector, that is the sum of the squared elements of the error vector, is given by the inner vector product as

$$e^T.e = (x - Yw)^T(x - Yw)$$
$$= x^Tx - x^TYw - w^TY^Tx + w^TY^TYw \tag{5.16}$$

The gradient of the squared error function with respect to the Wiener filter coefficients is obtained by differentiating Eq. (5.16) :

$$\frac{\partial e^T.e}{\partial w} = -2x^TY + 2w^TY^TY \tag{5.17}$$

The Wiener filter coefficients are obtained by setting the gradient of the squared error function of Eq. (5.17) to zero, this yields

$$(Y^TY)w = Y^Tx \tag{5.18}$$

or

$$w = (Y^TY)^{-1}Y^Tx \tag{5.19}$$

Note, that the signal matrix (Y^TY) is a time-averaged estimate of the autocorrelation matrix of the filter input R_{yy}, , and that the vector Y^Tx is a time-averaged estimate of r_{xy}. the cross correlation vector of the input and the desired signals. Theoretically, the Wiener filter is obtained from minimisation of the squared error across the ensemble of different realisations of a process as described in the previous section. For a correlation-ergodic process, as the signal length N approaches infinity the block-data Wiener filter of Eq. (5.19) approaches the Wiener filter Eq. (5.10) as

$$\lim_{N \to \infty} \left[w = (Y^TY)^{-1}Y^Tx \right] = R_{yy}^{-1}r_{xy} \tag{5.20}$$

Since the least squared error method, described in this section, requires a block of N samples of the input and the desired signals, it is also referred to as the block least squared (BLS) error estimation method. The block estimation method is appropriate for processing of signals which can be considered as time-invariant over the duration of the block.

An efficient and robust method for solving the least squared error Eq. (5.19) is the QR Decomposition (QRD) method. In QRD method the $N \times P$ signal matrix Y is decomposed into the product of an $N \times N$ orthonormal matrix Q and a $P \times P$ upper triangular matrix R as

$$QY = \begin{pmatrix} \mathcal{R} \\ \mathbf{0} \end{pmatrix} \tag{5.21}$$

where $\mathbf{0}$ is a $(N - P) \times P$ null matrix, $Q^T Q = Q Q^T = I$, and the upper triangular matrix R is in the form

$$\mathcal{R} = \begin{pmatrix} r_{00} & r_{01} & r_{02} & r_{03} & \cdots & r_{0P-1} \\ 0 & r_{11} & r_{12} & r_{13} & \cdots & r_{1P-1} \\ 0 & 0 & r_{22} & r_{23} & \cdots & r_{2P-1} \\ 0 & 0 & 0 & r_{33} & \cdots & r_{3P-1} \\ \vdots & \vdots & \vdots & \vdots & \ddots & \vdots \\ 0 & 0 & 0 & 0 & \cdots & r_{P-1P-1} \end{pmatrix} \tag{5.22}$$

Substitution of Eq. (5.21) in Eq. (5.18) yields

$$\begin{pmatrix} \mathcal{R} \\ \mathbf{0} \end{pmatrix}^T Q Q^T \begin{pmatrix} \mathcal{R} \\ \mathbf{0} \end{pmatrix} w = \begin{pmatrix} \mathcal{R} \\ \mathbf{0} \end{pmatrix}^T Q x \tag{5.23}$$

From Eq. (5.23) we have

$$\begin{pmatrix} \mathcal{R} \\ \mathbf{0} \end{pmatrix} w = Q x \tag{5.24}$$

From Eq. (5.24) we have

$$\mathcal{R} w = x_Q \tag{5.25}$$

where the vector x_Q in the right hand side of Eq. (5.25) is composed of the first P elements of the product Qx. Since the matrix R is upper triangular, the coefficients of the least squared error filter can be obtained easily through a process of back substitution from Eq. (5.25) starting with the coefficient $w_{P-1} = x_Q(P-1) / r_{P-1P-1}$.

The main computational steps in the QR decomposition is the determination of the orthonormal matrix Q and the upper triangular matrix R. The decomposition of a matrix into QR matrices can be achieved using a number of methods including Gram-Schmidt orthogonalisation method, Householder method and Givens rotation method.

5.3 Vector Space Interpretation of Wiener Filters

In this section Wiener filtering is visualised as the minimum distance, perpendicular, projection of the desired signal vector into the vector space of the input signal. In order to develop a vector space interpretation of the least squared error estimation, we rewrite the matrix Eq. (5.11) and express the filter output vector, $\hat{x}$, as a linear weighted combination of the vector columns of the input signal matrix :

$$
\begin{pmatrix} \hat{x}(0) \\ \hat{x}(1) \\ \hat{x}(2) \\ \vdots \\ \hat{x}(N-2) \\ \hat{x}(N-1) \end{pmatrix} = w_0 \begin{pmatrix} y(0) \\ y(1) \\ y(2) \\ \vdots \\ y(N-2) \\ y(N-1) \end{pmatrix} + w_1 \begin{pmatrix} y(-1) \\ y(0) \\ y(1) \\ \vdots \\ y(N-3) \\ y(N-2) \end{pmatrix} + \cdots + w_{P-1} \begin{pmatrix} y(1-P) \\ y(2-P) \\ y(3-P) \\ \vdots \\ y(N-1-P) \\ y(N-P) \end{pmatrix}
$$

(5.26)

In a compact notation Eq. (5.26) may be written as

$$
\hat{x} = w_0 y_0 + w_1 y_1 + \cdots + w_{P-1} y_{P-1}
$$

(5.27)

In Eq. (5.27) the signal estimate $\hat{x}$ is a linear combination of P basis vectors $[y_0, y_1, \ldots, y_{P-1}]$ and hence it can be said that $\hat{x}$ is in the vector space, or the subspace, formed by the input signal vectors $[y_0, y_1, \ldots, y_{P-1}]$. A vector space is the collection of an infinite number of vectors that can be obtained from linear combinations of P independent vectors.

In general, the P N-dimensional vectors $[y_0, y_1, \ldots, y_{P-1}]$ of the input signal in Eq. (5.27) define the *basis* vectors for a subspace in an N-dimensional signal space. If P, the number of basis vectors, is equal to N, the vector dimension, then the subspace defined by the input signal vectors encompasses the entire N-dimensional signal space and includes the desired signal vector x. In this case $\hat{x} = x$ and the estimation error is zero. However in practice $N > P$, and the input signal space is only a subspace of the N-dimensional signal space. In this case the estimation error is zero only if the desired signal x happens to be in the subspace of the input signal, otherwise the best estimate of x is the perpendicular projection of the vector x onto the vector space of $[y_0, y_1, \ldots, y_{P-1}]$, as explained in the following example.

Example 5.1 Figure 5.3 illustrates a vector space interpretation of a simple example of a least squared error estimation problem, where $y^T = [y(m), y(m-1), y(m-2)], y(m-3)]$ is the input signal, $x^T = [x(m), x(m-1), x(m-2)]$ is the desired signal, and $w^T=[w_0, w_1]$ is the filter coefficient vector. As in Eq. (5.26) the filter output can be written as

$$
\begin{pmatrix} \hat{x}(m) \\ \hat{x}(m-1) \\ \hat{x}(m-2) \end{pmatrix} = w_0 \begin{pmatrix} y(m) \\ y(m-1) \\ y(m-2) \end{pmatrix} + w_1 \begin{pmatrix} y(m-1) \\ y(m-2) \\ y(m-3) \end{pmatrix} \tag{5.28}
$$

In Eq. (5.28), the input signals $y_1^T=[y(m), y(m-1), y(m-2)]$ and $y_2^T=[y(m-1), y(m-2), y(m-3)]$ are three dimensional vectors. The subspace defined by the linear combinations of the two input vectors $[y_1, y_2]$ is a two-dimensional plane in a 3-dimensional signal space. The filter output is a linear combination of y_1 and y_2, and hence it is confined to the plane containing these two vectors. The least squared error estimate of x is the orthogonal projection of x on the plane of $[y_1, y_2]$ as shown by the shaded vector $\hat{x}$. If the desired vector happens to be in the plane defined by the vectors y_1 and y_2, then the estimation error would be zero, otherwise the estimation error will be the perpendicular distance of x from the plane containing y_1 and y_2.

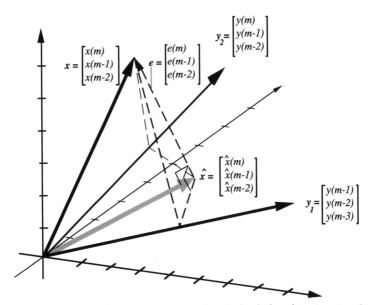

Figure 5.3 The least squared error projection of a desired signal vector x, onto the plane containing the input vectors y_1 and y_2, is the perpendicular projection of x shown as the shaded vector.

5.4 Analysis of the Least Mean Squared Error Signal

The optimality criterion in the formulation of the Wiener filter is the minimum mean squared distance between the filter output and a desired signal. In this section the variance of the error signal is analysed. Substituting the Wiener equation, $R_{yy}w=r_{yx}$, in Eq. (5.5) gives the least mean squared error as

$$
\begin{aligned}
\mathcal{E}[e^2(m)] &= r_{xx}(0) - w^T r_{yx} \\
&= r_{xx}(0) - w^T R_{yy} w
\end{aligned}
\tag{5.29}
$$

Now, for zero-mean signals, it is easy to show that in Eq. (5.29) the term $w^T R_{yy} w$ is the variance of the Wiener filter output $\hat{x}(m)$

$$
\sigma_{\hat{x}}^2 = \mathcal{E}[\hat{x}^2(m)] = w^T R_{yy} w
\tag{5.30}
$$

Therefore Eq. (5.29) may be written as

$$
\sigma_e^2 = \sigma_x^2 - \sigma_{\hat{x}}^2
\tag{5.31}
$$

Where $\sigma_x^2 = \mathcal{E}[x^2(m)], \sigma_{\hat{x}}^2 = \mathcal{E}[\hat{x}^2(m)], \sigma_e^2 = \mathcal{E}[e^2(m)]$ are the variances of the desired signal, the filter estimate of the desired signal, and the error signal respectively.

In general, the filter input $y(m)$ is composed of a signal component $x_c(m)$, and a random noise $n(m)$ as

$$
y(m) = x_c(m) + n(m)
\tag{5.32}
$$

where the signal $x_c(m)$ is correlated with the desired signal $x(m)$, and it is this part of the input signal which may be transformable, through a Wiener filter, to the desired signal. Using Eq. (5.32) the Wiener filter error may be decomposed into two distinct components as

$$
\begin{aligned}
e(m) &= x(m) - \sum_{k=0}^{P} w_k y(m-k) \\
&= \left(x(m) - \sum_{k=0}^{P} w_k x_c(m-k) \right) - \sum_{k=0}^{P} w_k n(m-k)
\end{aligned}
\tag{5.33}
$$

or

$$
e(m) = e_x(m) + e_n(m)
\tag{5.34}
$$

where $e_x(m)$ is the difference between the desired signal $x(m)$, and the filter output in response to the input signal component $x_c(m)$

$$e_x(m) = x(m) - \sum_{k=0}^{P-1} w_k x_c(m-k) \tag{5.35}$$

and $e_n(m)$ is the error in the output in response to the presence of noise $n(m)$ in the input signal

$$e_n(m) = - \sum_{k=0}^{P-1} w_k \, n(m-k) \tag{5.36}$$

The variance of filter error can be rewritten as

$$\sigma_e^2 = \sigma_{e_x}^2 + \sigma_{e_n}^2 \tag{5.37}$$

Ideally $e_x(m)=0$ and $e_n(m)=n(m)$, but this is possible only if the following conditions are satisfied : (a) the spectra of the signal and the noise are separable, (b) the signal component of the input, that is $x_c(m)$, is *linearly* transformable to $x(m)$, and (c) the filter length P is sufficiently large. The issue of signal and noise separability is addressed in Section 5.6.

5.5 Formulation of Wiener Filter in Frequency Domain

In the frequency domain, the Wiener filter output $\hat{X}(f)$ is the product of the input signal $Y(f)$, and the filter frequency response $W(f)$

$$\hat{X}(f) = W(f)Y(f) \tag{5.38}$$

The estimation error signal $E(f)$ is defined as the difference between the desired signal $X(f)$ and the filter output $\hat{X}(f)$ as

$$\begin{aligned} E(f) &= X(f) - \hat{X}(f) \\ &= X(f) - W(f)Y(f) \end{aligned} \tag{5.39}$$

and the mean squared error at a frequency f is given by

$$\mathcal{E}\left[|E(f)|^2\right] = \mathcal{E}\left[(X(f) - W(f)Y(f))^*(X(f) - W(f)Y(f))\right] \tag{5.40}$$

where $E[\,]$ is the expectation function, and the symbol * denotes the complex conjugate. Note that from Parsevals theorem the mean squared error in time and frequency are related as

$$\sum_{m=0}^{N-1} e^2(m) = \int_{-1/2}^{1/2} |E(f)|^2 \, df \tag{5.41}$$

To obtain the least mean squared error filter we set the complex derivative of Eq. (5.40) with respect to filter $W(f)$ to zero

$$\frac{\partial \mathcal{E}[|E(f)|^2]}{\partial W(f)} = 2W(f)P_{YY}(f) - 2P_{XY}(f) = 0 \tag{5.42}$$

where $P_{YY}(f)=E[Y(f)Y^*(f)]$ and $P_{XY}(f)=E[X(f)Y^*(f)]$ are the power spectrum of $Y(f)$, and the cross power spectrum of $Y(f)$ and $X(f)$ respectively. From Eq. (5.42) the least mean squared error Wiener filter in the frequency domain is given as

$$W(f) = \frac{P_{XY}(f)}{P_{YY}(f)} \tag{5.43}$$

Alternatively, the frequency domain Wiener filter Eq. (5.43) can be obtained by taking the Fourier transform of the time domain Wiener equation (5.9) as

$$\sum_{m}\sum_{k=0}^{P-1} w_k r_{yy}(m-k)e^{-j\omega m} = \sum_{m} r_{yx}(n)e^{-j\omega m} \tag{5.44}$$

From the Wiener-Kinchine relation, the correlation and the power spectral functions are Fourier transform pairs. Using this relation, and the Fourier transform property that convolution in time is equivalent to multiplication in frequency, it is easy to show that the Wiener filter is given by Eq. (5.43).

5.6 Some Applications of Wiener Filters

In this section we consider applications of Wiener filter in, suppression of broad band additive noise, time-alignment of signals in multi-channel or multi-sensor systems, and channel equalisation.

5.6.1 Wiener Filter for Additive Noise Reduction

Consider a signal $x(m)$ observed in a broad-band additive noise $n(m)$ and model as

$$y(m) = x(m) + n(m) \tag{5.45}$$

Assuming the signal and the noise are uncorrelated, it follows that the autocorrelation matrix of the noisy signal is the sum of the autocorrelation matrix of the signal $x(m)$, and the noise $n(m)$ as

$$\boldsymbol{R_{yy}} = \boldsymbol{R_{xx}} + \boldsymbol{R_{nn}} \tag{5.46}$$

and we also can write

$$\boldsymbol{r_{xy}} = \boldsymbol{r_{xx}} \tag{5.47}$$

Where $\boldsymbol{R_{yy}}$, $\boldsymbol{R_{xx}}$ and $\boldsymbol{R_{nn}}$ are the autocorrelation matrices of the noisy signal, the noise-free signal, and the noise respectively, and $\boldsymbol{r_{xy}}$ is the cross correlation vector of noisy signal and noise-free signal. Substitution of Eqs. (5.46) and (5.47) in the Wiener filter Eq. (5.10) yields

$$w = \left(\boldsymbol{R_{xx}} + \boldsymbol{R_{nn}}\right)^{-1}\boldsymbol{r_{xx}} \tag{5.48}$$

Eq. (5.48) is the optimal linear filter for the removal of additive noise. In the followings, a study of the frequency response of the Wiener filter provides useful insight into the operation of the Wiener filter. In frequency domain the noisy signal $Y(f)$ is given by

$$Y(f) = X(f) + N(f) \tag{5.49}$$

where $X(f)$ and $N(f)$ are the signal and noise spectra. For a signal observed in additive random noise, the frequency domain Wiener filter is obtained as

$$W(f) = \frac{P_{XX}(f)}{P_{XX}(f) + P_{NN}(f)} \tag{5.50}$$

where $P_{XX}(f)$ and $P_{NN}(f)$ are the signal and noise power spectra. Dividing the numerator and the denominator of Eq. (5.50) by the noise power spectra, $P_{NN}(f)$, and substituting a signal to noise ratio variable $SNR(f)=P_{XX}(f)/P_{NN}(f)$ yields

$$W(f) = \frac{SNR(f)}{SNR(f) + 1} \tag{5.51}$$

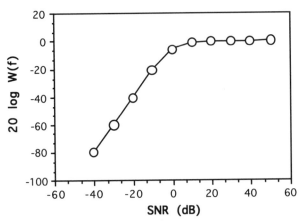

Figure 5.4 Variations of the gain of Wiener filter frequency response with SNR.

Note that the variable, *SNR(f)*, is expressed in terms of the power spectral ratio, and not in the more usual terms of log power ratio. Therefore *SNR(f)=0* corresponds to $-\infty$ *dB*.

From Eq(5.51), the following interpretation of the Wiener filter frequency response, *W(f)*, in terms of the signal to noise ratio can be deduced. For additive noise, the Wiener filter frequency response is a real positive number in the range $0 \leq W(f) \leq 1$. Now consider the two limiting cases of : (a) a noise-free signal $SNR(f)=\infty$, and (b) an extremely noisy signal *SNR(f)=0*. At very high SNR $W(f) \approx 1$, and the filter applies little or no attenuation to the noise-free frequency component. On the other extreme, when *SNR(f)=0*, *W(f)=0*. Therefore, *for additive noise, the Wiener filter attenuates each frequency component in proportion to an estimate of the signal to noise ratio.* Figure 5.4 shows the variation of the Wiener filter response *W(f)*, with the signal to noise ratio *SNR(f)*.

An alternative illustration of the variations of the Wiener filter frequency response with *SNR(f)* is shown in Figure 5.5. It illustrates the similarity between the Wiener filter frequency response and the signal spectrum for the case of an additive white noise disturbance. Note that at a spectral peak of the signal spectrum, where the *SNR(f)* is relatively high, the Wiener filter frequency response is also high, and the filter applies little attenuation. At a signal trough, the signal to noise ratio is low, and so is the Wiener filter response. Hence, for additive white noise the Wiener filter response broadly follows the signal spectrum.

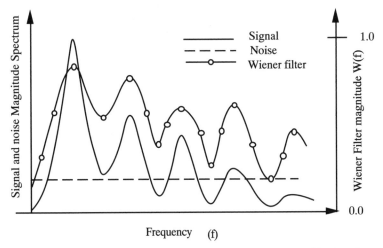

Figure 5.5 Illustration of variation of Wiener frequency response with signal spectrum for additive white noise. The Wiener filter response broadly follows the signal spectrum.

5.6.2   Wiener Filter and Separability of Signal and Noise

A signal is completely recoverable from noise if the spectra of the signal and the noise do not overlap. An example of a noisy signal with separable signal and noise spectra is shown in Figure 5.6.a. In this case the signal and the noise occupy different parts of the frequency spectrum, and can be separated with a lowpass, or a highpass, filter. Figure 5.6.b illustrates a more common example of a signal and

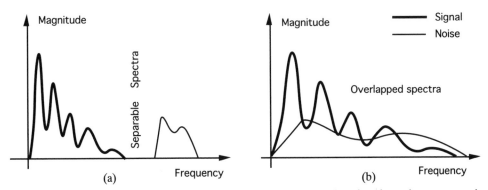

Figure 5.6 (a) The signal and noise spectra do not overlap, the signal can be recovered by a lowpass filter, (b) the signal and noise spectra overlap, the noise can be reduced but not completely removed.

noise process with overlapping spectra. .For this case it is not possible to completely separate the signal from the noise. However the effects of the noise can be reduced by using a Wiener filter that attenuates each noisy signal frequency in proportion to an estimate of the signal to noise ratio as described by Eq. (5.51).

5.6.3 Squared Root Wiener Filter

In the frequency domain, the Wiener filter output $\hat{X}(f)$ is the product of the input frequency $X(f)$ and the filter response $W(f)$ as expressed in Eq. (5.38). Taking the expectation of the squared magnitude frequency of both sides of Eq. (5.38) yields the power spectrum of the filtered signal as

$$
\begin{aligned}
\mathcal{E}[|\hat{X}(f)|^2] &= |W(f)|^2 \mathcal{E}[|Y(f)|^2] \\
&= |W(f)|^2 P_{YY}(f)
\end{aligned}
\tag{5.52}
$$

Substitution of $W(f)$ from Eq. (5.43) in Eq. (5.52) yields

$$
\mathcal{E}[|\hat{X}(f)|^2] = \frac{P_{XY}^2(f)}{P_{YY}(f)}
\tag{5.53}
$$

Now for a signal observed in an uncorrelated additive noise we have

$$
P_{YY}(f) = P_{XX}(f) + P_{NN}(f)
\tag{5.54}
$$

and

$$
P_{XY}(f) = P_{XX}(f)
\tag{5.55}
$$

Substitution of Eq. (5.54) and (5.55) in Eq. (5.53) yields

$$
\mathcal{E}[|\hat{X}(f)|^2] = \frac{P_{XX}^2(f)}{P_{XX}(f) + P_{NN}(f)}
\tag{5.56}
$$

Now in Eq. (5.38) if instead of the Wiener filter, the square root of the Wiener filter magnitude frequency response is used, the result will be

$$
\hat{X}(f) = |W(f)|^{1/2} Y(f)
\tag{5.57}
$$

and the power spectrum of the signal, filtered by the square-root Wiener filter, is given by

$$\mathcal{E}[|\hat{X}(f)|^2] = \left(|W(f)|^{1/2}\right)^2 \mathcal{E}[|Y(f)|^2]$$

$$= \frac{P_{XY}(f)}{P_{YY}(f)} P_{YY}(f) \qquad (5.58)$$

$$= P_{XY}(f)$$

Now for uncorrelated signal and noise Eq. (5.58) becomes

$$\mathcal{E}[|\hat{X}(f)|^2] = P_{XX}(f) \qquad (5.59)$$

Thus for additive noise the power spectrum of the output of the square-root Wiener filter is the same as the power spectrum of the desired signal.

5.6.4 Wiener Channel Equaliser

Communication channel distortions may be modelled by a combination of a linear filter and an additive random noise source as shown in Figure 5.7. The input/output signals of a linear time invariant channel can be modelled as

$$y(m) = \sum_{k=0}^{P-1} h_k \, x(m-k) + n(m) \qquad (5.60)$$

where $x(m)$ and $y(m)$ are the transmitted and received signals, $[h_k]$ is the impulse response of a linear filter model of the channel, and $n(m)$ models the channel noise. In the frequency domain Eq. (5.60) becomes

$$Y(f) = X(f)H(f) + N(f) \qquad (5.61)$$

where $X(f)$, $Y(f)$, $H(f)$ and $N(f)$ are the signal, noisy signal, channel and noise spectra respectively. To remove the channel distortions, the receiver is followed by an equaliser. The equaliser input is the distorted channel output, and the desired signal is the channel input.

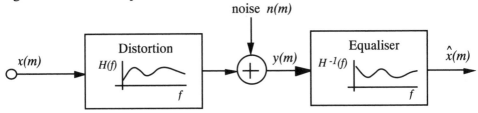

Figure 5.7 Illustration of a channel model followed by an equaliser.

Using Eq. (5.43) it can be shown that the Wiener equaliser in the frequency domain is given by

$$W(f) = \frac{P_{XX}(f)H^*(f)}{P_{XX}(f)|H(f)|^2 + P_{NN}(f)} \tag{5.62}$$

Where it is assumed that the channel noise and the signal are uncorrelated. In the absence of channel noise, $P_{NN}(f)=0$, and the Wiener filter is simply the inverse of the channel filter model $W(f)=H^{-1}(f)$. The equalisation problem is treated in detail in Chapter 14.

5.6.5 Time-alignment of Signals in Multi-Channel/Multi-Sensor Systems

In multi-channel/multi-sensor signal processing there are a number of noisy and distorted versions of a signal $x(m)$, and the objective is to use all the observations in estimating $x(m)$, as illustrated in Figure 5.8. As a simple example consider the problem of time-alignment of two noisy records of a signal given as

$$y_1(m) = x(m) + n_1(m) \tag{5.63}$$

$$y_2(m) = A x(m - D) + n_2(m) \tag{5.64}$$

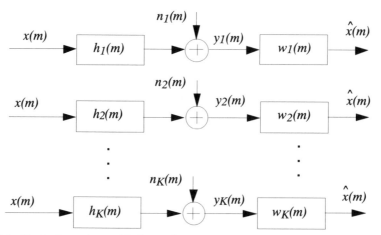

Figure 5.8 Illustration of a multi-channel system where Wiener filters are used to time align the signals from different channels.

where $y_1(m)$ and $y_2(m)$ are the noisy observations from channels *1* and *2*, $n_1(m)$ and $n_2(m)$ are uncorrelated noise in each channel, D is the time delay of arrival of the two signals, and A is an amplitude scaling factor. Now assume that $y_1(m)$ is used as the input to a Wiener filter and that, in the absence of the signal $x(m)$, $y_2(m)$ is used as the 'desired' signal. The error signal is given by

$$e(m) = y_2(m) - \sum_{k=0}^{P-1} w_k y_1(m)$$

$$= \left(A x(m-D) - \sum_{k=0}^{P-1} w_k x(m) \right) + \left(\sum_{k=0}^{P-1} w_k n_1(m) \right) + n_2(m)$$

(5.65)

The Wiener filter strives to minimise the terms shown inside the brackets in Eq. (5.65). Using the Wiener Eq. (5.10) we have

$$w = R_{y_1 y_1}^{-1} r_{y_1 y_2}$$

$$= \left(R_{xx} + R_{n_1 n_1} \right)^{-1} A r_{xx}(D)$$

(5.66)

where $r_{xx}(D) = E\ [x(PD)x(m)]$. The frequency domain equivalent of Eq. (5.65) can be derived as

$$W(f) = \frac{P_{XX}(f)}{P_{XX}(f) + P_{N_1 N_1}(f)}\, A e^{-j\omega D}$$

(5.67)

Note that in the absence of noise the Wiener filter becomes a pure phase (or a pure delay) filter with unity magnitude response

5.6.6 Implementation of Wiener Filters

The implementation of a Wiener filter for additive noise reduction, Eqs. (5.48) (5.50), requires the autocorrelation functions, or equivalently the power spectra, of the signal and noise. The noise power spectrum can be obtained from the signal inactive, noise only, periods. The assumption is that the noise is quasi-stationary, and that its power spectra remains relatively stationary in-between the update periods. This is a reasonable assumption for many noisy environments such as the noise inside a car emanating from the engine, aircraft noise, office noise from computer machines *etc.*

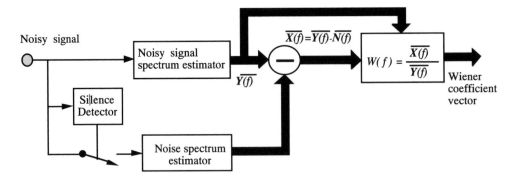

Figure 5.9 Configuration of a system for estimation of frequency Wiener filter coefficients.

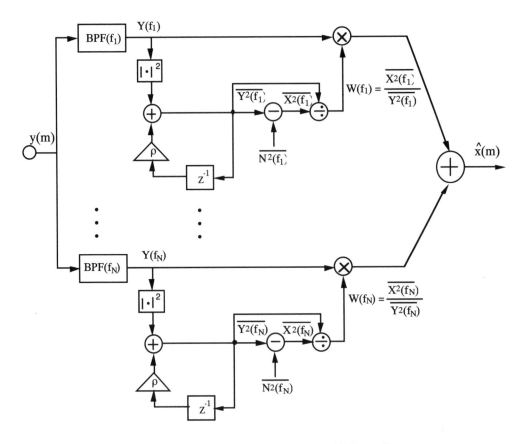

Figure 5.10 A filter-bank implementation of Wiener filter.

The main practical problem in implementation of a Wiener filter is that the desired signal is often observed in noise, and that the autocorrelation or power spectra of the desired signal are not readily available. Figure 5.9 illustrates the block diagram configuration of a system for implementation of a Wiener filter for additive noise reduction. An estimate of the desired signal power spectra is obtained by subtracting an estimate of the noise spectra from that of the noisy signal. A filter bank implementation of the Wiener filter is shown in Figure 5.10, where the incoming signal is divided into N bands of frequencies. A first order integrator, placed at the output of each bandpass filter, gives an estimate of the power spectra of noisy signal. The power spectrum of original signal is obtained by subtracting an estimate of the noise power spectrum from the noisy signal.

The Choice of Wiener Filter Order

The Choice of Wiener filter order affects : a) the ability of the filter to remove distortions and reduce the noise, b) the computational complexity of the filter, and c) the numerical stability of the of Wiener solution Eq. (5.10). The choice of the filter length also depends on the application and the method of implementation of the Wiener filter. For example in a filter bank implementation of the Wiener filter, for additive noise suppression, the number of filter coefficients is equal to the number of filter banks, and typically the number of filter banks are between 16 to 64. On the other hand for many applications, a direct implementation, of the time-domain Wiener filter requires a larger filter length say between, 64 to 256 taps.

A reduction in the required length of a time domain Wiener filter can be achieved by dividing the time domain signal into N subband signals. Each subband signal can then be decimated by a factor of N. The decimation results in a reduction, by a factor of N, in the required length of each subband Wiener filter. In Chapter 13 a subband echo canceller is described.

Summary

A Wiener filter is formulated to map an input signal to an output that is as close to a desired signal as possible. This chapter began with the derivation of the Wiener filter through minimisation of the mean squared error over the space of the input and desired signals. In Section 5.2 we derived the block least squared error Wiener filter for applications where only a finite length realisations of the input and the desired signals are available. In such cases the filter is obtained by minimising a time-averaged squared error function. In Section 5.3 we considered a vector space interpretation of the Wiener filters as the perpendicular projection of the desired signal onto the space of the input signal.

In Section 5.4 the least mean squared error signal was analysed. The mean squared error is zero only if the input signal is related to the desired signal through a linear and invertible filter. For most cases due to noise and/or nonlinear distortions of the input signal, the minimum mean squared error would be nonzero. In Section 5.5 we derived the Wiener filter in the frequency domain, and considered the issue of separability of signal and noise using a linear filter. Finally in Section 5.6 we considered some applications of Wiener filters in noise reduction, time delay estimation and channel equalisation.

Bibliography

AKAIKE H. (1974), A New Look at Statistical Model Identification, IEEE Trans. on Automatic Control, AC-19 Pages 716-23, Dec.

ALEXANDER S.T. (1986), Adaptive Signal Processing Theory and Applications. Springer-Verlag, New York.

ANDERSON B.D., MOOR J.B. (1979) Linear Optimal Control, Prentice-Hall, Englewood Cliffs, N. J.

BJORCK A. (1967), Solving Linear Least Squares Problems by Gram-Schmidt Orthogonalisation, BIT, Vol. 7, Pages 1-21.

DORNY C.N. (1975), A Vector Space Approach to Models and Optimisation, Wiley, New York.

DURBIN J. (1959), Efficient Estimation of Parameters in Moving Average Models, Biometrica Vol. 46, Pages 306-16.

GIORDANO A.A., HSU F.M. (1985), Least Square Estimation with Applications to Digital Signal Processing, Wiley, New York.

GIVENS W. (1958), Computation of Plane Unitary Rotations Transforming a General Matrix to Triangular Form, SIAM J. Appl. Math. Vol. 6, Pages 26-50.

GOLUB G.H., REINSCH (1970), Singular Value Decomposition and Least Squares Solutions, Numerical Mathematics, Vol. 14, Pages 403-20.

GOLUB G.H., VAN LOAN C.F. (1983). Matrix Computations, Johns Hopkins University Press, Baltimore, MD.

GOLUB G.H., VAN LOAN C.F. (1980). An Analysis of the Total Least Squares Problem, SIAM Journal of Numerical Analysis, Vol. 17, Pages 883-93.

HALMOS P. R. (1974), Finite-Dimensional Vector Spaces. Springer-Verlag, New York.

HAYKIN S. (1991), Adaptive Filter Theory, 2nd Edition, Prentice-Hall, Englewood Cliffs, N. J.

HOUSEHOLDER A.S.(1964), The Theory of Matrices in Numerical Analysis, Blaisdell, Waltham, Mass.

KAILATH T. (1974), A View of Three Decades of Linear Filtering Theory, IEEE Trans. Information Theory, Vol. IT-20, Pages 146-81.

KAILATH T. (1977), Linear Least Squares Estimation, Benchmark Papers in Electrical Engineering and Computer science, Dowden, Hutchinson &Ross.

KAILATH T. (1980), Linear Systems, Prentice-Hall, Englewood Cliffs, N. J.

KLEMA V.C., LAUB A. J. (1980), The Singular Value Decomposition: Its Computation and Some Applications, IEEE Trans. Automatic Control, Vol. AC-25, Pages 164-76.

KOLMOGROV A.N. (1939), Sur l' Interpolation et Extrapolation des Suites Stationaires, Comptes Rendus de l'Academie des Sciences, Vol. 208, Pages 2043-45.

LAWSON C.L., HANSON R.J. (1974), Solving Least Squares Problems, Prentice-Hall, Englewood Cliffs, N. J.

ORFANIDIS S. J. (1988), Optimum Signal Procesing : An introduction, 2nd Edition, Macmillan, New York.

SCHARF L.L. (1991), Statistical Signal Processing : Detection, Estimation, and Time Series Analysis, Addison Wesley, Reading, Mass.

STRANG G. (1976), Linear Algebra and Its Applications, 3rd ed., Harcourt Brace Jovanovich, San Diego, California.

WIENER N. (1949), Extrapolation, Interpolation and Smoothing of Stationary Time Series, MIT Press Cambridge, Mass.

WILKINSON J. H. (1965), The Algebraic Eigenvalue Problem, Oxford University Press, Oxford.

WHITTLE P.W. (1983), Prediction and Regulation by Linear Least-Squares Methods, University of Minnesota Press, Minneapolis, Minnesota.

WOLD H., (1954), The Analysis of Stationary Time Series, 2nd ed. Almquist and Wicksell, Uppsala, Sweden.

6

Kalman and Adaptive Least Squared Error Filters

6.1 State-space Kalman Filters
6.2 Sample-Adaptive Filters
6.3 Recursive Least Square (RLS) Filters
6.4 The Steepest Descent Method
6.5 The LMS Method

Adaptive least squared error filters are used with nonstationary signals and environments, or in applications where a low processing delay is required, such as multi-channel noise reduction, radar signal processing, channel equalisation, echo cancellation, and speech coding. We begin this chapter with a study of state-space Kalman filters. In Kalman theory a state equation models the dynamics of the signal generation process, and an observation equation models the channel distortions and additive noise. We then consider recursive least squared (RLS) error adaptive filters. The RLS filter is a sample adaptive formulation of the Wiener filter, and for stationary signals should converge to the same solution as the Wiener filter. In least squared error filtering, an alternative to using a Wiener-type closed form solution, is an iterative search for the optimal filter coefficients. The steepest descent search is a gradient-based method of searching the least squared error performance curve for the minimum error filter coefficients. We study the steepest descent method, and then consider the LMS method which is a computationally inexpensive gradient search algorithm.

6.1 State-space Kalman Filters

Kalman filter is a recursive least squared error method for estimation of a signal distorted in transmission through a channel and observed in noise. Kalman filters can be used with time-varying as well as time-invariant processes. The Kalman filter theory is based on a state-space approach in which, a state equation models the dynamics of the signal process, and an observation equation models the noisy observation signal. A state-equation model for the signal $x(m)$, and an observation model for the noisy observation signal $y(m)$, are defined as

$$x(m) = \Phi(m, m-1)x(m-1) + e(m) \qquad (6.1)$$

$$y(m) = H(m)x(m) + n(m) \qquad (6.2)$$

Where
$x(m)$ is the P dimensional signal, or the state, vector at time m,
$\Phi(m, m-1)$ is a $P \times P$ state transition matrix that relates the states of the process at
    times $m-1$ and m,
$e(m)$, is the P-dimensional uncorrelated excitation vector of the state equation,
$\Sigma_{ee}(m)$ is the $P \times P$ covariance matrix of $e(m)$
$y(m)$ is the M dimensional noisy and distorted observation vector,
$H(m)$ is the $M \times P$ channel distortion matrix,
$n(m)$ is the M dimensional additive noise process,
$\Sigma_{nn}(m)$ is the $M \times M$ covariance matrix of $n(m)$.

The Kalman filter is derived as a recursive minimum mean squared error predictor of a signal $x(m)$, given an observation signal $y(m)$. The filter derivation assumes that the state transition matrix $\Phi(m, m-1)$, the channel distortion matrix $H(m)$, the covariance matrix $\Sigma_{ee}(m)$ of the state equation input, and the covariance matrix $\Sigma_{nn}(m)$ of the additive noise are given.

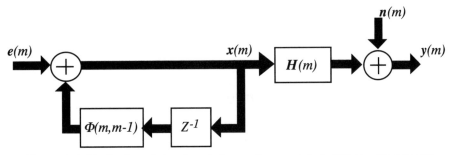

Figure 6.1 Illustration of signal and observation models in Kalman filter theory.

In this chapter we use the notation $\hat{y}(m| m - i)$ to denote a prediction of $y(m)$ based on the observation samples up to time m-i. Now assume that $\hat{y}(m| m - 1)$ is the least squared error prediction of $y(m)$ based on the observations $[y(0), ..., y(m-1)]$. Define a so called *innovation* signal as

$$v(m) = y(m) - \hat{y}(m|m - 1) \tag{6.3}$$

The innovation signal vector $v(m)$ contains all that is unpredictable from the past observations including both the noise and the unpredictable part of the signal. For an optimal linear minimum mean squared error estimate the innovation signal must be uncorrelated and orthogonal to the past observation vectors, hence we have

$$\mathcal{E}\left[v(m)y^T(m - k)\right] = 0 \qquad k > 0 \tag{6.4}$$

and

$$\mathcal{E}\left[v(m)v^T(k)\right] = 0 \qquad m \neq k \tag{6.5}$$

The concept of innovations is central to the derivation of the Kalman filter. The least squared error criterion is satisfied if the estimation error is orthogonal to the past samples. *In the following derivation of Kalman filter, the orthogonality condition of Eq. (6.4) is used as the starting point to derive an optimal linear filter whose innovations are orthogonal to the past observations.*
Substituting the observation Eq. (6.2) in Eq. (6.3) and using the relation

$$\begin{aligned}\hat{y}(m|m - 1) &= \mathcal{E}\left[y(m) \,|\, \hat{x}(m|m - 1)\right] \\ &= H(m)\hat{x}(m|m - 1)\end{aligned} \tag{6.6}$$

yields

$$\begin{aligned}v(m) &= H(m)x(m) + n(m) - H(m)\hat{x}(m|m - 1) \\ &= H(m)\tilde{x}(m) + n(m)\end{aligned} \tag{6.7}$$

where $\tilde{x}(m)$ is the signal prediction error vector defined as

$$\tilde{x}(m) = x(m) - \hat{x}(m|m - 1) \tag{6.8}$$

From Eq. (6.7) the covariance matrix of the innovation signal is given by

$$\begin{aligned}\Sigma_{vv}(m) &= \mathcal{E}\left[v(m)v^T(m)\right] \\ &= H(m)\Sigma_{\tilde{x}\tilde{x}}(m)H^T(m) + \Sigma_{nn}(m)\end{aligned} \tag{6.9}$$

where $\Sigma_{\bar{x}\bar{x}}(m)$ is the covariance matrix of $\bar{x}(m)$. Let $\hat{x}(m+1|m)$ denote the MMSE prediction of the signal $x(m+1)$. Now, the prediction of $x(m+1)$, based on the samples available up to the time m, can be expressed recursively as a linear combination of the prediction based on the samples available up to the time m-1 and the innovation signal at time m as

$$\hat{x}(m+1|m) = \hat{x}(m+1|m-1) + K(m)v(m) \tag{6.10}$$

Where the $P \times M$ matrix $K(m)$ is the Kalman gain matrix. Now from Eq. (6.1) we have

$$\hat{x}(m+1|m-1) = \Phi(m+1,m)\hat{x}(m|m-1) \tag{6.11}$$

Substituting Eq. (6.11) in (6.10) gives a recursive prediction equation as

$$\hat{x}(m+1|m) = \Phi(m+1,m)\hat{x}(m|m-1) + K(m)v(m) \tag{6.12}$$

To obtain a recursive relation for the computation and update of the Kalman gain matrix, multiply both sides of Eq. (6.12) by $v^T(m)$ and take the expectation of the results to yield

$$\mathcal{E}[\hat{x}(m+1|m)v^T(m)] = \mathcal{E}[\Phi(m+1,m)\hat{x}(m|m-1)v^T(m)] + K(m)\mathcal{E}[v(m)v^T(m)] \tag{6.13}$$

Due to the required orthogonality of the innovation sequence and the past samples we have

$$\mathcal{E}[\hat{x}(m|m-1)v^T(m)] = 0 \tag{6.14}$$

Hence from Eq. (6.13) and (6.14) the Kalman gain matrix is given by

$$K(m) = \mathcal{E}[\hat{x}(m+1|m)v^T(m)]\Sigma_{vv}^{-1}(m) \tag{6.15}$$

Now the first term in the right hand side of Eq. (6.15) can be expressed as

$$\begin{aligned} \mathcal{E}[\hat{x}(m+1|m)v^T(m)] &= \mathcal{E}[(x(m+1)-\bar{x}(m+1|m))v^T(m)] \\ &= \mathcal{E}[x(m+1)v^T(m)] \\ &= \mathcal{E}[(\Phi(m+1,m)x(m)+e(m+1))(y(m)-\hat{y}(m|m-1))^T] \\ &= \mathcal{E}[[\Phi(m+1,m)(\hat{x}(m|m-1)+\bar{x}(m|m-1))](H(m)\bar{x}(m|m-1)+n(m))^T] \\ &= \Phi(m+1,m)\mathcal{E}[\bar{x}(m|m-1)\bar{x}^T(m|m-1)]H^T(m) \end{aligned} \tag{6.16}$$

In developing successive lines of Eq. (6.16) we have used the following relations :

$$\mathcal{E}\left[\tilde{x}(m+1|m)v^T(m)\right] = 0 \tag{6.17}$$

$$\mathcal{E}\left[e(m+1)(y(m)-\hat{y}(m|m-1))^T\right]=0 \tag{6.18}$$

$$x(m) = \hat{x}(m|m-1) + \tilde{x}(m|m-1) \tag{6.19}$$

$$\mathcal{E}\left[\hat{x}(m|m-1)\ \tilde{x}(m|m-1)\right]=0 \tag{6.20}$$

and we also used the assumption that the signal and the noise are uncorrelated. Substitution of Eqs. (6.9) and (6.16) in Eq. (6.15) yields the following equation for the Kalman gain matrix

$$K(m) = \Phi(m+1,m)\Sigma_{\tilde{x}\tilde{x}}(m)H^T(m)\left[H(m)\Sigma_{\tilde{x}\tilde{x}}(m)H^T(m) + \Sigma_{nn}(m)\right]^{-1} \tag{6.21}$$

where $\Sigma_{\tilde{x}\tilde{x}}(m)$ is the covariance matrix of the signal prediction error $\tilde{x}(m|m-1)$. To derive a recursive relation for $\Sigma_{\tilde{x}\tilde{x}}(m)$ consider

$$\tilde{x}(m|m-1) = x(m) - \hat{x}(m|m-1) \tag{6.22}$$

Substitution of Eq. (6.1) and (6.12) in Eq. (6.22) and re-arrangement of the terms yields

$$\begin{aligned}
\tilde{x}(m|m-1) &= (\Phi(m,m-1)x(m-1) + e(m)) - (\Phi(m,m-1)\hat{x}(m-1|m-2) + K(m-1)v(m-1)) \\
&= \Phi(m,m-1)\tilde{x}(m-1) + e(m) - K(m-1)H(m-1)\tilde{x}(m-1) + K(m-1)n(m-1) \\
&= [\Phi(m,m-1) - K(m-1)H(m-1)]\tilde{x}(m-1) + e(m) + K(m-1)n(m-1)
\end{aligned} \tag{6.23}$$

From Eq. (6.23) we can derive the following recursive relation for the variance of the signal prediction error

$$\Sigma_{\tilde{x}\tilde{x}}(m) = L(m)\Sigma_{\tilde{x}\tilde{x}}(m-1)L^T(m) + \Sigma_{ee}(m) + K(m-1)\Sigma_{nn}(m-1)K^T(m-1) \tag{6.24}$$

where the $P \times P$ matrix $L(m)$ is defined as

$$L(m) = [\Phi(m,m-1) - K(m-1)H(m-1)] \tag{6.25}$$

Kalman Filtering Algorithm

Input, Observation vectors $\{y(m)\}$
Output, State or signal vectors $\{\hat{x}(m)\}$

Initial conditions
$$\Sigma_{\tilde{x}\tilde{x}}(0) = \delta I \tag{6.26}$$
$$\hat{x}(0|-1) = 0 \tag{6.27}$$

For $m = 0, 1, ...$
Innovation signal
$$v(m) = y(m) - H(m)\hat{x}(m|m-1) \tag{6.28}$$

Kalman gain
$$K(m) = \Phi(m+1,m)\Sigma_{\tilde{x}\tilde{x}}(m)H^T(m)\left[H(m)\Sigma_{\tilde{x}\tilde{x}}(m)H^T(m) + \Sigma_{nn}(m)\right]^{-1} \tag{6.29}$$

Prediction update
$$\hat{x}(m+1|m) = \Phi(m+1,m)\hat{x}(m|m-1) + K(m)v(m) \tag{6.30}$$

Prediction error correlation matrix update
$$L(m+1) = \left[\Phi(m+1,m) - K(m)H(m)\right] \tag{6.31}$$

$$\Sigma_{\tilde{x}\tilde{x}}(m+1) = L(m+1)\Sigma_{\tilde{x}\tilde{x}}(m)L(m+1)^T + \Sigma_{ee}(m+1) + K(m)\Sigma_{nn}(m)K(m) \tag{6.32}$$

Example 6.1 Consider the Kalman filtering of a first-order AR process $x(m)$ observed in an additive white Gaussian noise $n(m)$. Assume the signal generation and the observation equations are given as

$$x(m) = a(m)x(m-1) + e(m) \tag{6.33}$$

$$y(m) = x(m) + n(m) \tag{6.34}$$

Let $\sigma_e^2(m)$ and $\sigma_n^2(m)$ denote the variances of the excitation signal $e(m)$ and the noise $n(m)$ respectively. Substituting $\Phi(m+1,m) = a(m)$ and $H(m) = 1$ in the Kalman filter equations yields the following Kalman filter algorithm :

Initial Conditions
$$\sigma_{\tilde{x}}^2(0) = \delta \tag{6.35}$$
$$\hat{x}(0|-1) = 0 \tag{6.36}$$

For $m = 0, 1, \ldots$
Kalman gain

$$k(m) = \frac{a(m+1)\sigma_{\tilde{x}}^2(m)}{\sigma_{\tilde{x}}^2(m) + \sigma_n^2(m)} \tag{6.37}$$

Innovation signal

$$v(m) = y(m) - \hat{x}(m \mid m-1) \tag{6.38}$$

Prediction signal update

$$\hat{x}(m+1 \mid m) = a(m+1)\hat{x}(m \mid m-1) + k(m)v(m) \tag{6.39}$$

Prediction error update

$$\sigma_{\tilde{x}}^2(m+1) = [a(m+1) - k(m)]^2 \sigma_{\tilde{x}}^2(m) + \sigma_e^2(m+1) + k^2(m)\sigma_n^2(m) \tag{6.40}$$

where $\sigma_{\tilde{x}}^2(m)$ is the variance of prediction error signal.

Example 6.2 Recursive Estimation of a Constant Signal Observed in Noise

Consider the estimation of a constant signal observed in a random noise. The state and observation equations for this problem are given by

$$x(m) = x(m-1) = x \tag{6.41}$$

$$y(m) = x + n(m) \tag{6.42}$$

Note that $\Phi(m, m-1) = 1$, state excitation $e(m) = 0$, and $H(m) = 1$. Using the Kalman algorithm we have the following recursive solutions :

Initial Conditions
$$\sigma_{\tilde{x}}^2(0) = \delta \tag{6.43}$$
$$\hat{x}(0 \mid -1) = 0 \tag{6.44}$$

For $m = 0, 1, \ldots$
Kalman gain

$$k(m) = \frac{\sigma_{\tilde{x}}^2(m)}{\sigma_{\tilde{x}}^2(m) + \sigma_n^2(m)} \tag{6.45}$$

Innovation signal
$$v(m) = y(m) - \hat{x}(m \,|\, m - 1) \qquad\qquad (6.46)$$

Prediction signal update

$$\hat{x}(m + 1 | m) = \hat{x}(m | m - 1) + k(m)v(m) \qquad\qquad (6.47)$$

Prediction error update

$$\sigma_{\hat{x}}^2(m + 1) = [1 - k(m)]^2 \, \sigma_{\hat{x}}^2(m) + k^2(m) \, \sigma_n^2(m) \qquad\qquad (6.48)$$

6.2 Sample Adaptive Filters

Sample adaptive filters namely the RLS, the steepest descent and the LMS, are recursive, time-update, formulations of the least squared error Wiener filter. Sample-adaptive filters have a number of advantages over block-adaptive filters of Chapter 5 including lower delay, and better tracking of nonstationary signals. These are essential characteristics in applications such as echo cancellation, adaptive delay estimation, low delay predictive coding, noise cancellation etc., where low delay and fast tracking of time varying processes are important objectives.

Figure 6.2 illustrates the configuration of a least squared error adaptive filter. At each sampling time, an adaptation algorithm adjusts the filter coefficients to minimise the difference between the filter output and a desired, or target, signal. An adaptive filter starts at some initial state, and the filter coefficients are updated to minimise the difference between the filter output and a desired or target signal. The adaptation formula has the general recursive form :

Next Parameter Estimate = Previous Parameter Estimate + Update(error)

In adaptive filtering a number of decisions has to be made concerning the filter model and the adaptation algorithm. These are :

(a) Filter type : This can be a finite impulse response (FIR) filter, or an infinite impulse response (IIR) filter. In this chapter we only consider FIR filters as they have good stability and convergence properties and for this reason are the type most used in practice.

(b) Filter order : Often the correct number of filter taps is unknown. The filter order is either set using a *prior* knowledge of the input and the desired signals, or it may be obtained by monitoring the changes in the error signal as a function of the increasing filter order.

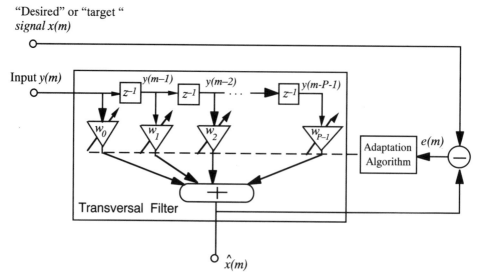

Figure 6.2 Illustration of the configuration of an adaptive filter.

(c) Adaptation algorithm : The two most widely used adaptation algorithms are the recursive least squared (RLS) error and the least mean squared error (LMS) methods. The factors that influence the choice of the adaptation algorithm are : the computational complexity, the speed of convergence to optimal operating condition, the minimum error at convergence, the numerical stability and the robustness of the algorithm to initial parameter states.

6.3 Recursive Least Squares (RLS) Adaptive Filters

Recursive least squared error (RLS) filter is a sample adaptive, time-update, version of the Wiener filter studied in Chapter 5. For stationary signals the RLS filter converges to the same optimal filter coefficients as the Wiener filter. For nonstationary signals the RLS filter tracks the time variations of the process. The RLS filter has a relatively fast rate of convergence to the optimal filter coefficients. This is useful in applications such as speech enhancement, channel equalisation and echo cancellation, where the filter has to track the changes in the signal process.

In recursive least squares algorithm, the adaptation starts with some initial filter state, and successive samples of the input signals are used to adapt the filter coefficients. As illustrated in Figure 6.2, let $y(m)$, $x(m)$ and $w(m)=[w_0(m),w_1(m), ..., w_{P-1}(m)]$ denote the filter input, the desired signal and the filter coefficient vector respectively. The filter output can be expressed as

$$\hat{x}(m) = w^T(m)y(m) \qquad (6.49)$$

where $\hat{x}(m)$ is an estimate of the desired signal $x(m)$. The filter error signal is defined as

$$e(m) = x(m) - \hat{x}(m)$$
$$= x(m) - w^T(m)y(m) \qquad (6.50)$$

The adaptation objective is to minimise the mean squared error defined as

$$\mathcal{E}[e^2(m)] = \mathcal{E}[(x(m) - w^T(m)y(m))^2]$$
$$= \mathcal{E}[x^2(m)] - 2w^T(m)\mathcal{E}[y(m)x(m)] + w^T(m)\mathcal{E}[y(m)y^T(m)]w(m)$$
$$= r_{xx}(0) - 2w^T(m)r_{yx}(m) + w^T(m)R_{yy}(m)w(m) \qquad (6.51)$$

The Wiener filter is obtained by minimising the mean squared error with respect to the filter coefficients. For stationary signals, the result of this minimisation is given in Chapter 5, Eq. (5.10), as

$$w = R_{yy}^{-1} r_{yx} \qquad (6.52)$$

Where R_{yy} is the autocorrelation matrix of the input signal and r_{yx} is the cross correlation vector of the input and the target signals. In the following we formulate a recursive, time-update, adaptive formulation of Eq. (6.52). From Section 5.2 for a block of N sample vectors the correlation matrix can be written as

$$R_{yy} = Y^T Y = \sum_{m=0}^{N-1} y(m)y^T(m) \qquad (6.53)$$

Where $y(m) = [y(m), ..., y(m-P)]^T$. Now, the sum vector product in Eq. (6.53) can be expressed in a recursive fashion as

$$R_{yy}(m) = R_{yy}(m-1) + y(m)y^T(m) \qquad (6.54)$$

To introduce adaptability to the time variations of the signal statistics, the autocorrelation estimate in Eq. (6.54) is windowed by an exponentially decaying window as

$$R_{yy}(m) = \lambda R_{yy}(m-1) + y(m)y^T(m) \qquad (6.55)$$

where λ is the so called adaptation or forgetting factor and is in the range $0 > \lambda > 1$. Similarly the cross correlation vector is given by

$$r_{yx} = \sum_{m=0}^{N-1} y(m)x(m) \qquad (6.56)$$

The sum products in Eq. (6.56) can be calculated in a recursive form as

$$r_{yx}(m) = r_{yx}(m-1) + y(m)x(m) \qquad (6.57)$$

And again this equation can be made adaptive using an exponentially decaying forgetting factor λ as

$$r_{yx}(m) = \lambda\, r_{yx}(m-1) + y(m)x(m) \qquad (6.58)$$

For a recursive solution of the least squared error Eq. (6.58), we need to obtain a recursive time-update formula for the inverse matrix in the form

$$R_{yy}^{-1}(m) = R_{yy}^{-1}(m-1) + Update(m) \qquad (6.59)$$

A recursive relation for the matrix inversion is obtained using the following lemma.

The Matrix Inversion Lemma : Let A and B be two positive-definite, $P \times P$ matrices related by

$$A = B^{-1} + CD^{-1}C^T \qquad (6.60)$$

where D is a positive-definite $N \times N$ matrix, C is a $P \times N$ matrix. The matrix inversion lemma states that the inverse of the matrix A can be expressed as

$$A^{-1} = B - BC(D + C^T BC)^{-1}C^T B \qquad (6.61)$$

This lemma can be proved by multiplying Eq. (6.60) and Eq. (6.61). The left and the right hand sides of the results of multiplication are the identity matrix.

The matrix inversion lemma can be used to obtain a recursive implementation for the inverse of the correlation matrix $R_{yy}^{-1}(m)$ as follows. Let

$$R_{yy}(m) = A \tag{6.62}$$

$$\lambda^{-1}R_{yy}^{-1}(m-1) = B \tag{6.63}$$

$$y(m) = C \tag{6.64}$$

$$D = Identity\ Matrix \tag{6.65}$$

Substituting Eqs. (6.62-63) in Eq. (6.61) we obtain

$$R_{yy}^{-1}(m) = \lambda^{-1}R_{yy}^{-1}(m-1) - \frac{\lambda^{-2}R_{yy}^{-1}(m-1)y(m)y^T(m)R_{yy}^{-1}(m-1)}{1+\lambda^{-1}y^T(m)R_{yy}^{-1}(m-1)y(m)} \tag{6.66}$$

Now define the variables $\Phi(m)$ and $k(m)$ as

$$\Phi_{yy}(m) = R_{yy}^{-1}(m) \tag{6.67}$$

and

$$k(m) = \frac{\lambda^{-1}R_{yy}^{-1}(m-1)y(m)}{1+\lambda^{-1}y^T(m)R_{yy}^{-1}(m-1)y(m)} \tag{6.68}$$

or

$$k(m) = \frac{\lambda^{-1}\Phi_{yy}(m-1)y(m)}{1+\lambda^{-1}y^T(m)\Phi_{yy}(m-1)y(m)} \tag{6.69}$$

Using Eq. (6.67) and (6.69), the recursive Eq. (6.66) for computing the inverse of the correlation matrix can be written as

$$\Phi_{yy}(m) = \lambda^{-1}\Phi_{yy}(m-1) - \lambda^{-1}k(m)y^T(m)\Phi_{yy}(m-1) \tag{6.70}$$

From Eqs. (6.69) and (6.70) we have

$$k(m) = \left(\lambda^{-1}\Phi_{yy}(m-1) - \lambda^{-1}k(m)y^T(m)\Phi_{yy}(m-1)\right)y(m)$$
$$= \Phi_{yy}(m)y(m) \tag{6.71}$$

Now Eqs. (6.70) and (6.71) are used in the following to derive the RLS adaptation algorithm.

Recursive Time-update of Filter Coefficients : The least squared error filter coefficients are

$$
\begin{aligned}
w(m) &= R_{yy}^{-1}(m)\, r_{yx}(m) \\
&= \Phi_{yy}(m) r_{yx}(m)
\end{aligned}
\tag{6.72}
$$

substituting the recursive form of the correlation vector in Eq. (6.72) yields

$$
\begin{aligned}
w(m) &= \Phi_{yy}(m)\left[\lambda r_{yx}(m-1) + y(m)x(m)\right] \\
&= \lambda \Phi_{yy}(m) r_{yx}(m-1) + \Phi_{yy}(m)y(m)x(m)
\end{aligned}
\tag{6.73}
$$

Now substitution of the recursive form of the matrix $\Phi_{yy}(m)$ from Eq. (6.70) and $k(m) = \Phi(m)y(m)$ from Eq(6.71) in the right hand side of Equation (6.73) yields

$$
w(m) = \left[\lambda^{-1} \Phi_{yy}(m-1) - \lambda^{-1}k(m)y^T(m)\Phi_{yy}(m-1)\right]\lambda\, r_{yx}(m-1) + k(m)x(m)
\tag{6.74}
$$

or

$$
w(m) = \Phi_{yy}(m-1)r_{yx}(m-1) - k(m)y^T(m)\Phi_{yy}(m-1)r_{yx}(m-1) + k(m)x(m)
\tag{6.75}
$$

Substitution of $w(m-1) = \Phi(m-1)r_{yx}(m-1)$ in Eq. (6.75) yields

$$
w(m) = w(m-1) - k(m)\left[x(m) - y^T(m)w(m-1)\right]
\tag{6.76}
$$

This equation can be rewritten in the following form

$$
w(m) = w(m-1) - k(m)e(m)
\tag{6.77}
$$

Eq. (6.77) is a recursive time-update implementation of the least squared error Wiener filter.

RLS Adaptation Algorithm:

Input signals $y(m)$ and $x(m)$

Initial values
$$
\begin{aligned}
\Phi_{yy}(m) &= \delta I \\
w(0) &= w_I
\end{aligned}
$$

For $m = 1, 2, ...$
Filter gain vector

$$k(m) = \frac{\lambda^{-1}\Phi_{yy}(m-1)y(m)}{1+\lambda^{-1}y^{T}(m)\Phi_{yy}(m-1)y(m)} \tag{6.78}$$

Error signal equation

$$e(m) = x(m) - w^{T}(m-1)y(m) \tag{6.79}$$

Filter coefficients

$$w(m) = w(m-1) - k(m)e(m) \tag{6.80}$$

Inverse correlation matrix update

$$\Phi_{yy}(m) = \lambda^{-1}\Phi_{yy}(m-1) - \lambda^{-1}k(m)y^{T}(m)\Phi_{yy}(m-1) \tag{6.81}$$

6.4   The Steepest Descent Method

The mean squared error surface with respect to the coefficients of an FIR filter, is a quadratic bowl-shaped curve, with a single global minimum that corresponds to the MMSE filter coefficients. Figure 6.3 illustrates the mean squared error curve for a single coefficient filter. The figure also illustrates the steepest descent search for the minimum mean squared error coefficient. The search is based on taking a number of successive downward steps in the direction of negative gradient of the error surface. Starting with a set of initial values, the filter coefficients are successively updated in the downward direction, until the minimum point, at which the gradient is zero, is reached. The steepest descent adaptation method can be expressed as

$$w(m+1) = w(m) + \mu\left(-\frac{\partial \mathcal{E}[e^{2}(m)]}{\partial w(m)}\right) \tag{6.82}$$

where μ is the adaptation step size. From Eq. (5.7) the gradient of mean squared error function is given by

$$\frac{\partial \mathcal{E}[e^{2}(m)]}{\partial w(m)} = -2r_{yx} + 2R_{yy}w(m) \tag{6.83}$$

Substituting Eq. (6.83) in Eq. (6.82) yields

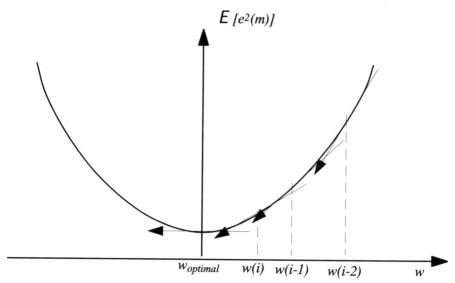

Figure 6.3 Illustration of gradient search of the mean squared error surface for the minimum error point.

$$w(m+1) = w(m) + \mu \left(r_{yx} - R_{yy}w(m) \right) \tag{6.84}$$

where the factor of 2 in Eq. (6.83) has been absorbed in the adaptation step size μ. Let w_o denote the optimal MMSE filter coefficient vector, we define a filter coefficients error vector $\tilde{w}(m)$ as

$$\tilde{w}(m) = w(m) - w_o \tag{6.85}$$

For a stationary process, the optimal MMSE filter w_o is obtained from the Wiener filter Eq. (5.10) as

$$w_o = R_{yy}^{-1} r_{yx} \tag{6.86}$$

Subtracting w_o from both sides of Eq. (6.84), and then substituting $R_{yy}w_o$ for r_{yx}, and using Eq. (6.85) yields

$$\tilde{w}(m+1) = \left[I - \mu R_{yy} \right] \tilde{w}(m) \tag{6.87}$$

It is desirable that the filter error vector $\tilde{w}(m)$ vanishes as quickly as possible. The parameter μ, the adaptation step size, controls the stability and the rate of convergence of the adaptive filter. Too big a value for μ causes instability, too small a value gives a low convergence rate. The stability of the parameter estimation method depends on the choice of the adaptation parameter μ and the autocorrelation matrix. From Eq. (6.87) a recursive equation for the error in each individual filter coefficient can be obtained as follows. The correlation matrix can be expressed in terms of the matrices of eigenvectors and eigenvalues as

$$R_{yy} = Q\Lambda Q^T \tag{6.88}$$

where Q is an orthonormal matrix of the eigenvectors of R_{yy}, and Λ is a diagonal matrix with its diagonal elements corresponding to the eigenvalues of R_{yy}. Substituting R_{yy} from Eq. (6.88) in Eq. (6.87) gives

$$\tilde{w}(m+1) = [I - \mu\, Q\Lambda Q^T]\tilde{w}(m) \tag{6.89}$$

Multiplying both sides of Eq. (6.89) by Q^T and using the relation $Q^TQ=QQ^T=I$ gives

$$Q^T\tilde{w}(m+1) = [I - \mu\Lambda]Q^T\tilde{w}(m) \tag{6.90}$$

Let

$$v(m) = Q^T\,\tilde{w}(m) \tag{6.91}$$

then

$$v(m+1) = [I - \mu\Lambda]\,v(m) \tag{6.92}$$

As Λ and I are both diagonal matrices, Eq. (6.92) can be expressed in terms of the equations for the individual elements of the error vector $v(m)$ as

$$v_k(m+1) = [1 - \mu\lambda_k]v_k(m) \tag{6.93}$$

where λ_k is the kth eigenvalue of the autocorrelation matrix of the filter input $y(m)$. Figure 6.4 is a feedback network model of the time-variations of the error vector. From Eq. (6.93), the condition for the stability of the adaptation process and the decay of the coefficient error vector is

$$-1 < 1 - \mu\lambda_k < 1 \tag{6.94}$$

Let λ_{max} denote the maximum eigenvalue of the autocorrelation matrix of $y(m)$, then from Eq. (6.94) the limits on μ for stable adaptation is given by

$$0 < \mu < \frac{2}{\lambda_{max}} \tag{6.95}$$

Convergence Rate : The convergence rate of the filter coefficients depends on the choice of the adaptation step size μ where $0<\mu<1/\lambda_{max}$. When the eigenvalues of the correlation matrix are unevenly spread, the filter coefficients converge at different speeds : the smaller the kth eigenvalue the slower the speed of convergence of the kth coefficients. The filter coefficients with the maximum and minimum eigenvalues, λ_{max} and λ_{min} converge according to the following equations

$$v_{max}(m+1) = (1 - \mu\,\lambda_{max})v_{max}(m) \tag{6.96}$$

$$v_{min}(m+1) = (1 - \mu\,\lambda_{min})v_{min}(m) \tag{6.97}$$

The ratio of the maximum to the minimum eigenvalue of a correlation matrix is called the eigenvalue spread of the correlation matrix

$$Eigenvalue\ spread = \frac{\lambda_{max}}{\lambda_{min}} \tag{6.98}$$

Note that the spread in the speed of convergence of filter coefficients is proportional to the spread in eigenvalue of the autocorrelation matrix of the input signal.

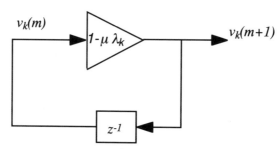

Figure 6.4 A feedback model of the variation of coefficient error with time.

6.5 The LMS Adaptation Method

The steepest descent method employs the gradient of the *averaged* squared error to search for the MMSE filter coefficients. A computationally simpler version of the gradient search method is the least mean squared (LMS) algorithm, in which the gradient of the *mean* squared error is substituted with the gradient of *instantaneous* squared error function. The LMS adaptation method is defined as

$$w(m+1) = w(m) + \mu \left(-\frac{\partial e^2(m)}{\partial w(m)} \right) \tag{6.99}$$

where the error signal $e(m)$ is given by

$$e(m) = x(m) - w^T(m)x(m) \tag{6.100}$$

The instantaneous gradient of the squared error can be re-expressed as

$$\begin{aligned}
\frac{\partial e^2(m)}{\partial w(m)} &= \frac{\partial}{\partial w(m)}[x(m) - w^T(m)y(m)]^2 \\
&= -2y(m)[x(m) - w^T(m)y(m)]^2 \\
&= -2y(m)e(m)
\end{aligned} \tag{6.101}$$

Substituting Eq. (6.101) into the recursion update equation of the filter parameters (6.99) yields the LMS adaptation equation :

$$w(m+1) = w(m) + \mu \, (y(m)e(m)) \tag{6.102}$$

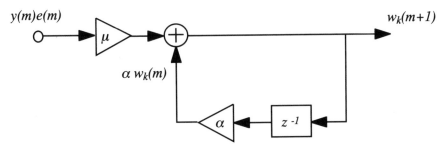

Figure 6.5 Illustration of LMS adaptation of a filter coefficient.

It can seen that the filter update equation is very simple. The LMS filter is widely used in adaptive filter applications such as adaptive equalisation, echo cancellation etc. The main advantage of the LMS algorithm is its simplicity both in terms of the memory requirement and the computational complexity which is in the order of $O(P)$, where P is the filter length.

Leaky LMS Algorithm : The stability and the adaptability of the recursive LMS adaptation Eq. (6.86) can improved by introducing a so called leakage factor α as

$$w(m+1) \ = \ \alpha w(m) \ + \ \mu \ (y(m)e(m)) \tag{6.103}$$

where α is the leakage factor. Note that the feedback equation for the time update of the filter coefficients is essentially a recursive (infinite impulse response) system with input $\mu y(m)e(m)$ and its poles at α.. When the parameter $\alpha<1$, the effect is to introduce more stability and accelerate the filter adaptation to the changes in input signal characteristics.

Steady State Error : The optimal minimum mean squared error, E_{min}, is achieved when the coefficients of the filter approach the optimum value defined by the block least square error equation $w_o = R_{yy}^{-1} r_{yx}$ derived in Chapter 5. The steepest decent method uses the average of the gradient of the error surface for incremental updates of the filter coefficients towards the optimal value. Hence when it reaches the minimum point of mean squared error curve, the *averaged* gradient is zero, and it will remain zero so long as the error surface is stationary. In contrast, examination of the LMS equation shows that for applications in which the MMSE is nonzero such as noise reduction, the incremental update term $\mu e(m)y(m)$ would remain nonzero even when the optimal point is reached. Thus at the convergence the LMS filter will randomly vary about the MMSE point with the result that the MMSE for the LMS will be in excess of the MMSE for Wiener or steepest descent methods. Note that at, or near, convergence a gradual decrease in μ would decrease the excess MMSE at the expense of some loss of adaptability to further changes in the signal characteristics.

Summary

This chapter began with an introduction to Kalman filter theory. The Kalman filter was derived using the orthogonality principle that : for the optimal filter the innovation sequence must be an uncorrelated process and orthogonal to the past observations. Note that the same principle can also be used to derive the Wiener filter coefficients. Although, like Wiener filter, the derivation of the Kalman filter is based on the least squared error criterion, the Kalman filter differs from the Wiener filter in two respects. Firstly, Kalman filter can be applied to nonstationary processes, and secondly, the Kalman theory employs a model of the signal generation process in the form of the state equation. This is an important advantage in the sense that the Kalman filter can be used for the explicit modelling of the dynamics of a signal process.

For many practical applications such as echo cancellation, channel equalisation, adaptive noise cancellation, time-delay estimation *etc.*, the RLS and the LMS filters provide a suitable alternative to Kalman filter. The RLS filter is a recursive implementation of the Wiener filter, and for stationary processes it should converge to the same solution as the Wiener filter. The main advantage of the LMS is the relative simplicity of the algorithm. However, for signals with a large spectral dynamic range, or equivalently a large eigenvalue spread, the LMS has an uneven and slow rate of convergence. If in addition to having a large eigenvalue spread a signal is also nonstationary (e.g. speech and audio signals), then the LMS can be an unsuitable adaptation method, and the RLS method, with its better convergence rate and less sensitivity to the eigenvalue spread, becomes a more attractive alternative.

Bibliography

ALEXANDER S. T. (1986) Adaptive Signal Processing : Theory and Applications, Springer-Verlag, New York.

BELLANGER M. G. (1988), Adaptive Filters and Signal Analysis, Marcel-Dekker, New York.

BERSHAD N. J. (1986), Analysis of the Normalised LMS Algorithm with Gaussian Inputs, IEEE Trans. Acoustic Speech Signal Processing ASSP-34, Pages 793-806.

BERSHAD N. J., QU L. Z. (1989), On the Probability Density Function of the LMS Adaptive Filter Weights, IEEE Trans. Acoustic Speech Signal Processing ASSP-37, Pages 43-56.

CIOFFI J. M., KAILATH T., Fast Recursive Least Squares transversal filters for adaptive filtering, IEEE Trans. Acoustic Speech and Signal Processing, Vol. ASSP-32, Pages 304-337.

CLASSEN T. A., MECKLANBRAUKER W. F., (1985), Adaptive Techniques for signal Processing in Communications, IEEE Communications, Vol-23, Pages 8-19.

COWAN C. F., GRANT P. M. (1985), Adaptive Filters, Prentice-Hall, Englewood Cliffs, N.J.

EWEDA E., MACCHI O., (1985), Tracking Error Bounds of Adaptive Nonsationary Filtering, Automatica, Vol. 21, Pages 293-302.

GABOR D., WILBY W. P., WOODCOCK R. (1960), A Universal Non-linear Filter, Predictor and Simulator which Optimises Itself by a Learning Process, IEE Proc. Vol. 108 Pages 422-38.

GABRIEL W. F. (1976), Adaptive Arrays : An Introduction, Proc. IEEE Vol. 64 Pages 239-272.

HAYKIN S.(1991), Adaptive Filter Theory, Prentice Hall, Englewood Cliffs, N.J..

HODGKISS W. S., ALEXANDROU D (1983), Applications of Adaptive Least Squares Lattice Structures to Problems in Under Water Acoustics, Proc. SPIE Vol. 431 Real Time Signal Processing Pages 45-48.

HONIG M.L., MESSERSCHMITT D. G. (1984), Adaptive filters : Structures, Algorithms and Applications, Kluwer Boston, Hingham, Mass.

KAILATH T. (1970), The Innovations Approach to Detection and Estimation Theory, Proc. IEEE, Vol. 58, Pages 680-95.

KALMAN R. E. (1960), A New Approach to Linear Filtering and Prediction Problems, Trans. of the ASME, Series D, Journal of Basic Engineering, Vol. 82 Pages 34-45.

KALMAN R. E., BUCY R. S. (1961), New Results in Linear Filtering and Prediction Theory, Trans. ASME J. Basic Eng., Vol. 83, Pages 95-108.

WIDROW B. (1990), 30 Years of Adaptive Neural Networks: Perceptron, Madaline, and Back Propagation, Proc. IEEE, Special Issue on Neural Networks I, Vol. 78.

WIDROW B., STERNS S.D. (1985), Adaptive Signal Processing Prentice Hall, Englewood Cliffs, N.J..

WILKINSON J. H. (1965), The Algebraic Eigenvalue Problem, Oxford University Press, Oxford.

ZADEH L. A., DESOER C. A. (1963), Linear System Theory : The State-Space Approach, McGraw-Hill.

7

Linear Prediction Models

7.1 Linear Prediction Coding
7.2 Forward, Backward and Lattice Predictors
7.3 Short-term and Long-Term Linear Prediction
7.4 MAP Estimation of Predictor Coefficients
7.5 Signal Restoration Using Linear Prediction Models

L inear prediction modelling is used in a diverse area of applications such as data forecasting, speech recognition, low bit rate coding, model-based spectral analysis, interpolation, signal restoration etc. In statistical literature, linear prediction models are referred to as autoregressive (AR) processes. In this chapter we introduce the theory of linear prediction, and consider efficient methods for computation of the predictor coefficients. We study the forward, the backward and the lattice predictors, and consider various methods of formulation of the least squared error predictor coefficients. For the modelling of signals with a quasi-periodic structure, such as voiced speech, an extended linear predictor, that simultaneously utilises both the short and the long term correlation structures, is introduced. Finally the application of linear prediction in enhancement of noisy speech is considered. Further applications of linear prediction models, in this book, are in Chapter 11 on the interpolation of a sequence of lost samples, and in Chapters 12 and 13 on the detection and removal of impulsive noise and transient noise pulses.

7.1 Linear Prediction Coding

The degree to which a signal can be predicted from its past samples, depends on the autocorrelation, or equivalently the bandwidth and the power spectrum, of the signal. In the time domain a predictable signal has smooth and correlated fluctuations, and in the frequency domain the energy of a predictable signal is concentrated in relatively narrow bands of frequencies. In contrast, the energy of an unpredictable signal, such as a white noise, is spread in a wide band of frequencies.

For a signal to have a capacity to convey information it must have a degree of randomness. Most signals, such as speech and music, are partially predictable and partially random. These signals can be modelled as the output of a filter excited by an uncorrelated input. The random input models the unpredictable part of the signal, whereas the filter models the predictable structure of the signal. The aim of linear prediction is to model the mechanism that introduces the correlation in a signal.

Linear prediction models are extensively used in speech processing, in low bit rate speech coders, and in speech recognition systems. Speech is generated by exhaling air through the glottis and the vocal tract. The noise-like air, from the lung, is modulated and shaped by the vibrations of the glottal cords and the resonances of the vocal tract. Figure 7.1 illustrates a source-filter model of speech. The source models the lung, and emits a random excitation signal which is filtered by a pitch filter. The pitch filter models the vibrations of the glottal cords, it is also termed "the long term predictor" as it models the correlation of each sample with the samples a pitch-period away. The main source of correlation in speech is the vocal tract. The vocal tract is modelled by a linear prediction model, also termed the short-term predictor because it models the correlation of each sample with the few immediately preceding samples. In this section we study short-term linear prediction models. In Section 7.3 the predictor model is extended to include pitch period correlations.

A linear predictor model forecasts the amplitude of a signal at time m, $x(m)$, using a linearly weighted combination of P past samples $[x(m-1), x(m-2), ..., x(m-P)]$ as

$$\hat{x}(m) = \sum_{k=1}^{P} a_k x(m - k) \tag{7.1}$$

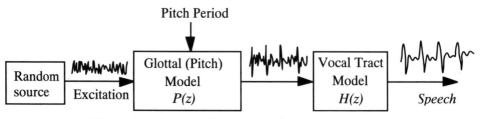

Figure 7.1 A source-filter model of speech production.

Where the integer variable m is the discrete-time index, $\hat{x}(m)$ is the prediction of $x(m)$, and a_k are the predictor coefficients. A block diagram implementation of Eq. (7.1) is illustrated in Figure 7.2.

The prediction error $e(m)$, defined as the difference between the actual sample $x(m)$ and its predicted value $\hat{x}(m)$, is given by

$$e(m) = x(m) - \hat{x}(m)$$
$$= x(m) - \sum_{k=1}^{P} a_k x(m-k) \tag{7.2}$$

For information bearing signals, the prediction error $e(m)$ may be regarded as the information, or the innovation, content of the sample $x(m)$. From Eq. (7.2) a signal generated, or modelled, by a linear predictor can be described by the following feedback equation

$$x(m) = \sum_{k=1}^{P} a_k x(m-k) + e(m) \tag{7.3}$$

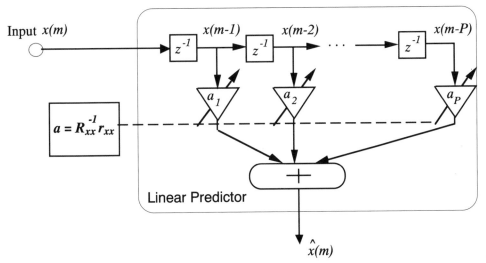

Figure 7.2 Block diagram illustration of a linear prediction system.

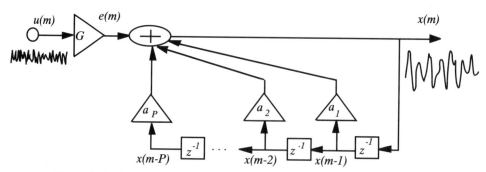

Figure 7.3 Illustration of a signal generated by a linear prediction model.

Figure 7.3 illustrates a linear predictor model of a signal $x(m)$. In this Figure the prediction error is $e(m)=G\,u(m)$, where $u(m)$ is a zero-mean unit variance random signal, and G, a gain term, is the square root of the variance of $e(m)$

$$G = (\mathcal{E}\,[e^2(m)])^{1/2} \qquad (7.4)$$

where $E[.]$ is an averaging operator. Taking the z-transform of Eq. (7.3) shows that the linear prediction model is an all-pole digital filter with the z-transfer function

$$H(z) \;=\; \frac{X(z)}{U(z)} \;=\; \frac{G}{1 - \displaystyle\sum_{k=1}^{P} a_k z^{-k}} \qquad (7.5)$$

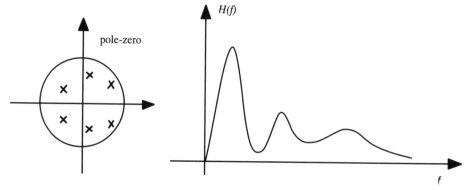

Figure 7.4 Pole-zero position, and the frequency response, of a linear predictor model.

In general, a linear predictor of order P has $P/2$ complex pole pairs, and can model up to $P/2$ resonances of the signal spectrum as illustrated in Figure 7.4. Spectral analysis using linear prediction models is discussed in Chapter 8.

7.1.1   Least Mean Squared Error Predictor

The mean squared prediction error function can be written as

$$\mathcal{E}[e^2(m)] = \mathcal{E}\left[\left(x(m) - \sum_{k=1}^{P} a_k x(m-k)\right)^2\right]$$

$$= \mathcal{E}[x^2(m)] - 2\sum_{k=1}^{P} a_k \mathcal{E}[x(m)x(m-k)] + \sum_{k=1}^{P} a_k \sum_{j=1}^{P} a_j \mathcal{E}[x(m-k)x(m-j)]$$

$$= r_{xx}(0) - 2r_{xx}^T a + a^T R_{xx} a$$

$$(7.6)$$

Where $R_{xx} = E[xx^T]$ is the autocorrelation matrix of the input vector $x^T = [x(m-1), x(m-2), \ldots, x(m-P)]$, $r_{xx} = E[x(m)x]$ is the autocorrelation vector and $a^T = [a_1, a_2, \ldots, a_P]$ is the predictor coefficients vector. From Eq. (7.6), the gradient of the mean squared prediction error with respect to the predictor coefficient vector is given by

$$\frac{\partial}{\partial a}\mathcal{E}[e^2(m)] = -2r_{xx}^T + 2a^T R_{xx} \qquad (7.7)$$

Where the gradient vector is defined as

$$\frac{\partial}{\partial a} = \left(\frac{\partial}{\partial a_1}, \frac{\partial}{\partial a_2}, \ldots, \frac{\partial}{\partial a_P}\right)^T \qquad (7.8)$$

The minimum mean squared prediction error solution, obtained by setting Eq. (7.7) to zero, is given by

$$R_{xx} a = r_{xx} \qquad (7.9)$$

The least squared error predictor coefficient vector is given by

$$a = R_{xx}^{-1} r_{xx} \qquad (7.10)$$

Eq(7.10) may also be written in an expanded form as

$$
\begin{pmatrix} a_1 \\ a_2 \\ a_3 \\ \vdots \\ a_P \end{pmatrix} = \begin{pmatrix} r_{xx}(0) & r_{xx}(1) & r_{xx}(2) & \cdots & r_{xx}(P-1) \\ r_{xx}(1) & r_{xx}(0) & r_{xx}(1) & \cdots & r_{xx}(P-2) \\ r_{xx}(2) & r_{xx}(1) & r_{xx}(0) & \cdots & r_{xx}(P-3) \\ \vdots & \vdots & \vdots & \ddots & \vdots \\ r_{xx}(P-1) & r_{xx}(P-2) & r_{xx}(P-3) & \cdots & r_{xx}(0) \end{pmatrix}^{-1} \begin{pmatrix} r_{xx}(1) \\ r_{xx}(2) \\ r_{xx}(3) \\ \vdots \\ r_{xx}(P) \end{pmatrix} \tag{7.11}
$$

An alternative formulation of the least squared error problem is as follows. For a signal block of N samples $[x(0), ..., x(N-1)]$, we can write a set of N linear prediction error equations as

$$
\begin{pmatrix} e(0) \\ e(1) \\ e(2) \\ \vdots \\ e(N-1) \end{pmatrix} = \begin{pmatrix} x(0) \\ x(1) \\ x(2) \\ \vdots \\ x(N-1) \end{pmatrix} - \begin{pmatrix} x(-1) & x(-2) & x(-3) & \cdots & x(-P) \\ x(0) & x(-1) & x(-2) & \cdots & x(1-P) \\ x(1) & x(0) & x(-1) & \cdots & x(2-P) \\ \vdots & \vdots & \vdots & \ddots & \vdots \\ x(N-2) & x(N-3) & x(N-4) & \cdots & x(N-P-1) \end{pmatrix} \begin{pmatrix} a_1 \\ a_2 \\ a_3 \\ \vdots \\ a_P \end{pmatrix} \tag{7.12}
$$

where $x_I = [x(-1), ..., x(-P)]$ is the initial vector. In a compact vector/matrix notation Eq. (7.12) can be written as

$$
e = x - Xa \tag{7.13}
$$

Using Eq. (7.13), the sum of squared prediction error over a block of N samples can be expressed as

$$
e^T e = x^T x - 2x^T Xa - a^T X^T Xa \tag{7.14}
$$

The least squared error predictor is obtained by setting the derivative of Eq. (7.14) with respect to the parameter vector a to zero as

$$
\frac{\partial e^T e}{\partial a} = -2x^T X - a^T X^T X = 0 \tag{7.15}
$$

From Eq. (7.15) the least squared error predictor is given by

$$
a = (X^T X)^{-1} (X^T x) \tag{7.16}
$$

A comparison of Eqs. (7.10) and (7.16) shows that in (7.16) the autocorrelation matrix and vector of Eq. (7.10) are replaced by the time-averaged estimates as

$$\hat{r}_{xx}(m) = \frac{1}{N} \sum_{k=0}^{N-1} x(k)x(k-m) \tag{7.17}$$

Eqs. (7.11) (7.16) may be solved efficiently by utilising the regular Toeplitz structure of the correlation matrix $\boldsymbol{R}_{xx}$. In a Toeplitz matrix, all the elements on a left-right diagonal are equal. The correlation matrix is also cross-diagonal symmetric. Note that altogether there are only $P+1$ unique elements, $[r_{xx}(0), r_{xx}(1), \ldots, r_{xx}(P)]$, in the correlation matrix and cross correlation vector. An efficient method for solution of Eq. (7.10) is the Levinson/Durbin algorithm, introduced in Section 7.2.2.

7.1.2 The Inverse Filter : Spectral Whitening

The all-pole linear predictor model, in Figure 7.3, transforms an uncorrelated excitation signal, $u(m)$, to a correlated signal $x(m)$. In the frequency domain the input-output relation of the all-pole filter of Figure 7.3 is given by

$$X(f) = \frac{GU(f)}{A(f)} = \frac{E(f)}{1 - \sum_{k=1}^{P} a_k \, e^{-j2\pi fk}} \tag{7.18}$$

where $X(f)$, $E(f)$, $U(f)$ are the spectra of $x(m)$, $e(m)$ and $u(m)$ respectively, G is the input gain factor, and $A(f)$ is the frequency response of the inverse predictor. As the

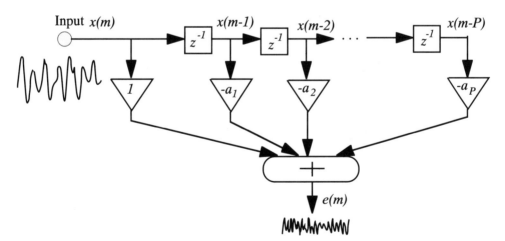

Figure 7.4 Illustration of the inverse (or whitening) filter.

excitation signal $e(m)$ is assumed to have a flat spectrum, it follows that the shape of $X(f)$ is due to the frequency response, $1/A(f)$, of the all-pole predictor model.

excitation signal $e(m)$ is assumed to have a flat spectrum, it follows that the shape of $X(f)$ is due to the frequency response, $1/A(f)$, of the all-pole predictor model.

The inverse linear predictor, as the name implies, transforms a correlated signal $x(m)$, back to an uncorrelated signal $e(m)$. The inverse filter, also known as the prediction error filter, is an all-zero, finite impulse response, filter defined as

$$e(m) = x(m) - \hat{x}(m)$$

$$= x(m) - \sum_{k=1}^{P} a_k x(m-k) \tag{7.19}$$

$$= (a^{inv})^T x$$

where the inverse filter $(a^{inv})^T = [1, -a_1, \ldots, -a_P] = [1, -a]$ and $x^T = [x(m), \ldots, x(m-P)]$. The z-transfer function of the inverse predictor model is given by

$$A(z) = 1 - \sum_{k=1}^{P} a_k z^{-k} \tag{7.20}$$

A linear predictor model is an all-pole filter, where the poles model the resonances of the signal spectrum. The inverse of an all-pole filter is an-all zero filter, with the zeros situated at the same angular frequencies as the poles of the all-pole filter, as illustrated in Figure 7.5. Consequently, the zeros of the inverse filter introduce anti-resonances

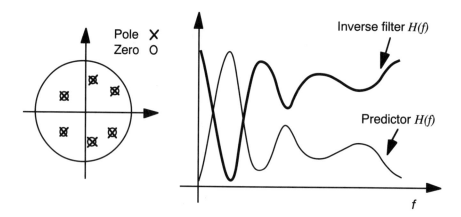

Figure 7.5 Illustration of the pole-zero diagram, and the frequency responses of : an all-pole predictor, and it's all-zero inverse filter.

that cancel out the resonances of the poles of the predictor. The inverse filter has the effect of flattening the spectrum of the input signal and is also known as a spectral whitening, or decorrelation filter.

7.1.3 The Prediction Error Signal

The prediction error signal is in general composed of three components: (a) the excitation signal, (b) the errors due to the modelling inaccuracies, and (c) the noise. The mean squared prediction error becomes zero only if, the signal is deterministic, correctly modelled by a predictor of order P, and noise-free. For example a mixture of $P/2$ sine waves can be modelled by a predictor of order P, and the prediction error is zero. However, in practice, the prediction error is nonzero because information bearing signals are random, often only approximately modelled by a linear system, and usually observed in noise. The minimum mean squared prediction error, obtained from substitution of Eq. (7.9) in (7.6), is

$$E^{(P)} = \mathcal{E}[e^2(m)] = r_{xx}(0) - \sum_{k=1}^{P} a_k r_{xx}(k) \tag{7.21}$$

Where $E^{(P)}$ denotes the prediction error for a predictor of order P. The prediction error decreases, initially rapidly and then slowly, with the increasing predictor order up to the correct model order. For the correct model order, the signal $e(m)$ is an uncorrelated zero mean random process with an autocorrelation function defined as

$$\mathcal{E}[e(m)e(m-k)] = \begin{cases} \sigma_e^2 = G^2 & \text{if } m = k \\ 0 & \text{if } m \neq k \end{cases} \tag{7.22}$$

Where σ_e^2 is the variance of $e(m)$.

7.2 Forward, Backward and Lattice Predictors

The forward predictor model of Eq. (7.1), predicts a sample, $x(m)$ from a linear combination of P past samples $x(m-1)$, $x(m-2)$, . . .,$x(m-P)$. Similarly we can define a backward predictor, that predicts a sample $x(m-P)$ from P future samples $x(m-P+1)$, . . ., $x(m)$ as

$$\hat{x}(m-P) = \sum_{k=1}^{P} c_k\, x(m-k+1) \tag{7.23}$$

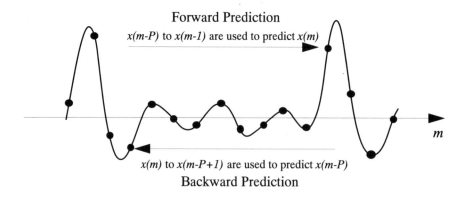

Figure 7.6 Illustration of forward and backward predictor.

The backward prediction error is defined as the difference between the actual sample and its predicted value as

$$b(m) = x(m - P) - \hat{x}(m - P)$$

$$= x(m - P) - \sum_{k=1}^{P} c_k x(m - k + 1) \qquad (7.24)$$

From Eq. (7.24), a signal generated by a backward predictor is given by

$$x(m - P) = \sum_{k=1}^{P} c_k x(m - k + 1) + b(m) \qquad (7.25)$$

The coefficients of the least squared error backward predictor, obtained in a similar method to that of the forward predictor in Section 7.1.1, are give by

$$
\begin{pmatrix}
r_{xx}(0) & r_{xx}(1) & r_{xx}(2) & \cdots & r_{xx}(P-1) \\
r_{xx}(1) & r_{xx}(0) & r_{xx}(1) & \cdots & r_{xx}(P-2) \\
r_{xx}(2) & r_{xx}(1) & r_{xx}(0) & \cdots & r_{xx}(P-3) \\
\vdots & \vdots & \vdots & \ddots & \vdots \\
r_{xx}(P-1) & r_{xx}(P-2) & r_{xx}(P-3) & \cdots & r_{xx}(0)
\end{pmatrix}
\begin{pmatrix}
c_1 \\ c_2 \\ c_3 \\ \vdots \\ c_P
\end{pmatrix}
=
\begin{pmatrix}
r_{xx}(P) \\ r_{xx}(P-1) \\ r_{xx}(P-2) \\ \vdots \\ r_{xx}(1)
\end{pmatrix}
\qquad (7.26)
$$

Note that the main difference between Eqs. (7.26) and (7.11) is that the correlation vector on the right hand side of the backward predictor Eq. (7.26) is up-side-down compared to the forward predictor (Eq. (7.11)). Since the correlation matrix is Toeplitz and symmetric, Eq. (7.11) for the forward predictor may be rearranged and rewritten in the following form :

$$
\begin{pmatrix}
r_{xx}(0) & r_{xx}(1) & r_{xx}(2) & \cdots & r_{xx}(P-1) \\
r_{xx}(1) & r_{xx}(0) & r_{xx}(1) & \cdots & r_{xx}(P-2) \\
r_{xx}(2) & r_{xx}(1) & r_{xx}(0) & \cdots & r_{xx}(P-3) \\
\vdots & \vdots & \vdots & \ddots & \vdots \\
r_{xx}(P-1) & r_{xx}(P-2) & r_{xx}(P-3) & \cdots & r_{xx}(0)
\end{pmatrix}
\begin{pmatrix}
a_P \\
a_{P-1} \\
a_{P-2} \\
\vdots \\
a_1
\end{pmatrix}
=
\begin{pmatrix}
r_{xx}(P) \\
r_{xx}(P-1) \\
r_{xx}(P-2) \\
\vdots \\
r_{xx}(1)
\end{pmatrix}
\tag{7.27}
$$

A comparison of Eqs. (7.27) and (7.26) shows that the coefficients of the backward predictor are the time-reversed version of those of the forward predictor

$$
\mathbf{c} =
\begin{pmatrix}
c_1 \\
c_2 \\
c_3 \\
\vdots \\
c_P
\end{pmatrix}
=
\begin{pmatrix}
a_P \\
a_{P-1} \\
a_{P-2} \\
\vdots \\
a_1
\end{pmatrix}
= \mathbf{a}^B
\tag{7.28}
$$

where the vector $\mathbf{a}^B$ is the reversed version of the vector $\mathbf{a}$. The relation between the backward and the forward predictors is employed in the Levinson-Durbins algorithm to derive an efficient method of calculation of the predictor coefficients as described in Section 7.2.2.

7.2.1 Augmented Equations for Forward and Backward Predictors

The inverse forward predictor coefficient vector is $[1, -a_1, ..., -a_P] = [1, -\mathbf{a}^T]$. Eqs. (7.11) and (7.21) may be combined to yield a matrix equation for the inverse forward predictor coefficients as

$$
\begin{pmatrix}
r(0) & \mathbf{r}_{xx}^T \\
\mathbf{r}_{xx} & \mathbf{R}_{xx}
\end{pmatrix}
\begin{pmatrix}
1 \\
-\mathbf{a}
\end{pmatrix}
=
\begin{pmatrix}
E^{(P)} \\
\mathbf{0}
\end{pmatrix}
\tag{7.29}
$$

Eq(7.29) is called the augmented forward predictor equation. Similarly, for the inverse backward predictor we can define an augmented backward predictor equation as

$$\begin{pmatrix} R_{xx} & r_{xx}^B \\ r_{xx}^{BT} & r(0) \end{pmatrix} \begin{pmatrix} -a^B \\ 1 \end{pmatrix} = \begin{pmatrix} 0 \\ E(P) \end{pmatrix} \tag{7.30}$$

Where $r_{xx}^T = [r_{xx}(1), \cdots, r_{xx}(P)]$ and $r_{xx}^{BT} = [r_{xx}(P), \cdots, r_{xx}(1)]$. The augmented forward and backward matrices Eqs. (7.29) and (7.30) are used to derive an order-update solution for the linear predictor coefficients as follows.

7.2.2 Levinson-Durbin Recursive Solution

The Levinson-Durbin algorithm is a recursive, order-update, method for calculation of linear predictor coefficients. A forward prediction error filter of order i can be described in terms of the forward and backward prediction error filters of order $i-1$ as

$$\begin{pmatrix} 1 \\ -a_1^{(i)} \\ \vdots \\ -a_{i-1}^{(i)} \\ -a_i^{(i)} \end{pmatrix} = \begin{pmatrix} 1 \\ -a_1^{(i-1)} \\ \vdots \\ -a_{i-1}^{(i-1)} \\ 0 \end{pmatrix} + k_i \begin{pmatrix} 0 \\ -a_{i-1}^{(i-1)} \\ \vdots \\ -a_1^{(i-1)} \\ 1 \end{pmatrix} \tag{7.31}$$

or in a more compact vector notation as

$$\begin{pmatrix} 1 \\ -a^{(i)} \end{pmatrix} = \begin{pmatrix} 1 \\ -a^{(i-1)} \\ 0 \end{pmatrix} + k_i \begin{pmatrix} 0 \\ -a^{(i-1)B} \\ 1 \end{pmatrix} \tag{7.32}$$

where k_i is called the reflection coefficient. The proof of Eq. (7.32) and the derivation of the value of reflection coefficient for k_i follows shortly. Similarly, a backward prediction error filter of order i is described in terms of the forward and backward prediction error filters of order $i-1$ as

$$\begin{pmatrix} -a^{(i-1)B} \\ 1 \end{pmatrix} = \begin{pmatrix} 0 \\ -a^{(i-1)B} \\ 1 \end{pmatrix} + k_i \begin{pmatrix} 1 \\ -a^{(i-1)} \\ 0 \end{pmatrix} \tag{7.33}$$

To prove the order-update Eq. (7.32), (or alternatively Eq. (7.33)), we multiply both sides of the equation by the $(i+1) \times (i+1)$ augmented matrix $R_{xx}^{(i+1)}$ and use the equality

$$R_{xx}^{(i+1)} = \begin{pmatrix} R_{xx}^{(i)} & r_{xx}^{(i)B} \\ r_{xx}^{(i)BT} & r_{xx}(0) \end{pmatrix} = \begin{pmatrix} r_{xx}(0) & r_{xx}^{(i)T} \\ r_{xx}^{(i)} & R_{xx}^{(i)} \end{pmatrix} \tag{7.34}$$

to obtain

$$\begin{pmatrix} R_{xx}^{(i)} & r_{xx}^{(i)B} \\ r_{xx}^{(i)BT} & r_{xx}(0) \end{pmatrix} \begin{pmatrix} 1 \\ -a^{(i)} \end{pmatrix} = \begin{pmatrix} R_{xx}^{(i)} & r_{xx}^{(i)B} \\ r_{xx}^{(i)BT} & r_{xx}(0) \end{pmatrix} \begin{pmatrix} 1 \\ -a^{(i-1)} \\ 0 \end{pmatrix} + k_i \begin{pmatrix} r_{xx}(0) & r_{xx}^{(i)T} \\ r_{xx}^{(i)} & R_{xx}^{(i)} \end{pmatrix} \begin{pmatrix} 0 \\ -a^{(i-1)B} \\ 1 \end{pmatrix} \tag{7.35}$$

Where in Eq. (7.34) (7.35) $r_{xx}^{(i)T} = [r_{xx}(1), \cdots, r_{xx}(i)]$, and $r_{xx}^{(i)BT} = [r_{xx}(i), \cdots, r_{xx}(1)]$ is the reversed version of $r_{xx}^{(i)T}$. Matrix-vector multiplication of both sides of Eq. (7.35) and the use of the augmented Eqs. (7.29) and (7.30) yields

$$\begin{pmatrix} E^{(i)} \\ 0^{(i)} \end{pmatrix} = \begin{pmatrix} E^{(i-1)} \\ 0^{(i-1)} \\ \Delta^{(i-1)} \end{pmatrix} + k_i \begin{pmatrix} \Delta^{(i-1)} \\ 0^{(i-1)} \\ E^{(i-1)} \end{pmatrix} \tag{7.36}$$

where

$$\Delta^{(i-1)} = \begin{bmatrix} 1 & -a^{(i-1)} \end{bmatrix}^T r_{xx}^{(i)B}$$

$$= r_{xx}(i) - \sum_{k=1}^{i-1} a_k^{(i-1)} r_{xx}(i-k) \tag{7.37}$$

If Eq. (7.36) is true it follows that Eq. (7.32) must also be true. The conditions for Eq. (7.36) to be true are

$$E^{(i)} = E^{(i-1)} + k_i \; \Delta^{(i-1)} \tag{7.38}$$

and

$$0 = \Delta^{(i-1)} + k_i \; E^{(i-1)} \tag{7.39}$$

From (7.39)

$$k_i = -\frac{\Delta^{(i-1)}}{E^{(i-1)}} \tag{7.40}$$

Substitution of $\Delta^{(i-1)}$ from Eq. (7.40) into Eq. (7.38) yields

$$E^{(i)} = E^{(i-1)}(1 - k_i^2)$$

$$= E^{(0)} \prod_{j=1}^{i} (1 - k_j^2) \tag{7.41}$$

Note that it can be shown that $\Delta^{(i)}$ is the cross correlation of the forward and backward prediction errors

$$\Delta^{(i-1)} = \mathcal{E}[b^{(i-1)}(m-1)e^{(i-1)}(m)] \tag{7.42}$$

The parameter $\Delta^{(i-1)}$ is known as partial correlation.

Durbins algorithm:

$$E^{(0)} = r_{xx}(0) \tag{7.43}$$

$$For\ i = 1, ... P$$

$$\Delta^{(i-1)} = r_{xx}(i) - \sum_{k=1}^{i-1} a_k^{(i-1)} r_{xx}(i-k) \tag{7.44}$$

$$k_i = -\frac{\Delta^{(i-1)}}{E^{(i-1)}} \tag{7.45}$$

$$a_i^{(i)} = k_i \tag{7.46}$$

$$a_j^{(i)} = a_j^{(i-1)} - k_i\ a_{i-j}^{(i-1)} \qquad 1 \le j \le i\text{-}1 \tag{7.47}$$

$$E^{(i)} = (1 - k_i^2)E^{(i-1)} \tag{7.48}$$

Equations(7.43-48) are solved recursively for $i = 1, ..., P$. The Durbin algorithm starts with a predictor of order zero for which $E^{(0)}=r_{xx}(0)$. The algorithm computes the coefficients of a predictor of order i, using the coefficients of a predictor of order i-1. In the process of solving for the coefficients of a predictor of order P, the solutions for the predictor coefficients of all orders less than P are also obtained.

7.2.3   Lattice Predictors

The lattice structure, shown in Figure 7.7, is a cascade connection of similar units, with each unit specified by a single parameter k_i, known as the reflection coefficient. A major attraction of a lattice structure is the relative ease with which the model order can be extended. A further advantage is that, for a stable model, the magnitude of k_i

is bounded by unity ($|k_i|<1$), and therefore it is easy to check a lattice structure for stability. The lattice structure is derived from the forward and backward prediction errors as follows. An order-update recursive equation can be obtained for the forward prediction error by multiplying both sides of Eq. (7.32) by the input vector $[x(m), x(m-1), \ldots, x(m-i)]$ as

$$e^{(i)}(m) \; = \; e^{(i-1)}(m) \; - \; k_i \, b^{(i-1)}(m-1) \tag{7.49}$$

Similarly, we can obtain an order-update recursive equation for the backward prediction error by multiplying both sides of Eq. (7.33) by the input vector $[x(m-i), x(m-i+1), \ldots, x(m)]$ as

$$b^{(i)}(m) \; = \; b^{(i-1)}(m-1) \; - \; k_i \, e^{(i-1)}(m) \tag{7.50}$$

Equations (7.49) and (7.50) are interrelated and may be implemented by a lattice network as shown in Figure 7.7. Minimisation of the squared forward prediction error of Eq (7.49) over N samples yields

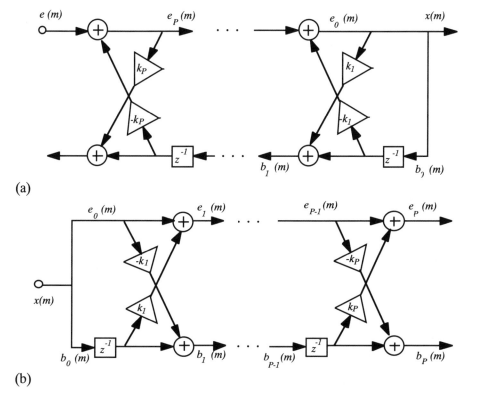

Figure 7.7 Configuration of (a) a lattice predictor, and (b) inverse lattice predictor.

$$k_i = \frac{\sum_{m=0}^{N-1} e^{(i-1)}(m)b^{(i-1)}(m-1)}{\sum_{m=0}^{N-1} \left(e^{(i-1)}(m)\right)^2} \tag{7.51}$$

Note that a similar relation for k_i can be obtained through minimisation of the squared backward prediction error of Eq (7.50) over N samples. The reflection coefficients are also known as the normalised partial correlation (PARCOR) coefficients.

7.2.4   Alternative Formulations of Least Squared Error Prediction

The methods described above, for derivation of the predictor coefficients, are based on minimisation of either the forward or the backward prediction error. In this section we consider alternative methods based on the minimisation of the sum of forward and backward prediction error.

Burg's Method

The Burg method is based on minimisation of the sum of the forward and backward squared prediction errors. The squared error function is defined as

$$E^{(i)}_{fb} = \sum_{m=0}^{N-1} \left(\left(e^{(i)}(m)\right)^2 + \left(b^{(i)}(m)\right)^2 \right) \tag{7.52}$$

Substitution of Eqs. (7.49) and (7.50) in Eq. (7.52) yields

$$E^{(i)}_{fb} = \sum_{m=0}^{N-1} \left(\left(e^{(i-1)}(m) - k_i\, b^{(i-1)}(m-1)\right)^2 + \left(b^{(i-1)}(m-1) - k_i\, e^{(i-1)}(m)\right)^2 \right) \tag{7.53}$$

Minimisation of $E^{(i)}_{fb}$ with respect to the reflection coefficients k_i yields

$$k_i = \frac{2 \sum_{m=0}^{N-1} e^{(i-1)}(m)\, b^{(i-1)}(m-1)}{\sum_{m=0}^{N-1} \left(\left(e^{(i-1)}(m)\right)^2 + \left(b^{(i-1)}(m-1)\right)^2 \right)} \tag{7.54}$$

Direct Minimisation of Backward and Forward Prediction Error

From Eq. (7.28) we have that the backward predictor coefficients vector is the reversed version of the forward predictor coefficients vector. Hence a predictor of order P can be obtained through simultaneous minimisation of the sum of the squared backward and forward prediction errors defined by the following equation

$$
\begin{aligned}
E_{fb}^{(P)} &= \sum_{m=0}^{N-1}\left((e^{(P)}(m))^2 + (b^{(P)}(m))^2\right) \\
&= \sum_{m=0}^{N-1}\left(\left(x(m) - \sum_{k=1}^{P} a_k x(m-k)\right)^2 + \left(x(m-P) - \sum_{k=1}^{P} a_k x(m-P+k)\right)^2\right) \\
&= (x - Xa)^T(x - Xa) + (x_b - X_b a)^T(x_b - X_b a)
\end{aligned}
$$

$$(7.55)$$

where X and x are the signal matrix and vector defined by Eqs. (7.12) and (7.13) respectively, and similarly X_b and x_b are the signal matrix and vector for the backward predictor. Using an approach similar to that used in derivation of Eq. (7.16), the minimisation of the mean squared error function of Eq. (7.54) yields

$$a = \left(X^T X + X_b^T X_b\right)^{-1}\left(X^T x + X_b^T x_b\right) \tag{7.56}$$

Note that for an ergodic signal as the signal length N increases Eq. (7.56) converges to the so called normal Eq. (7.10).

7.2.5   Model Order Selection

One procedure for the determination of the correct model order is to increment the model order, and monitor the differential change in the error power, until the change levels off. The incremental change in error power with the increasing model order from i-1 to i is defined as

$$\Delta E^{(i)} = E^{(i-1)} - E^{(i)} \tag{7.57}$$

The order P beyond which the decrease in the error power $\Delta E^{(P)}$ becomes less than a threshold is taken as the model order.

In linear prediction two coefficients are required for modelling each spectral peak of the signal spectrum. For example the modelling of a signal with K dominant resonances in the spectrum needs $P=2K$ coefficients. Hence a procedure for model selection is to examine the power spectrum of the signal process, and to set the model order to twice the number of significant spectral peaks in the spectrum.

When the model order is less than the correct order, the signal is under-modelled. In this case the prediction error is not well decorrelated and will be more than the optimal minimum. A further consequence of under-modelling is a decrease in spectral resolution of the model; adjacent spectral peaks of the signal could be merged and appear as a single spectral peak when the model order is too small. When the model order is larger than the correct order the signal is over-modelled. An over-modelled problem can result in an ill-conditioned matrix equation, unreliable numerical solutions, and appearance of spurious spectral peaks in the model.

7.3 Short-term and Long-term Predictors

For quasi-periodic signals, such as voiced speech, there are two types of correlation structures which can be utilised for a more accurate prediction, these are : (a) the short term correlation, which is the correlation of each sample with the P immediate past samples $x(m-1), \ldots, x(m-P)$, and (b) the long term correlation, which is the correlation of a sample $x(m)$ with say $2Q+1$ similar samples a pitch period T away $x(m-T+Q), \ldots, x(m-T-Q)$. The short-term correlation of a signal may be modelled by the linear prediction Eq. (7.3). The remaining correlation, in the prediction error signal $e(m)$, is called the long term correlation. The long-term correlation in the prediction error signal may be modelled by a pitch predictor defined as

$$\hat{e}(m) = \sum_{k=-Q}^{Q} p_k\, e(m - T - k) \tag{7.58}$$

where p_k are the coefficients of a long-term predictor of order $2Q+1$. The pitch period T can be obtained from the correlation function of $x(m)$ or $e(m)$, it is the first nonzero time lag where the correlation function attains a maximum. Assuming that the long term correlation is correctly modelled, the prediction error of the long term filter is a completely random signal with a white spectrum, and is given by

$$\varepsilon(m) = e(m) - \hat{e}(m)$$

$$= e(m) - \sum_{k=-Q}^{Q} p_k\, e(m - T - k) \tag{7.59}$$

Minimisation of $\mathcal{E}[\varepsilon^2(m)]$ results in the following solution for the pitch predictor

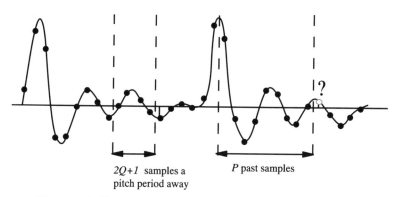

2Q+1 samples a   P past samples
pitch period away

Figure 7.8 Illustration of short-term and long-term prediction.

$$\begin{pmatrix} P_{-Q} \\ P_{-Q+1} \\ \vdots \\ P_{Q-1} \\ P_Q \end{pmatrix} = \begin{pmatrix} r_{xx}(0) & r_{xx}(1) & r_{xx}(2) & \cdots & r_{xx}(2Q) \\ r_{xx}(1) & r_{xx}(0) & r_{xx}(1) & \cdots & r_{xx}(2Q-1) \\ r_{xx}(2) & r_{xx}(1) & r_{xx}(0) & \cdots & r_{xx}(2Q-2) \\ \vdots & \vdots & \vdots & \ddots & \vdots \\ r_{xx}(2Q) & r_{xx}(2Q-1) & r_{xx}(2Q-2) & \cdots & r_{xx}(0) \end{pmatrix}^{-1} \begin{pmatrix} r_{xx}(T-Q) \\ r_{xx}(T-Q+1) \\ \vdots \\ r_{xx}(T+Q-1) \\ r_{xx}(T+Q) \end{pmatrix}$$

$$(7.60)$$

An alternative to separate modelling of the short-term and long-term correlations, is to combine the short and long term predictors into a single model described as

$$x(m) = \underbrace{\sum_{k=1}^{P} a_k x(m-k)}_{short\ term\ prediction} + \underbrace{\sum_{k=-Q}^{Q} p_k x(m-k-T)}_{long\ term\ prediction} + \varepsilon(m) \qquad (7.61)$$

In Eq. (7.61) each sample is expressed as a linear combination of P immediate past samples and $2Q+1$ samples a pitch period away. Minimisation of $\mathcal{E}[\varepsilon^2(m)]$ results in the following solution for the pitch predictor

$$
\begin{pmatrix} a_1 \\ a_2 \\ a_3 \\ \vdots \\ a_P \\ P_{-Q} \\ P_{-Q+1} \\ \vdots \\ P_{+Q} \end{pmatrix} = \begin{pmatrix} r(0) & r(1) & \cdots & r(P-1) & r(T+Q-1) & r(T+Q) & \cdots & r(T-Q-1) \\ r(1) & r(0) & \cdots & r(P-2) & r(T+Q-2) & r(T+Q-1) & \cdots & r(T+Q-2) \\ r(2) & r(1) & \cdots & r(P-3) & r(T+Q-3) & r(T+Q-2) & \cdots & r(T+Q-3) \\ \vdots & \vdots & \ddots & \vdots & \vdots & \vdots & \ddots & \vdots \\ r(P-1) & r(P-2) & \cdots & r(0) & r(T+Q-P) & r(T+Q-P+1) & \cdots & r(T+Q-P) \\ r(T+Q-1) & r(T+Q-2) & \cdots & r(T+Q-P) & r(0) & r(1) & \cdots & r(2Q) \\ r(T+Q) & r(T+Q-1) & \cdots & r(T+Q-P+1) & r(1) & r(0) & \cdots & r(2Q-1) \\ \vdots & \vdots & \ddots & \vdots & \vdots & \vdots & \ddots & \vdots \\ r(T-Q-1) & r(T-Q-2) & \cdots & r(T-Q-P) & r(2Q) & r(2Q-1) & \cdots & r(0) \end{pmatrix}^{-1} \begin{pmatrix} r(1) \\ r(2) \\ r(3) \\ \vdots \\ r(P) \\ r(T+Q) \\ r(T+Q-1) \\ \vdots \\ r(T-Q) \end{pmatrix}
$$

$$(7.62)$$

In Eq. (7.62), for simplicity the subscript xx of $r_{xx}(k)$ has been omitted. In Chapter 10 the predictor model of Eq. (7.61) is used for interpolation of a sequence of missing samples.

7.4 MAP Estimation of Predictor Coefficients

The posterior probability density function of a predictor coefficient vector a, given a signal x and the initial samples x_I, can be expressed, using the Bayes rule, as

$$
f_{A|X, X_I}(a|x, x_I) = \frac{f_{X|A, X_I}(x|a, x_I) f_{A|X_I}(a|x_I)}{f_{X|X_I}(x|x_I)} \tag{7.63}
$$

In Eq. (7.63) the pdfs are conditioned on P initial samples $x_I = [x(-P), x(-P+1), ..., x(-1)]$. Note that for a given set of samples $[x, x_I]$, $f_{X|X_I}(x|x_I)$ is a constant, and it is reasonable to assume that $f_{A|X_I}(a|x_I) = f_A(a)$

Probability Density Function of Predictor Output $f_{X|A,X_I}(x|a,x_I)$

The pdf $f_{X|A,X_I}(x| a,x_I)$ of the signal x, given the predictor coefficient vector a and the initial samples x_I, is equal to the pdf of the input signal e

$$
f_{X|A, X_I}(x|a, x_I) = f_E(x - Xa) \tag{7.64}
$$

where the input signal vector is given by

$$
e = x - Xa \tag{7.65}
$$

and $f_E(e)$ is the pdf of e. Eq. (7.64) can be expanded as

$$
\begin{pmatrix} e(0) \\ e(1) \\ e(2) \\ \vdots \\ e(N-1) \end{pmatrix} = \begin{pmatrix} x(0) \\ x(1) \\ x(2) \\ \vdots \\ x(N-1) \end{pmatrix} - \begin{pmatrix} x(-1) & x(-2) & x(-3) & \cdots & x(-P) \\ x(0) & x(-1) & x(-2) & \cdots & x(1-P) \\ x(1) & x(0) & x(-1) & \cdots & x(2-P) \\ \vdots & \vdots & \vdots & \ddots & \vdots \\ x(N-2) & x(N-3) & x(N-4) & \cdots & x(N-P-1) \end{pmatrix} \begin{pmatrix} a_1 \\ a_2 \\ a_3 \\ \vdots \\ a_P \end{pmatrix} \quad (7.66)
$$

Assuming that the input excitation signal $e(m)$ is a zero mean, uncorrelated, Gaussian process with variance σ_e^2, the likelihood function in Eq. (7.64) becomes

$$
f_{X|A,X_I}(x|a,x_I) = f_E(x - Xa)
$$

$$
= \frac{1}{(2\pi\sigma_e^2)^{N/2}} \exp\left(\frac{1}{2\sigma_e^2}(x - Xa)^T(x - Xa)\right) \quad (7.67)
$$

An alternative form of Eq. (7.67) can be obtained by rewriting Eq. (7.66) in the following form :

$$
\begin{pmatrix} e_0 \\ e_1 \\ e_3 \\ e_4 \\ \vdots \\ e_{N-1} \end{pmatrix} = \begin{pmatrix} -a_P & \cdots & -a_2 & -a_1 & 1 & 0 & 0 & 0 & 0 & 0 \\ 0 & -a_P & \cdots & -a_2 & -a_1 & 1 & 0 & 0 & 0 & 0 \\ 0 & 0 & -a_P & \cdots & -a_2 & -a_1 & 1 & 0 & 0 & 0 \\ 0 & 0 & 0 & -a_P & \cdots & -a_2 & -a_1 & 1 & 0 & 0 \\ \vdots & \vdots & \vdots & \vdots & \vdots & \ddots & \vdots & \vdots & \vdots & \vdots \\ 0 & 0 & 0 & 0 & 0 & -a_P & \cdots & -a_2 & -a_1 & 1 \end{pmatrix} \begin{pmatrix} x_{-P} \\ x_{-P+1} \\ x_{-P+2} \\ x_{-P+3} \\ \vdots \\ x_{N-1} \end{pmatrix}
$$

$$(7.68)$$

in a compact notation Eq. (7.68) can be written as

$$
e = Ax \quad (7.69)
$$

Using Eq. (7.69), and assuming that the excitation signal $e(m)$ is a zero mean, uncorrelated process with variance σ_e^2, the likelihood function of Eq. (7.67) can be written as

$$
f_{X|A,X_I}(x|a,x_I) = \frac{1}{(2\pi\sigma_e^2)^{N/2}} \exp\left(-\frac{1}{2\sigma_e^2} x^T A^T A x\right) \quad (7.70)
$$

Using the Prior pdf of the Predictor Coefficients $f_A(a)$.

The prior pdf of the predictor coefficient vector is assumed to have a Gaussian distribution with a mean vector μ_a and a covariance matrix Σ_{aa} :

$$f_A(a) = \frac{1}{(2\pi)^{P/2}|\Sigma_{aa}|^{1/2}} \exp\left(-\frac{1}{2}(a-\mu_a)^T \Sigma_{aa}^{-1}(a-\mu_a)\right) \qquad (7.71)$$

Substituting Eqs.(7.67) and (7.71) in Eq. (7.63), the posterior pdf of the predictor coefficient vector $f_{A|X,X_I}(a|x,x_I)$ can be expressed as

$$f_{A|X,X_I}(a|x,x_I) = \frac{1}{f_{X|X_I}(x|x_I)} \frac{1}{(2\pi)^{(N+P)/2} \sigma_e^N |\Sigma_{aa}|^{1/2}} \times$$

$$\exp\left(-\frac{1}{2}\left(\frac{1}{\sigma_e^2}(x-Xa)^T(x-Xa) + (a-\mu_a)^T \Sigma_{aa}^{-1}(a-\mu_a)\right)\right) \qquad (7.72)$$

The maximum a posterior estimate is obtained by maximising the log likelihood function :

$$\frac{\partial}{\partial a}\left(\ln f_{A|X,X_I}(a|x,x_I)\right) = \frac{\partial}{\partial a}\left(\frac{1}{\sigma_e^2}(x-Xa)^T(x-Xa) + (a-\mu_a)^T \Sigma_{aa}^{-1}(a-\mu_a)\right) = 0 \qquad (7.73)$$

this yields

$$\hat{a}^{MAP} = \left(\Sigma_{aa}X^TX + \sigma_e^2 I\right)^{-1}\Sigma_{aa}X^Tx + \sigma_e^2\left(\Sigma_{aa}X^TX + \sigma_e^2 I\right)^{-1}\mu_a \qquad (7.74)$$

Note that as the covariance matrix Σ_{aa} of the Gaussian prior increases, the Gaussian prior tends to a uniform prior, and the MAP solution tends to the least squared error solution :

$$\hat{a}^{LS} = (X^TX)^{-1}(X^Tx) \qquad (7.75)$$

In the following section we consider the use of linear prediction models in signal restoration.

7.5 Signal Restoration Using Linear Prediction Models

For a noisy signal, linear prediction analysis models the combined spectra of the signal and the noise. For example the frequency spectrum of a linear prediction model of speech, observed in additive white noise, would be flatter than the spectrum of the noise-free speech, due to the effects of the flat spectrum of white noise. In this section we consider the estimation of the coefficients of a predictor model from noisy observations, and the use of linear prediction model in signal restoration. The noisy signal $y(m)$ is modelled as

$$y(m) = x(m) + n(m)$$

$$= \sum_{k=1}^{P} a_k x(m-k) + e(m) + n(m) \qquad (7.76)$$

where the signal $x(m)$ is modelled by a linear prediction model with coefficients a_k and random input $e(m)$, and it is assumed that the noise $n(m)$ is additive. The least squared error predictor model of the noisy signal $y(m)$ is given by

$$\boldsymbol{R}_{yy}\hat{\boldsymbol{a}} = \boldsymbol{r}_{yy} \qquad (7.77)$$

where $\boldsymbol{R}_{yy}$ and $\boldsymbol{r}_{yy}$ are the autocorrelation matrix and vector of the noisy signal $y(m)$. For additive noise model, Eq. (7.77) can be written as

$$(\boldsymbol{R}_{xx} + \boldsymbol{R}_{nn})(\boldsymbol{a} + \tilde{\boldsymbol{a}}) = (\boldsymbol{r}_{xx} + \boldsymbol{r}_{nn}) \qquad (7.78)$$

where $\tilde{\boldsymbol{a}}$ is the error in the predictor coefficients due to noise. A simple method for removing the effects of noise is to subtract an estimate of the autocorrelation of the noise from that of the noisy signal. The drawback of this approach is that, due to random variations of noise, correlation subtraction can cause numerical instability in Eq. (7.78) and result in spurious solutions. In the following we formulate the pdf of the noisy signal and describe an iterative signal-restoration/parameter-estimation procedure proposed by Lee and Oppenheim.

From Bayes rule the MAP estimate of the predictor coefficient vector $\boldsymbol{a}$, given an observation signal $y=[y(0), y(1), ..., y(N-1)]$, and the initial samples x_I is

$$f_{A|Y,X_I}(\boldsymbol{a}|y,x_I) = \frac{f_{Y|A,X_I}(y|\boldsymbol{a},x_I)f_{A,X_I}(\boldsymbol{a},x_I)}{f_{Y,X_I}(y,x_I)} \qquad (7.79)$$

Now consider the variance of the signal y in the argument of the term $f_{Y|A,X_I}(y|a,x_I)$ in Eq. (7.79). The innovation of $y(m)$ can be defined as

$$\varepsilon(m) = y(m) - \sum_{k=1}^{P} a_k y(m-k)$$

$$= e(m) + n(m) - \sum_{k=1}^{P} a_k n(m-k)$$

(7.80)

The variance of $y(m)$, given the previous P samples and the coefficient vector a, is the variance of the innovation signal $\varepsilon(m)$ given by

$$Var[y(m)|y(m-1),\ldots,y(m-P),a] = \sigma_\varepsilon^2 = \sigma_e^2 + \sigma_n^2 - \sigma_n^2 \sum_{k=1}^{P} a_k^2 \qquad (7.81)$$

where σ_e^2 and σ_n^2 are the variance of the excitation signal and noise respectively. From Eq. (7.81) the variance of $y(m)$ is a function of the coefficient vector a. Consequently maximisation of $f_{Y|A,X_I}(y|a,x_I)$ with respect to a is a nonlinear and nontrivial exercise.

Lim and Oppenheim (1979) proposed the following iterative process in which an estimate $\hat{a}$, of the predictor coefficient vector, is used to make an estimate $\hat{x}$ of the signal vector, and the signal estimate $\hat{x}$ is then used to improve the estimate of the parameter vector $\hat{a}$, and the process is iterated until convergence.

The posterior pdf of the noise-free signal x given the noisy signal y and an estimate of the parameter vector $\hat{a}$ is

$$f_{X|A,Y}(x|\hat{a},y) = \frac{f_{Y|A,X}(y|\hat{a},x) \, f_{X|A}(x|\hat{a})}{f_{Y|A}(y|\hat{a})} \qquad (7.82)$$

Consider the likelihood term $f_{Y|A,X}(y|\hat{a},x)$. Since the noise is additive we have

$$f_{Y|A,X}(y|\hat{a},x) = f_N(y-x)$$

$$= \frac{1}{(2\pi\sigma_n^2)^{N/2}} \exp\left(-\frac{1}{2\sigma_n^2}(y-x)^T(y-x)\right) \qquad (7.83)$$

Assuming that the input of the predictor model is a zero mean Gaussian process with variance σ_e^2, the pdf of the signal x given an estimate of the predictor coefficient vector a is

$$f_{Y|A,X}(x|\hat{a}) = \frac{1}{\left(2\pi\sigma_e^2\right)^{N/2}} \exp\left(-\frac{1}{2\sigma_e^2}e^T e\right)$$

$$= \frac{1}{\left(2\pi\sigma_e^2\right)^{N/2}} \exp\left(-\frac{1}{2\sigma_e^2}x^T\hat{A}^T\hat{A}x\right)$$
(7.84)

Where $e = \hat{A}x$ as in Eq. (7.69). Substitution of Eq. (7.83) and (7.84) in Eq. (7.82) yields

$$f_{X|A,Y}(x|\hat{a},y) = \frac{1}{f_{Y|A}(y|\hat{a})}\frac{1}{\left(2\pi\sigma_n\sigma_e\right)^N} \exp\left(-\frac{1}{2\sigma_n^2}(y-x)^T(y-x)-\frac{1}{2\sigma_e^2}x^T\hat{A}^T\hat{A}x\right)$$
(7.85)

In Eq. (7.85), for a given signal y and coefficient vector $\hat{a}$, $f_{Y|A}(y|\hat{a})$ is a constant. From Eq. (7.85) the ML signal estimate is obtained by maximising the log likelihood function as

$$\frac{\partial}{\partial a}\left(\ln f_{X|A,Y}(x|\hat{a},y)\right) = \frac{\partial}{\partial x}\left(-\frac{1}{2\sigma_e^2}x^T\hat{A}^T\hat{A}x - \frac{1}{2\sigma_n^2}(y-x)^T(y-x)\right) = 0 \quad (7.86)$$

which gives

$$\hat{x} = \sigma_e^2\left(\sigma_n^2\hat{A}^T\hat{A} + \sigma_e^2 I\right)^{-1} y$$
(7.87)

The signal estimate of Eq. (7.87) can be used to obtain an updated estimate of the predictor parameter. Assuming that the signal is a zero mean Gaussian process, the estimate of a is given by

$$\hat{a}(\hat{x}) = \left(\hat{X}^T\hat{X}\right)^{-1}\left(\hat{X}^T\hat{x}\right)$$
(7.88)

Equations form the basis for an iterative signal restoration/parameter estimation method.

7.5.1 Frequency Domain Signal Restoration Using Prediction Models

The following algorithm is a frequency domain implementation of the linear prediction model-based restoration of a signal observed in additive white noise.

Initialisation : Set the initial signal estimate to noisy signal $\hat{x}_0 = y$,

For Iterations $i = 0, 1, ...$

Step 1 Estimate the predictor parameter vector $\hat{a}_i$:

$$\hat{a}_i(\hat{x}_i) = \left(\hat{X}_i^T \hat{X}_i\right)^{-1}\left(\hat{X}_i^T \hat{x}_i\right) \tag{7.89}$$

Step 2 Calculate an estimate of the model gain G using the Parseval's theorem :

$$\frac{1}{N}\sum_{f=0}^{N-1}\frac{\hat{G}^2}{\left|1-\sum_{k=1}^{P}\hat{a}_{k,i}\,e^{-j2\pi fk/N}\right|^2} = \sum_{m=0}^{N-1}y^2(m) - N\hat{\sigma}_n^2 \tag{7.90}$$

where $\hat{a}_{k,i}$ are the coefficient estimates at iteration i, and $N\hat{\sigma}_n^2$ is the energy of white noise over N samples.

Step 3 Calculate an estimate of the power spectrum of speech model :

$$\hat{P}_{X_iX_i}(f) = \frac{\hat{G}^2}{\left|1-\sum_{k=1}^{P}\hat{a}_{k,i}\,e^{-j2\pi fk/N}\right|^2} \tag{7.91}$$

Step 4 Calculate the Wiener filter frequency response :

$$\hat{W}_i(f) = \frac{\hat{P}_{X_iX_i}(f)}{\hat{P}_{X_iX_i}(f) + \hat{P}_{N_iN_i}(f)} \tag{7.92}$$

where $\hat{P}_{N_iN_i}(f) = \hat{\sigma}_n^2$ is an estimate of the noise power spectrum.

Step 5 Filter the magnitude spectrum of the noisy speech as

$$\hat{X}_{i+1}(f) = \hat{W}_i(f)\,Y(f) \tag{7.93}$$

Restore the time domain signal $\hat{x}_{i+1}$ by combining $\hat{X}_{i+1}(f)$ with the phase of noisy signal and the complex signal to time domain.

Step 6 Goto step 1 and repeat until convergence, or for a specified number of iterations.

Figure 7.9 illustrates a block diagram configuration of a Wiener filter using a linear prediction estimate of the signal spectrum. Figure 7.10 illustrates an iterative restoration of the spectrum of a noisy speech signal.

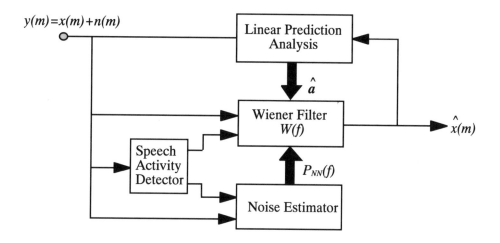

Figure 7.9 Illustration of an iterative implementation of a signal restoration system based on a linear prediction model of speech.

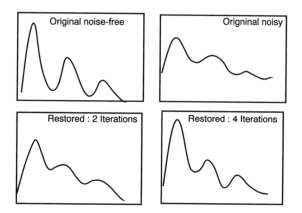

Figure 7.10 Illustration of restoration of a noisy signal with iterative linear prediction based method.

Summary

Linear prediction models are used in a wide range of signal processing applications from low bit rate speech coding to model-based spectral analysis. We began this chapter with an introduction to linear prediction theory, and considered different methods of formulation of the prediction problem and derivations of the predictor coefficients. A main attraction of the linear prediction method is the closed form solution of the predictor coefficients, and the availability of a number of efficient and relatively robust methods for solving the prediction equation; such as the Levinson-Durbin method. In Section 7.2 we considered the forward, the backward and the lattice predictors. Although the direct form implementation of the linear predictor is the most convenient method, for many applications such as transmission of the predictor coefficients in speech coding it is advantageous to use the lattice form of the predictor. This is because the lattice form can be conveniently checked for stability, and furthermore a perturbation of the parameter of any section of the lattice structure has a limited and more localised effect. In Section 7.3 we considered a modified form of linear prediction that models the short-term and the long-term correlations of the signal. This method can be used for the modelling of signals with a quasi-periodic structure such as voiced speech. In Section 7.4 we considered MAP estimation and the use of a prior pdf for derivation of the predictor coefficients. Finally in Section 7.5 linear prediction model was applied to restoration of a signal observed in additive noise.

Bibliography

AKAIKE H. (1970), Statistical Predictor Identification, Annals of the Institute of Statistical Mathematics, Vol. 22, Pages 203-217.

AKAIKE H. (1974), A New Look at Statistical Model Identification, IEEE Trans. on Automatic Control, Vol. AC-19, Pages 716-723, Dec.

ANDERSON O.D. (1976), Time Series Analysis and Forecasting, The Box-Jenkins Approach, Butterworth, London.

AYRE A.J. (1972), Probability and Evidence Columbia University Press.

BOX G.E.P, JENKINS G.M. (1976), Time Series Analysis: Forecasting and Control, Holden-Day, San Francisco, California.

BURG J.P., (1975) Maximum Entropy Spectral Analysis, P.h.D. thesis, Stanford University, Stanford, California.

DURBIN J. (1959), Efficient Estimation of Parameters in Moving Average Models, Biometrica, Vol. 46, Pages 306-317.

DURBIN J. (1960), "The Fitting of Time Series Models", Rev. Int. Stat. Inst., Vol. 28 Pages 233-244.

FULLER W.A,(1976), Introduction to Statistical Time Series, Wiley, New York.

HANSEN J. H., CLEMENTS M. A. (1987). " Iterative Speech Enhancement with Spectral Constrains", IEEE Proc. Int. Conf. on Acoustics, Speech and Signal Processing ICASSP-87, Vol. 1, Pages 189-192, Dallas, April.

HANSEN J. H., CLEMENTS M. A. (1988). "Constrained Iterative Speech Enhancement with Application to Automatic Speech Recognition", IEEE Proc. Int. Conf. on Acoustics, Speech and Signal Processing, ICASSP-88, Vol. 1, Pages 561-564, New York, April.

KOBATAKE H., INARI J., KAKUTA S. (1978), "Linear prediction Coding of Speech Signals in a High Ambient Noise Environment", IEEE Proc. Int. Conf. on Acoustics, Speech and Signal Processing, Pages 472-475, April.

LEVINSON N. (1947),"The Wiener RMS (Root Mean Square) Error Criterion in Filter Design and Prediction", J. Math Phys., Vol. 25, Pages 261-278.

LIM J. S., OPPENHEIM A. V. (1978), " All-Pole Modelling of Degraded Speech", IEEE Trans. Acoustics, Speech and Signal Processing, Vol. ASSP-26, No 3, pages 197-210, June.

LIM J. S., OPPENHEIM A. V. (1979), " Enhancement and Bandwidth Compression of Noisy Speech", Proc. IEEE, No 67 Pages 1586-1604.

MAKOUL J.(1975), Linear Prediction : A Tutorial review. Proceedings of the IEEE Vol. 63, Pages 561-580.

MARKEL J.D., GRAY A.H. (1976), Linear Prediction of Speech, Springer Verlag, New York.

RABINER L.R., SCHAFER R.W. (1976), Digital Processing of Speech Signals Prentice-Hall, Englewood Cliffs, N. J .

TONG H. (1975),"Autoregressive Model Fitting with Noisy Data by Akaike's Information Criterion", IEEE Trans. Information Theory, Vol. IT-23, Pages 409-410.

STOCKHAM T.G, CANNON T.M, INGEBRETSEN R.B (1975), "Blind Deconvolution Through Digital Signal Processing", IEEE Proc. Vol. 63, No. 4, Pages 678-692.

8

Power Spectrum Estimation

8.1 Fourier Transform, Power Spectrum and Correlation
8.2 Non-Parametric Spectral Estimation
8.3 Model-based Spectral Estimation
8.4 High Resolution Spectral Estimation Based on Subspace Eigen Analysis

Power spectrum reveals the existence, or the absence, of repetitive patterns and correlation structures in a signal process. These structural patterns are important in a wide range of applications such as data forecasting, signal coding, signal detection, radar, pattern recognition, and decision making systems. The most common method of spectral estimation is based on the fast Fourier transform (FFT). For many applications, FFT-based methods produce sufficiently good results. However, more advanced methods of spectral estimation can offer better frequency resolution, and less variance. This chapter begins with an introduction to the Fourier transform and the basic principles of spectral estimation. The classical methods for power spectrum estimation are based on periodograms. Various methods of averaging periodograms, and their effects on the variance of spectral estimates, are considered. We then study the maximum entropy, and the model based spectral estimation methods. We also consider several high-resolution spectral estimation methods, based on eigen analysis, for estimation of sinusoids observed in additive white noise.

8.1 Fourier Transform, Power Spectrum and Correlation

Power spectrum of a signal gives the distribution of the signal power among various frequencies. The power spectrum is the Fourier transform of the correlation function, and reveals information on the correlation structure of the signal. In general the more correlated or predictable a signal the more concentrated its power spectrum, and conversely the more random or unpredictable a signal the more spread its power spectrum. For example, the power spectrum of a white noise is a flat line and contains all frequencies with equal intensity, whereas the power spectrum of a periodic waveform is concentrated in narrow bands of fundamental and harmonic frequencies as shown in Figure 8.1. Therefore the power spectrum of a signal can be used to deduce the existence of repetitive structures or correlated patterns in the signal process. Such information is crucial in detection, decision making and estimation problems and in system analysis.

8.1.1 Fourier Transform, Frequency Spectrum

The objective of a signal transform is to express the signal in terms of a set of relatively simple basic (or "building block") signals, known as the basis functions, so that the transformed signal lends itself to a more convenient interpretation and manipulation. The basis function of the Fourier transform is the complex exponential $e^{-j2\pi ft}=\cos(2\pi ft)+j\sin(2\pi ft)$. Cosine and sine basis functions have the advantages of being orthogonal, infinitely differentiable and related to each other through a phase shift of $\pi/4$. The Fourier transform of a continuous-time signal $x(t)$ is defined as

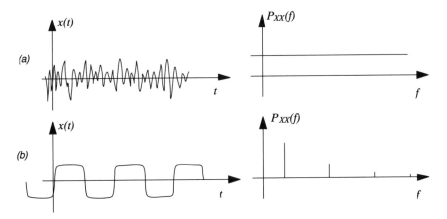

Figure 8.1 Concentration or spread of power in frequency spectrum indicates the correlated or random character of a signal : (a) a random signal, (b) a predictable signal.

$$X(f) = \int_{-\infty}^{\infty} x(t)e^{-j2\pi ft} dt \tag{8.1}$$

where $X(f)$ is a complex number that conveys the magnitude and the phase of the signal at frequency f. The inverse Fourier transform is given by

$$x(t) = \int_{-\infty}^{\infty} X(f)e^{j2\pi ft} df \tag{8.2}$$

Note from Eq. (8.1), that $X(f)$ *may be interpreted as a measure of correlation of the signal x(t) and the complex sinusoid* $e^{-j2\pi ft}$. The condition for existence of the Fourier transform integral of a signal $x(t)$ is that the signal must have finite energy

$$\int_{-\infty}^{\infty} |x(t)|^2 dt < \infty \tag{8.3}$$

The Fourier transform of a sampled signal $x(m)$ can be obtained from Eq. (8.1) as

$$X(f) = \sum_{m=-\infty}^{\infty} x(m)e^{-j2\pi fm} \tag{8.4}$$

where m denotes the discrete time variable. Using Eq. (8.4) it is easy to show that the spectrum of a sampled signal is periodic with a period equal to the sampling frequency f_s. The inverse Fourier transform of a sampled signal is defined as

$$x(m) = \int_{-1/2}^{1/2} X(f)\, e^{j2\pi fm}\, df \tag{8.5}$$

Note that the time domain signal $x(m)$ and its frequency spectrum $X(f)$ are equivalent in that they contain the same information in different domains. In particular, as expressed in Parsevals theorem, the energy of the signal may be computed either in the time or in the frequency domain as

$$Signal\ Energy = \sum_{m=-\infty}^{\infty} x^2(m) = \int_{-1/2}^{1/2} |X(f)|^2 \, \tag{8.6}$$

The function $|X(f)|^2$ is called the energy density spectrum.

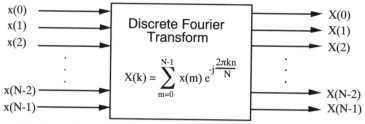

Figure 8.2 Illustration of DFT as a parallel input parallel output signal processor.

.8.1.2 Discrete Fourier Transform (DFT)

For a finite duration signal of length N samples, the discrete Fourier transform (DFT) is defined as

$$X(k) = \sum_{m=0}^{N-1} x(m)\, e^{-j\frac{2\pi}{N}km} \qquad k = 0, \ldots, N\text{-}1 \qquad (8.7)$$

And, the inverse discrete Fourier transform (IDFT) is given by

$$x(m) = \frac{1}{N} \sum_{k=0}^{N-1} X(k)\, e^{j\frac{2\pi}{N}km} \qquad m = 0, \ldots, N\text{–}1 \qquad (8.8)$$

A periodic signal has a discrete spectrum, and conversely any discrete frequency spectrum must be of a periodic signal. Hence *the implicit assumption in DFT, is that the signal x(m) is periodic with a period of N samples.*

8.1.3 Frequency Resolution and Spectral Smoothing

As illustrated in Figure 8.2, the DFT takes as the input N uniformly spaced time domain samples of $x(m)$, and outputs N uniformly spaced samples in frequency domain. The frequency resolution of the DFT spectrum is proportional to the signal length N, and is given by $\Delta f = 2\pi/N$. For a short length record the spectral resolution is low. However the spectrum of a short length signal can be interpolated to obtain a smoother spectrum. Interpolation of the frequency spectrum $X(k)$ is achieved by *zero-padding* of the time domain signal $x(m)$. Consider a signal of N samples [$(x(0)$, . . ., $x(N–1)$]. Increase the signal length from N to $2N$ samples by padding N zeros to obtain the padded sequence [$(x(0)$, . . ., $x(N–1)$, 0, . . ., 0].

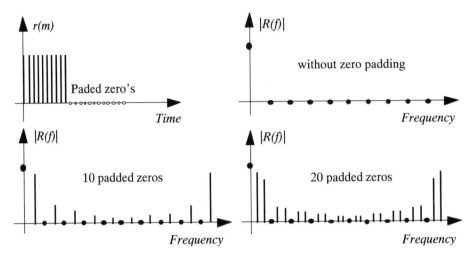

Figure 8.3 Illustration of the interpolating effect, in the frequency domain, of zero padding a signal in the time domain.

The DFT of the padded signal is given by

$$X(k) = \sum_{m=0}^{2N-1} x(m)\, e^{-j\frac{2\pi}{2N}km}$$

$$= \sum_{m=0}^{N-1} x(m)\, e^{-j\frac{\pi}{N}km}$$

$$k = 0, \ldots, 2N-1 \qquad (8.9)$$

The spectrum of the zero-padded signal, Eq. (8.9), is composed of $2N$ spectral samples; N of which, $\{X(0), X(2), X(4), X(6), \ldots X(2N-2)\}$ are the same as those that would be obtained from a DFT of the original N time domain samples, and the other N samples $\{X(1), X(3), X(5), X(6), \ldots X(2N-1)\}$ are the interpolated spectral lines that result from zero padding. Note that zero padding does not increase the spectral resolution, it merely has an interpolating, or smoothing, effect in the frequency domain.

8.1.4 Energy Spectral Density and Power Spectral Density

Energy, or power, spectrum analysis is concerned with the distribution of the signal energy or power in frequency domain. For a deterministic discrete-time signal, the energy spectral density is defined as

$$|X(f)|^2 = \left| \sum_{m=-\infty}^{\infty} x(m)e^{-j2\pi fm} \right|^2 \tag{8.10}$$

The energy spectrum of $x(m)$ may be expressed as the Fourier transform of the autocorrelation function of $x(m)$ as

$$\begin{aligned} |X(f)|^2 &= X(f)X^*(f) \\ &= \sum_{m=-\infty}^{\infty} r_{xx}(m)e^{-j2\pi fm} \end{aligned} \tag{8.11}$$

Where the variable $r_{xx}(m)$ is the autocorrelation function of $x(m)$. The Fourier transform exists only for finite energy signals. An important theoretical class of signals consists of stationary stochastic signals which, since they are stationary, are infinitely long, have infinite energy, and therefore do not possess a Fourier transform. For stochastic signals the quantity of interest is the power spectral density defined as the Fourier transform of the autocorrelation function

$$P_{XX}(f) = \sum_{m=-\infty}^{\infty} r_{xx}(m)e^{-j2\pi fm} \tag{8.12}$$

Where the autocorrelation function of $r_{xx}(m)$ is defined as

$$r_{xx}(m) = \mathcal{E}\left[x(m)x(m+k) \right] \tag{8.13}$$

In practice, the autocorrelation function is estimated from a signal record of length N samples as

$$\hat{r}_{xx}(m) = \frac{1}{N-|m|} \sum_{k=0}^{N-|m|-1} x(k)\, x(k+m) \quad k = 0, \ldots, N-1 \tag{8.14}$$

In Eq. (8.14) as the correlation lag m approaches the record length N, the estimate of $\hat{r}_{xx}(m)$ is obtained from the average of fewer samples and has a higher variance. A triangular window may be used to down-weight the correlation estimates for larger values of lag m. The triangular window has the form

$$w(m) = \begin{cases} 1 - \dfrac{|m|}{N} & |m| \leq N-1 \\ 0 & otherwise \end{cases} \tag{8.15}$$

Multiplication of Eq. (8.14) by the window of Eq. (8.15) yields

$$\hat{r}_{xx}(m) = \frac{1}{N} \sum_{k=0}^{N-|m|-1} x(k) \, x(k+m) \tag{8.16}$$

The expectation of the windowed correlation estimate $\hat{r}_{xx}(m)$ is given by

$$\mathcal{E}[\hat{r}_{xx}(m)] = \frac{1}{N} \sum_{k=0}^{N-|m|-1} \mathcal{E}[x(k)\, x(k+m)]$$
$$= \left(1 - \frac{|m|}{N}\right) r_{xx}(m) \tag{8.17}$$

In Jenkins it is shown that the variance of $\hat{r}_{xx}(m)$ is given by

$$Var[\hat{r}_{xx}(m)] \approx \frac{1}{N} \sum_{k=-\infty}^{\infty} \left[r_{xx}^2(k) + r_{xx}(k-m) r_{xx}(k+m) \right] \tag{8.18}$$

From Eqs. (8.17) and (8.18) $\hat{r}_{xx}(m)$ is an asymptotically unbiased and consistent estimate.

8.2 Non-Parametric Power Spectrum Estimation

The classic method for estimation of the power density spectrum of an N sample record is the periodogram introduced by Sir Arthur Schuster in 1898. The periodogram is defined as

$$\hat{P}_{XX}(f) = \frac{1}{N} \left| \sum_{m=0}^{N-1} x(m) e^{-j2\pi fm} \right|^2$$
$$= \frac{1}{N} |X(f)|^2 \tag{8.19}$$

The power spectral density function, or power spectrum for short, defined in Eq. (8.19) is the basis of non-parametric methods of spectral estimation. Due to finite length and the random nature of most signals, the spectra obtained from different records of a signal vary randomly about the average spectrum. A number of methods have been developed which attempt to reduce the variance of the periodogram.

8.2.1   The Mean and Variance of Periodograms

The mean of the periodogram is obtained by taking the expectation of Eq. (8.19) as

$$
\mathcal{E}[\hat{P}_{XX}(f)] = \frac{1}{N}\mathcal{E}\left[|X(f)|^2\right]
$$

$$
= \frac{1}{N}\mathcal{E}\left(\sum_{m=0}^{N-1}x(m)e^{-j2\pi fm}\sum_{n=0}^{N-1}x(n)e^{j2\pi fn}\right) \tag{8.20}
$$

$$
= \sum_{m=-(N-1)}^{N-1}\left(1-\frac{|m|}{N}\right)r_{xx}(m)e^{-j2\pi fm}
$$

As the number of signal samples N increases we have

$$
\lim_{N\to\infty}\mathcal{E}[\hat{P}_{XX}(f)] = \sum_{m=-\infty}^{\infty}r_{xx}(m)e^{-j2\pi fm} = P_{XX}(f) \tag{8.21}
$$

For a Gaussian random sequence, the variance of the periodogram can be obtained as

$$
Var[\hat{P}_{XX}(f)] = P_{XX}^2(f)\left[1 + \left(\frac{\sin 2\pi fN}{N\sin 2\pi f}\right)^2\right] \tag{8.22}
$$

As the record length N increases the expectation of the periodogram converges to the power spectrum $P_{XX}(f)$ and the variance of $\hat{P}_{XX}(f)$ converges to $P_{XX}^2(f)$. Hence the periodogram is an unbiased but not a consistent estimate. The periodograms can be calculated from a DFT of the signal $x(m)$, or from a DFT of the autocorrelation estimates $\hat{r}_{xx}(m)$. In addition the signal from which the periodogram, or the autocorrelation samples, are obtained can be segmented into overlapping blocks to result in a larger number of periodograms which can then be averaged. These methods and their effects on the variance of periodograms are considered in the following.

8.2.2   Averaging Periodograms (Bartlett Method)

In this method several periodograms, from different segments of a signal, are averaged in order to reduce the variance of the periodogram. The Bartlett periodogram is obtained as the average of K periodograms as

$$\hat{P}_{XX}^{B}(f) = \frac{1}{K} \sum_{i=1}^{K} \hat{P}_{XX}^{(i)}(f) \tag{8.23}$$

where $\hat{P}_{XX}^{(i)}(f)$ is the periodogram of the ith segment of the signal. The expectation of the Bartlett periodogram $\hat{P}_{XX}^{B}(f)$ is given by

$$\mathcal{E}[\hat{P}_{XX}^{B}(f)] = \mathcal{E}[\hat{P}_{XX}^{(i)}(f)]$$

$$= \sum_{m=-(N-1)}^{N-1} \left(1 - \frac{|m|}{N}\right) r_{xx}(m) e^{-j2\pi fm} \, v \tag{8.24}$$

$$= \frac{1}{N} \int_{-1/2}^{1/2} P_{XX}(v) \left(\frac{\sin \pi (f - v) N}{\sin \pi (f - v)}\right)^2 dv$$

Where $(\sin \pi f N / \sin \pi f)^2 / N$ is the frequency response of the triangular window $(1 - |m|/N)$. From Eq. (8.24) the Bartlett periodogram is asymptotically unbiased. The variance of $\hat{P}_{XX}^{B}(f)$ is $1/K$ of the variance of the periodogram and is given by

$$Var\left[\hat{P}_{XX}^{B}(f)\right] = \frac{1}{K} P_{XX}^2(f)\left(1 + \left(\frac{\sin 2\pi f N}{N \sin 2\pi f}\right)^2\right) \tag{8.25}$$

8.2.3 Welch Method: Averaging Periodograms from Overlapped and Windowed Segments

In this method a signal $x(m)$, of length M samples, is divided into K overlapping segments of length N, and each segment is windowed prior to computing the periodogram. The ith segment is defined as

$$x_i(m) = x(m + iD) \qquad m=0, \ldots, N-1, \ i=0, \ldots, K-1 \tag{8.26}$$

Where D is the overlap. For half overlap $D=N/2$, and $D=N$ corresponds to no overlap. For the ith windowed segment the periodogram is given by

$$\hat{P}_{XX}^{(i)}(f)] = \frac{1}{NU} \left|\sum_{m=0}^{N-1} w(m) x_i(m) \, e^{-j2\pi fm}\right|^2 \tag{8.27}$$

Where $w(m)$ is the window function and U is the power in the window function given by

$$U = \frac{1}{N} \sum_{m=0}^{N-1} w^2(m) \tag{8.28}$$

The spectrum of a finite length signal typically exhibits side lobes due to discontinuities at the endpoints. The window function $w(m)$ alleviates the discontinuities and reduces the spread of the spectral energy into the side lobes of the spectrum. The Welch power spectrum is the average of K periodograms obtained from overlapped and windowed segments of a signal :

$$\hat{P}_{XX}^W(f) = \frac{1}{K} \sum_{i=0}^{K-1} \hat{P}_{XX}^{(i)}(f) \tag{8.29}$$

Using Eqs. (8.27) and (8.29) the expectation of $\hat{P}_{XX}^W(f)$ can be obtained as

$$
\begin{aligned}
\mathcal{E}[P_{XX}^W(f)] &= \mathcal{E}[\hat{P}_{XX}^{(i)}(f)] \\
&= \frac{1}{NU} \sum_{n=0}^{N-1} \sum_{m=0}^{N-1} w(n)w(m)\mathcal{E}[x_i(m)x_i(n)]e^{-j2\pi f(n-m)} \\
&= \frac{1}{NU} \sum_{n=0}^{N-1} \sum_{m=0}^{N-1} w(n)w(m)r_{xx}(n-m)e^{-j2\pi f(n-m)} \tag{8.30} \\
&= \int_{-1/2}^{1/2} P_{XX}(v)W(v-f)dv
\end{aligned}
$$

where

$$W(f) = \frac{1}{NU} \left| \sum_{m=0}^{N-1} w(n)e^{-j2\pi fn} \right|^2 \tag{8.31}$$

and the variance of the Welch estimate is given by

$$Var[\hat{P}_{XX}^W(f)] = \frac{1}{K^2} \sum_{i=0}^{K-1} \sum_{j=0}^{K-1} \mathcal{E}\left[\hat{P}_{XX}^{(i)}(f)\hat{P}_{XX}^{(j)}(f)\right] - \left(\mathcal{E}\left[\hat{P}_{XX}^W(f)\right]\right)^2 \tag{8.32}$$

Welch has shown that for the case when there is no overlap, $D=N$,

$$Var[P_{XX}^W(f)] = \frac{Var[P_{XX}^{(i)}(f)]}{K_1} \approx \frac{P_{XX}^2(f)}{K_1} \tag{8.33}$$

and for half overlap, $D=N/2$,

$$Var[\hat{P}_{XX}^W(f)] = \frac{9}{8K_2} P_{XX}^2(f)] \tag{8.34}$$

8.2.4   Blackman-Tukey Method

In this method an estimate of a signal power spectrum is obtained from the Fourier transform of the windowed estimate of the autocorrelation function as

$$\hat{P}_{XX}^{BT}(f) = \sum_{m=-(N-1)}^{N-1} w(m)\, \hat{r}_{xx}(m)\, e^{-j2\pi fm} \tag{8.35}$$

For a signal of N samples, the number of samples available for estimation of the autocorrelation value at the lag m, $\hat{r}_{xx}(m)$, decrease as m approaches N. Therefore for large m the variance of the autocorrelation estimate increases, and the estimate becomes less reliable. The window $w(m)$ has the effect of down-weighting the high variance coefficients at and around the end-points. The mean of the Blackman-Tukey power spectrum estimate is

$$\mathcal{E}[\hat{P}_{XX}^{BT}(f)] = \sum_{m=-(N-1)}^{N-1} \mathcal{E}[\hat{r}_{xx}(m)]w(m)\, e^{-j2\pi fm} \tag{8.36}$$

Now $\mathcal{E}[\hat{r}_{xx}(m)] = r_{xx}(m)\, w_B(m)$, where $w_B(m)$ is the Bartlett, or triangular window. Eq. (8.36) may be written as

$$\mathcal{E}[\hat{P}_{XX}^{BT}(f)] = \sum_{m=-(N-1)}^{N-1} r_{xx}(m)w_c(m)\, e^{-j2\pi fm} \tag{8.37}$$

where $w_c(m) = w_B(m)w(m)$. The right hand side of Eq. (8.37) can be written in terms of the Fourier transform of the autocorrelation and the window functions as

$$\mathcal{E}[\hat{P}_{XX}^{BT}(f)] = \int_{-1/2}^{1/2} P_{XX}(v)\, W_c(f-v)dv \tag{8.38}$$

where $W_c(f)$ is the Fourier transform of $w_c(m)$. The variance of Blackman-Tukey estimate is given by

$$Var[\hat{P}_{XX}^{BT}(f)] \approx \frac{U}{N} P_{XX}^2(f) \qquad (8.39)$$

where U is the energy of the window $w_c(m)$.

8.2.5 Power Spectrum Estimation from Autocorrelation of Overlapped Segments

In Blackman-Tukey method, in calculating a correlation sequence of length N from a signal record of length N, progressively less samples are admitted in estimation of $\hat{r}_{xx}(m)$ as the lag m approaches N. Hence the variance of $\hat{r}_{xx}(m)$ increases with the lag m. This problem can be solved by using a signal of length $2N$ samples for calculation of N correlation values. In a generalisation of this method the signal record, $x(m)$, of length M samples, is divided into a number K of overlapping segments of length $2N$. The ith segment is defined as

$$x_i(m) = x(m + iD) \qquad m = 0, 1, \ldots, 2N-1 \quad (8.40)$$
$$i = 0, 1, \ldots, K-1$$

Where D is the overlap. For each segment of length $2N$ the correlation function in the range $0 \geq m \geq N$ is given by

$$\hat{r}_{xx}(m) = \frac{1}{N} \sum_{k=0}^{N-1} x_i(k)x_i(k+m) \qquad m = 0, 1, \ldots, N-1 \quad (8.41)$$

In Eq. (8.41) the estimate of each correlation value is obtained as the averaged sum of N products, and all correlation estimates have the same average variance.

8.3 Model-based Power Spectrum Estimation

In nonparametric power spectrum estimation, the autocorrelation function is assumed to be zero for lags $|m| \geq N$ beyond which no estimates are available. In parametric or model-based methods, a model of the signal process is used to extrapolate the autocorrelation function beyond the range $|m| \geq N$ for which data is available. Model based spectral estimators have a better resolution than the periodograms, mainly because they do not assume that the correlation sequence is zero-valued for the range of lags for which no measurements are available.

In linear model-based spectral estimation, it is assumed that the signal, $x(m)$, can be modelled as the output of a linear time-invariant system excited with a random, flat spectrum, excitation. The assumption that the input has a flat spectrum implies that the power spectrum of the model output is *shaped* entirely by the frequency response of the model. The input-output relation of a generalised discrete linear time-invariant model is given by

$$x(m) = \sum_{k=1}^{P} a_k x(m-k) + \sum_{k=0}^{Q} b_k e(m-k) \tag{8.42}$$

where $x(m)$ is the model output, $e(m)$ is the input, and a_k's and b_k's are the parameters of the model. Eq. (8.42) is known as an Auto-Regressive-Moving-Average (ARMA) model. The system function, $H(z)$, of the discrete linear time-invariant model of Eq. (8.42) is given by

$$H(z) = \frac{B(z)}{A(z)} = \frac{\sum_{k=0}^{Q} b_k z^{-k}}{1 - \sum_{k=1}^{P} a_k z^{-k}} \tag{8.43}$$

Where $1/A(z)$ and $B(z)$ are the autoregressive and moving-average parts of $H(z)$ respectively. The power spectrum of the signal, $x(m)$, is given as the product of the power spectrum of the input signal and the squared magnitude frequency response of the model as

$$P_{XX}(f) = P_{EE}(f)|H(f)|^2 \tag{8.44}$$

where $H(f)$ is the frequency response of the model and $P_{EE}(f)$ is the input power spectrum. Assuming that the input is a white noise process with unit variance, i.e. $P_{EE}(f)=1$, Eq. (8.44) becomes

$$P_{XX}(f) = |H(f)|^2 \tag{8.45}$$

Thus the power spectrum of the model output is the squared magnitude of the frequency response of the model. An important aspect of model-based spectral estimation is the choice of the model. The model may be an auto regressive (all-pole), a moving average (all-zero), or an ARMA (pole-zero) model.

8.3.1   Maximum Entropy Spectral Estimation

The power spectrum of a stationary signal is defined as the Fourier transform of the autocorrelation sequence as

$$P_{XX}(f) = \sum_{n=-\infty}^{\infty} r_{xx}(m) e^{-j2\pi fm} \qquad (8.46)$$

Eq. (8.46) requires the autocorrelation $r_{xx}(m)$ for the lags m in the range $\pm\infty$. In practice, an estimate of the autocorrelation $r_{xx}(m)$ is available only for the values of m in a finite range of say $\pm P$. In general, there are an infinite number of different correlation sequences that have the same values in the range $|m| \leq P$ as the measured values. The particular estimate used in the nonparametric methods assumes the correlation values are zero for the lags beyond $\pm P$ for which no estimates are available. This arbitrary assumption results in spectral leakage and loss of frequency resolution. *The maximum entropy estimate is based on the principle that the estimate of the autocorrelation sequence must correspond to the most random signal whose correlation values in the range $|m| \leq P$ coincide with the measured values.* The maximum entropy principle is appealing because it assumes no more structure in the correlation sequence than that indicated by the measured data. The randomness or the entropy of a signal is defined as

$$H[P_{XX}(f)] = \int_{-1/2}^{1/2} \ln P_{XX}(f) \, df \qquad (8.47)$$

To obtain the maximum entropy correlation estimate, we differentiate Eq. (8.47) with respect to the unknown values of the correlation coefficients, and set the derivative to zero as

$$\frac{\partial H[P_{XX}(f)]}{\partial r_{xx}(m)} = \int_{-1/2}^{1/2} \frac{\partial \ln P_{XX}(f)}{\partial r_{xx}(m)} \, df = 0 \qquad for \ |m| > P \quad (8.48)$$

Now from Eq. (8.46), the derivative of the power spectrum with respect to the autocorrelation values is given by

$$\frac{\partial P_{XX}(f)}{\partial r_{xx}(m)} = e^{-j2\pi fm} \qquad (8.49)$$

From Eq. (8.49), for the derivative of the logarithm of power spectrum we have

$$\frac{\partial \ln P_{XX}(f)}{\partial r_{xx}(m)} = P_{XX}^{-1}(f)\, e^{-j2\pi fm} \tag{8.50}$$

Substitution of Eq. (8.50) in Eq. (8.48) gives

$$\int_{-1/2}^{1/2} P_{XX}^{-1}(f)\, e^{-j2\pi fm}\, df = 0 \qquad \text{for } |m| > P \tag{8.51}$$

Assuming that $P_{XX}^{-1}(f)$ is integrable, it may be associated with an autocorrelation sequence $c(m)$ as

$$P_{XX}^{-1}(f) = \sum_{m=-\infty}^{\infty} c(m)\, e^{-j2\pi fm} \tag{8.52}$$

where

$$c(m) = \int_{-1/2}^{1/2} P_{XX}^{-1}(f)\, e^{j2\pi fm}\, df \tag{8.53}$$

From Eq. (8.51) and Eq. (8.53) we have $c(m)=0$ for $|m| > P$. Hence from Eq. (8.53), the inverse of the maximum entropy power spectrum may be obtained from the Fourier transform of a finite length autocorrelation sequence as

$$P_{XX}^{-1}(f) = \sum_{m=-P}^{P} c(m) e^{-j2\pi fm} \tag{8.54}$$

and the maximum entropy power spectrum is given by

$$\hat{P}_{XX}^{ME}(f) = \frac{1}{\displaystyle\sum_{m=-P}^{P} c(m) e^{-j2\pi fm}} \tag{8.55}$$

Since the denominator polynomial in Eq. (8.55) is symmetric it follows that for every zero of this polynomial situated at a radius r there is a zero at radius $1/r$. Hence this symmetric polynomial can be factorised and expressed as

$$\sum_{m=-P}^{P} c(m)\, z^{-m} = \frac{1}{\sigma^2} A(z)A(z^{-1}) \tag{8.56}$$

where $1/\sigma^2$ is a gain term, and $A(z)$ is a polynomial of order P defined as

$$A(z) = 1 + a_1 z^{-1} + \cdots + a_p z^{-P} \qquad (8.57)$$

From Eqs. (8.55), and (8.56) the maximum entropy power spectrum may be expressed as

$$\hat{P}_{XX}^{ME}(f) = \frac{\sigma^2}{A(z)A(z^{-1})} \qquad (8.58)$$

Eq. (8.58) shows that the maximum entropy power spectrum estimate is the power spectrum of an auto regressive (AR) model. Eq. (8.58) was obtained by maximising the entropy of the power spectrum with respect to the unknown autocorrelation values. The known values of the autocorrelation function can be used to obtain the coefficients of the AR model of Eq. (8.58) as discussed in the next section.

8.3.2 Autoregressive Power Spectrum Estimation

In the preceding section it was shown that the maximum entropy spectrum is equivalent to the spectrum of an autoregressive model of the signal. An autoregressive, or linear prediction model, considered in detail in Chapter 7, is defined as

$$x(m) = \sum_{k=1}^{P} a_k x(m - k) + e(m) \qquad (8.59)$$

where $e(m)$ is a random signal of variance σ_e^2. The power spectrum of an auto regressive process is given by

$$P_{XX}^{AR}(f) = \frac{\sigma_e^2}{\left| 1 - \sum_{k=1}^{P} a_k e^{-j 2\pi f k} \right|^2} \qquad (8.60)$$

An AR model extrapolates the correlation sequence beyond the range for which estimates are available. The relation between the autocorrelation values and the AR model parameters is obtained by multiplying both sides of Eq. (8.59) by $x(m-j)$ and taking the expectation as

$$\mathcal{E}[x(m)x(m-j)] = \sum_{k=1}^{P} a_k \mathcal{E}[x(m-k)x(m-j)] + \mathcal{E}[e(m)x(m-j)] \quad (8.61)$$

As the random input $e(m)$ is orthogonal to past samples Eq. (8.61) becomes

$$r_{xx}(j) = \sum_{k=1}^{P} a_k r_{xx}(j-k) \qquad\qquad j=1, 2, \ldots \quad (8.62)$$

Given P correlation values, Eq. (8.62) can be used to solve for the AR coefficients. Eq. (8.62) can also be used to extrapolate the correlation sequence. The methods of solving the AR model coefficients are discussed in Chapter 7.

8.3.3 Moving Average Power Spectral Estimation

A moving average model is also known as an all-zero or a finite impulse response (FIR) filter. A signal $x(m)$, modelled as a moving average process is described as

$$x(m) = \sum_{k=0}^{Q} b_k e(m-k) \qquad (8.63)$$

where $e(m)$ is a zero mean random input and Q is the model order. The cross correlation of the input and output of a moving average process is given by

$$r_{xe}(m) = \mathcal{E}[x(j)e(j-m)]$$
$$= \mathcal{E}\left[\sum_{k=0}^{Q} b_k e(j-k)\, e(j-m)\right] = \sigma_e^2 b_m \qquad (8.64)$$

and the autocorrelation function of a moving average process is

$$r_{xx}(m) = \begin{cases} \sigma_e^2 \sum_{k=0}^{Q-|m|} b_k b_{k+m} & |m| \le Q \\ 0 & |m| > Q \end{cases} \qquad (8.65)$$

From Eq. (8.65) the power spectrum obtained from the Fourier transform of the autocorrelation sequence is the same as the power spectrum of a moving average model of the signal. Hence the power spectrum of a moving average process may be obtained directly from the Fourier transform of the autocorrelation function as

$$P_{XX}^{MA} = \sum_{m=-Q}^{Q} r_{xx}(m)\, e^{-j2\pi fm} \tag{8.66}$$

Note also that moving average spectral estimation is identical to Blackman-Tukey method of estimating periodograms from autocorrelation sequence.

8.3.4   Autoregressive Moving Average Power Spectral Estimation

The ARMA, or the pole-zero, model is described by Eq. (8.42). The relationship between the ARMA parameters and the autocorrelation sequence can be obtained, by multiplying both sides of Eq. (8.42) by $x(m\text{-}j)$ and taking the expectation, as

$$r_{xx}(j) = -\sum_{k=1}^{P} a_k r_{xx}(j-k) + \sum_{k=0}^{Q} b_k r_{xe}(j-k) \tag{8.67}$$

The moving average part of Eq. (8.67) influences the autocorrelation values only up to lag Q. Hence for the autoregressive part of Eq. (8.67) we have

$$r_{xx}(m) = -\sum_{k=1}^{P} a_k r_{xx}(m-k) \qquad for\ m > Q \tag{8.68}$$

Hence Eq. (8.68) can be used to obtain the coefficients a_k which may then be substituted in Eq. (8.67) for solving the coefficients b_k. Once the coefficients of an ARMA model are identified the spectral estimate is given by

$$P_{XX}^{ARMA}(f) = \sigma_e^2 \frac{\left| \sum_{k=0}^{Q} b_k e^{-j2\pi fk} \right|^2}{\left| 1 + \sum_{k=1}^{P} a_k e^{-j2\pi fk} \right|^2} \tag{8.69}$$

Where σ_e^2 is the variance of the input of the ARMA model. In general the poles model the resonances of the signal spectrum, whereas the zeros model the anti-resonances of the spectrum.

8.4 High Resolution Spectral Estimation Based on Subspace Eigen Analysis

The eigen-based methods considered in this section are primarily used for estimation of the parameters of sinusoidal signals observed in an additive white noise. Eigen analysis is used for partitioning the eigenvectors and the eigenvalues, of the autocorrelation matrix of a noisy signal, into two subspaces : a) the signal subspace composed of the *principle* eigenvectors associated with the largest eigenvalues, and b) the noise subspace represented by the smallest eigenvalues. The decomposition of a noisy signal into a signal subspace and a noise subspace forms the basis of eigen analysis methods considered in this section.

8.4.1 Pisarenko Harmonic Decomposition

A real-valued sinewave can be modelled by a second order autoregressive (AR) model, with its poles on the unit circle at the angular frequency of the sinusoid as shown in Figure 8.4. The AR model for a sinusoid of frequency f_i at a sample rate of f_s is given by

$$x(m) = 2\cos(2\pi f_i / f_s)x(m-1) - x(m-2) + A\delta(m-t_0) \qquad (8.70)$$

where $A\delta(m\text{-}t_0)$ is the initial impulse for a sine wave of amplitude A. In general a signal composed of P real sinusoids can be modelled by an AR model of order $2P$ as

$$x(m) = \sum_{k=1}^{2P} a_k x(m-k) + A\delta(m-t_0) \qquad (8.71)$$

The transfer function of the AR model is given by

$$H(z) = \frac{A}{1 - \sum_{k=1}^{2P} a_k z^{-k}} = \frac{A}{\prod_{k=1}^{P}(1 - e^{-j2\pi f_k}z^{-1})(1 - e^{+j2\pi f_k}z^{-1})} \qquad (8.72)$$

Where the angular positions of the poles on the unit circle, $e^{\pm j2\pi f_k}$, correspond to the angular frequencies of the sinusoids. For P real sinusoids observed in an additive white noise we can write

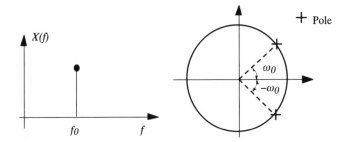

Figure 8.4 A 2nd order all pole model of a sinusoidal signal.

$$y(m) = x(m) + n(m)$$

$$= \sum_{k=1}^{2P} a_k x(m - k) + n(m) \tag{8.73}$$

Substituting $[y(m-k)-n(m-k)]$ for $x(m-k)$ in Eq. (8.73) yields

$$y(m) - \sum_{k=1}^{2P} a_k y(m - k) = n(m) - \sum_{k=1}^{2P} a_k n(m - k) \tag{8.74}$$

From Eq. (8.74) the noisy sinusoidal signal $y(m)$ can be modelled by an ARMA process in which the AR and the MA sections are identical, and the input is the noise process. Eq. (8.74) can also be expressed in a vector notation as

$$y^T a = n^T a \tag{8.75}$$

where $y^T=[y(m), \ldots, y(m-2P)]$, $a^T=[1, a_1, \ldots, a_{2P}]$ and $n^T=[n(m), \ldots, n(m-2P)]$. To obtain the parameter vector a, we multiply both sides of Eq. (8.75) by the vector y and take the expectation as

$$\mathcal{E}[yy^T] a = \mathcal{E}[yn^T] a \tag{8.76}$$

or

$$R_{yy} a = R_{yn} a \tag{8.77}$$

Where $\mathcal{E}[yy^T] = R_{yy}$, and $\mathcal{E}[yn^T] = R_{yn}$ can be written as

$$R_{yn} = \mathcal{E}[(x+n)n^T]$$
$$= \mathcal{E}[nn^T] = R_{nn} = \sigma_n^2 I \tag{8.78}$$

where σ_n^2 is the noise variance. Using Eq. (8.77), Eq. (8.76) becomes

$$R_{yy}\, a = \sigma_n^2\, a \tag{8.79}$$

Eq. (8.79) is in the form of an eigen equation. If the dimension of the matrix R_{yy} is greater than $2P \times 2P$, then the largest $2P$ eigenvalues are associated with the eigenvectors of the noisy sinusoids and the minimum eigenvalue corresponds to the noise variance σ_n^2. The parameter vector a is obtained as the eigenvector of R_{yy}, with its first element unity and associated with the minimum eigenvalue. From the AR parameter vector a we can obtain the frequencies of the sinusoids by first calculating the roots of the polynomial

$$1 + a_1 z^{-1} + a_2 z^{-2} + \cdots + a_2 z^{-2P+2} + a_1 z^{-2P+1} - z^{-2P} = 0 \tag{8.80}$$

Note that for sinusoids the AR parameters form a symmetric polynomial, that is $a_k = a_{2P-k}$. The frequencies f_k of the sinusoids can be obtained from the roots z_k of Eq. (8.80) using the relation

$$z_k = e^{j2\pi f_k} \tag{8.81}$$

The powers of the sinusoids are calculated as follows. For P sinusoids observed in additive white noise the autocorrelation function is given by

$$r_{yy}(k) = \sum_{i=1}^{P} P_i \cos 2k\pi f_i + \sigma_n^2\, \delta(k) \tag{8.82}$$

Where $P_i = A_i^2/2$ is the power of the sinusoid $A_i \sin(2\pi f_i)$, and white noise affects only the correlation at lag zero $r_{yy}(0)$. Hence Eq. (8.82), for the correlation lags $k=1$, ..., P can be written as

$$
\begin{pmatrix}
\cos 2\pi f_1 & \cos 2\pi f_2 & \cdots & \cos 2\pi f_P \\
\cos 4\pi f_1 & \cos 4\pi f_2 & \cdots & \cos 4\pi f_P \\
\vdots & \vdots & \ddots & \vdots \\
\cos 2P\pi f_1 & \cos 2P\pi f_2 & \cdots & \cos 2P\pi f_P
\end{pmatrix}
\begin{pmatrix}
P_1 \\ P_2 \\ \vdots \\ P_P
\end{pmatrix}
=
\begin{pmatrix}
r_{yy}(1) \\ r_{yy}(2) \\ \vdots \\ r_{yy}(P)
\end{pmatrix}
\tag{8.83}
$$

Given an estimate of the frequencies f_i from Eqs. (8.79-80), and an estimate of the autocorrelation function $\hat{r}_{yy}(k)$, Eq. (8.83) can be solved to obtain the powers of the sinusoids P_i. The noise variance can then be obtained from Eq. (8.82) as

$$\sigma_n^2 = r_{yy}(0) - \sum_{i=1}^{P} P_i \tag{8.84}$$

8.4.2 Multiple Signal Classification (MUSIC) Spectral Estimation

The MUSIC algorithm is an eigen-based subspace decomposition method for estimation of the frequencies of complex sinusoids observed in additive white noise. Consider a signal $y(m)$ modelled as

$$y(m) = \sum_{k=1}^{P} A_k \, e^{-j(2\pi f_k m + \phi_k)} + n(m) \tag{8.85}$$

An N sample vector $y=[y(m), \ldots, y(m+N-1)]$ of the noisy signal can be written as

$$
\begin{aligned}
y &= x + n \\
&= Sa + n
\end{aligned}
\tag{8.86}
$$

where the signal vector $x=Sa$ is defined as

$$
\begin{pmatrix} x(m) \\ x(m+1) \\ \vdots \\ x(m+N-1) \end{pmatrix}
=
\begin{pmatrix}
e^{j2\pi f_1 m} & e^{j2\pi f_2 m} & \cdots & e^{j2\pi f_P m} \\
e^{j2\pi f_1 (m+1)} & e^{j2\pi f_2 (m+1)} & \cdots & e^{j2\pi f_P (m+1)} \\
\vdots & \vdots & \ddots & \vdots \\
e^{j2\pi f_1 (m+N-1)} & e^{j2\pi f_2 (m+N-1)} & \cdots & e^{j2\pi f_P (m+N-1)}
\end{pmatrix}
\begin{pmatrix} A_1 e^{j2\pi\phi_1} \\ A_2 e^{j2\pi\phi_2} \\ \vdots \\ A_P e^{j2\pi\phi_P} \end{pmatrix}
\tag{8.87}
$$

The matrix S and the vector a are defined in the right hand side of Eq. (8.87). The autocorrelation matrix of the noisy signal y can be written as the sum of the autocorrelation matrices of the signal x and the noise as

$$
\begin{aligned}
R_{yy} &= R_{xx} + R_{nn} \\
&= SPS^H + \sigma_n^2 I
\end{aligned}
\tag{8.88}
$$

where $R_{xx}=SPS^H$ and $R_{nn}=\sigma_n^2 I$ are the autocorrelation matrices of the signal and noise processes, the exponent H denotes the Hermitian transpose, and the diagonal matrix P defines the power of the sinusoids as

$$P = aa^H = diag[P_1, P_2, \cdots, P_P]$$ (8.89)

where $P_i = A_i^2$ is the power of the complex sinusoid $e^{-j2\pi f_i}$. The correlation matrix of the signal can also be expressed in the form

$$R_{xx} = \sum_{k=1}^{P} P_k s_k s_k^H$$ (8.90)

where $s_k^H = [1, e^{j2\pi f_k}, \cdots, e^{j2\pi(N-1)f_k}]$. Now consider an eigen-decomposition of the $N \times N$ correlation matrix R_{xx}

$$\begin{aligned} R_{xx} &= \sum_{k=1}^{N} \lambda_k v_k v_k^H \\ &= \sum_{k=1}^{P} \lambda_k v_k v_k^H \end{aligned}$$ (8.91)

where λ_k and v_k are the eigenvalues and eigenvectors of the matrix R_{xx} respectively. We have also used the fact that the autocorrelation matrix R_{xx} of P complex sinusoids has only P nonzero eigenvalues; $\lambda_{P+1}=\lambda_{P+2}$, ..., $\lambda_N=0$. Since the sum of the cross products of the eigenvectors forms an identity matrix we can also express the diagonal autocorrelation matrix of the noise in terms of the eigenvectors of R_{xx} as

$$R_{nn} = \sigma_n^2 I = \sigma_n^2 \sum_{k=1}^{N} v_k v_k^H$$ (8.92)

The correlation matrix of the noisy signal may be expressed in terms of its eigenvectors and the associated eigenvalues of the noisy signal as

$$\begin{aligned} R_{yy} &= \sum_{k=1}^{P} \lambda_k v_k v_k^H + \sigma_n^2 \sum_{k=1}^{N} v_k v_k^H \\ &= \sum_{k=1}^{P} (\lambda_k + \sigma_n^2) v_k v_k^H + \sigma_n^2 \sum_{k=P+1}^{N} v_k v_k^H \end{aligned}$$ (8.93)

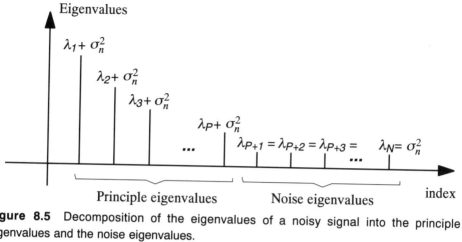

Figure 8.5 Decomposition of the eigenvalues of a noisy signal into the principle eigenvalues and the noise eigenvalues.

From Eq. (8.93), the eigenvectors and the eigenvalues of the correlation matrix of the noisy signal can be partitioned into two disjoint subsets. The set of eigenvectors $\{v_1, \cdots, v_P\}$, associated with the P largest eigenvalues span the *signal subspace* and are called the *principal eigenvectors*. The signal vectors s_i can be expressed as linear combinations of the principle eigenvectors. The second subset of eigenvectors $\{v_{P+1}, \cdots, v_N\}$ span the *noise subspace* and have σ_n^2 as their eigenvalue. Since the signal and noise eigenvectors are orthogonal it follows that the signal subspace and the noise subspace are orthogonal. Hence the sinusoidal signal vectors s_i which are in the signal subspace are orthogonal to the noise subspace and we have

$$s_i^H(f)v_k = \sum_{j=0}^{N-1} v_k(j)e^{-j2\pi f_i j} = 0 \quad i = 1,\cdots,P \;\; k = P+1,\cdots,N \quad (8.94)$$

Eq. (8.94) implies that the frequencies of the P sinusoids can be obtained by solving for the zeros of the following polynomial function of f

$$\sum_{k=P+1}^{N} s^H(f)v_k \qquad (8.95)$$

In MUSIC algorithm the power spectral estimate is defined as

$$P_{XX}(f) = \sum_{k=P+1}^{N} \left| s^H(f)v_k \right|^2 \qquad (8.96)$$

where $s(f) = [1, e^{j2\pi f}, \ldots, e^{j2\pi(N-1)f}]$ is the complex sinusoidal vector, and $\{v_{P+1}, \ldots, v_N\}$ are the eigenvectors in the noise subspace. From Eq. (8.94) and Eq. (8.96) we have that at the

$$P_{XX}(f_i) = 0 \qquad i = 1, \ldots, P \qquad (8.97)$$

Since $P_{XX}(f)$ has its zeros at the frequencies of the sinusoids, it follows that the reciprocal of $P_{XX}(f)$ has its poles at the sinusoids frequencies. The MUSIC spectrum is defined as

$$P_{XX}^{MUSIC}(f) = \frac{1}{\displaystyle\sum_{k=P+1}^{N} |s^H(f)v_k|^2} = \frac{1}{s^H(f)V(f)V^H(f)s(f)} \qquad (8.98)$$

Where $V = [v_{P+1}, \ldots, v_N]$ is the matrix of eigenvectors of the noise subspace. $P_{MUSIC}(f)$ is sharply peaked at the frequencies of the sinusoidal components of the signal, and hence the frequencies of its peaks are taken as the MUSIC estimates.

8.4.3   Estimation of Signal Parameters via Rotational Invariance Techniques (ESPIRIT)

The ESPIRIT algorithm is an eigen-decomposition approach for estimating the frequencies of a number of complex sinusoids observed in additive white noise. Consider a signal, $y(m)$, composed of P complex-valued sinusoids and additive white noise

$$y(m) = \sum_{k=1}^{P} A_k e^{-j(2\pi f_k m + \phi_k)} + n(m) \qquad (8.99)$$

The ESPIRIT algorithm exploits the deterministic relation between sinusoidal component of the signal vector $y(m)=[y(m), \ldots, y(m+N-1)]^T$ and that of the time-shifted vector $y(m+1)=[y(m+1), \ldots, y(m+N)]^T$. The signal component of the noisy vector $y(m)$ may be expressed as

$$x(m) = S\,a \qquad (8.100)$$

where S is the complex sinusoidal matrix and a is the vector containing the amplitude and phase of the sinusoids as in Eqs. (8.86) (8.87). A complex sinusoid $e^{j2\pi f_i m}$ can be time shifted through multiplication by a phase term $e^{j2\pi f_i}$. Hence the time shifted sinusoidal signal vector $x(m+1)$ may be obtained from $x(m)$ by phase shifting each complex sinusoidal component of $x(m)$ as

$$x(m+1) = S \Phi a \qquad (8.101)$$

where Φ is a $P \times P$ phase matrix defined as

$$\Phi = diag[e^{j2\pi f_1}, e^{j2\pi f_2}, \cdots, e^{j2\pi f_P}] \qquad (8.102)$$

The diagonal elements of Φ are the relative phase between the adjacent samples of the sinusoids. The matrix Φ is a unitary matrix an is known as a *rotation matrix* as it relates the time shifted vectors $x(m)$ and $x(m+1)$. The autocorrelation matrix of the noisy signal vector $y(m)$ can be written as

$$R_{y(m)y(m)} = SPS^H + \sigma_n^2 I \qquad (8.103)$$

where the matrix P is diagonal, and its diagonal elements are the powers of the complex sinusoids $P = diag[A_1^2, \cdots, A_P^2] = aa^H$. The cross covariance matrix of the vectors $y(m)$ and $y(m+1)$ is

$$R_{y(m)y(m+1)} = SP\Phi^H S^H + R_{n(m)n(m+1)} \qquad (8.104)$$

Where the autocovariance matrices $R_{y(m)y(m+1)}$ and $R_{n(m)n(m+1)}$ are defined as

$$R_{y(m)y(m+1)} = \begin{pmatrix} r_{yy}(1) & r_{yy}(2) & r_{yy}(3) & \cdots & r_{yy}(N) \\ r_{yy}(0) & r_{yy}(1) & r_{yy}(2) & \cdots & r_{yy}(N-1) \\ r_{yy}(1) & r_{yy}(0) & r_{yy}(1) & \cdots & r_{yy}(N-2) \\ \vdots & \vdots & \vdots & \ddots & \vdots \\ r_{yy}(N-2) & r_{yy}(N-3) & r_{yy}(N-4) & \cdots & r_{yy}(1) \end{pmatrix} \qquad (8.105)$$

and

$$R_{n(m)n(m+1)} = \begin{pmatrix} 0 & 0 & \cdots & 0 & 0 \\ \sigma_n^2 & 0 & \cdots & 0 & 0 \\ 0 & \sigma_n^2 & \cdots & 0 & 0 \\ \vdots & \vdots & \ddots & \vdots & \vdots \\ 0 & 0 & \cdots & \sigma_n^2 & 0 \end{pmatrix} \qquad (8.106)$$

The correlation matrix of the signal vector $x(m)$ can be estimated as

$$R_{x(m)x(m)} = R_{y(m)y(m)} - R_{n(m)n(m)} = SPS^H \qquad (8.107)$$

and the cross correlation matrix of the signal vector $x(m)$ with its time-shifted version $x(m+1)$ is obtained as

$$R_{x(m)x(m+1)} = R_{y(m)y(m+1)} - R_{n(m)n(m+1)} = SP\Phi^H S^H \qquad (8.108)$$

Subtraction of a fraction $\lambda_i = e^{-j2\pi f_i}$ of Eq. (8.107) from Eq. (8.108) yields

$$R_{x(m)x(m)} - \lambda_i R_{x(m)x(m)} = SP(I - \lambda_i \Phi^H)S^H \qquad (8.109)$$

From Eqs. (8.102) and (8.109), the frequencies of the sinusoids can be estimated as the roots of Eq. (8.109).

Summary

Power spectrum estimation is perhaps the most widely used method of signal analysis. The main objective of any transformation is to express a signal in a form that lends itself to a more convenient analysis and manipulation. The power spectrum is related to the correlation function through the Fourier transform. The power spectrum reveals the periodic and correlated patterns of a signal, which are important in detection, estimation, data forecasting, and decision making systems. We began this chapter with Section 8.1 on basic definitions of Fourier transform, energy spectrum and power spectrum. In Section 8.2 we considered nonparametric DFT-based methods of spectral analysis. These methods do not offer the high resolution of parametric and eigen-based methods. However, they have attractions in that they are computationally less expensive than model based methods and are relatively robust. In Section 8.3 we considered the maximum entropy and the model based spectral estimation methods. These method can extrapolate the correlation values beyond the range for which data is available and hence offer a higher resolution and less sidelobes. In section 8.4 we considered the eigen-based spectral estimation of noisy signals. These methods decomposed the eigen variables of the noisy signal into a signal subspace and noise subspace. The orthogonality of the signal and noise subspaces is used to estimate the signal and noise parameters. In the next chapter we use DFT-based spectral estimation for restoration of signals observed in noise.

Bibliography

BARTLETT M. S. (1950), Periodogram Analysis and Continuous Spectra, Biometrica, Vol. 37, Pages 1-16.

BLACKMAN R. B., TUKEY J. W. (1958), The Measurement of Power Spectra from the Point of View of Communication Engineering, Dover Publications, New York.

BURG J. P. (1975), Maximum Entropy Spectral Analysis, PhD Thesis, Department of Geophysics, Stanford University, California.

CHILDERS D. G., Editor (1978), Modern Spectrum Analysis, IEEE Press New York.

COHEN L. (1989), Time-Frequency Distributions - A review, Proceedings of the IEEE, Vol. 77(7), Pages 941-981.

CADZOW J. A. (1979), ARMA Spectral Estimation: An Efficient Closed-form Procedure, Proc. RADC Spectrum estimation Workshop, Pages 81-97.

CAPON J. (1969), High Resolution Frequency-Wavenumber Spectrum Analysis, proc. IEEE, Vol. 57, Pages 1408-1418.

HAYKIN S. (1985), Array Signal Processing, Prentice-Hall, Englewood Cliffs, N. J.

JENKINS G. M., WATTS D. G. (1968), Spectral Analysis and Its Applications, Holden-Day, San Francisco, California.

KAY S. M., MARPLE S. L., (1981) Spectrum Analysis : A Modern Perspective Proceedings of the IEEE, Vol. 69, Pages 1380-1419.

KAY S. M. (1988) Modern Spectral Estimation : Theory and Application. Prentice Hall-Englewood Cliffs, N. J.

LACOSS R. T. (1971), Data Adaptive Spectral Analysis Methods, Geophysics, Vol. 36, Pages 661-675.

MARPLE S. L. (1987) Digital Spectral Analysis with Applications. Prentice Hall-Englewood Cliffs, N. J.

PARZEN E. (1957),On Consistent Estimates of the Spectrum of a Stationary Time series , Am. Math. Stat., Vol. 28, Pages 329-348.

PISARENKO V. F. (1973), The Retrieval of Harmonics from a Covariance Function , Geophy. J. R. Astron. Soc., Vol. 33, Pages 347-366

ROY R. H. (1987), ESPRIT-Estimation of Signal Parameters via Rotational Invariance Techniques. PhD Thesis, Stanford University, California.

SCHMIDT R. O. (1981), A signal Subspace Approach to Multiple Emitter Location and Spectral Estimation, PhD Thesis, Stanford University, California.

STANISLAV B. K., Editor (1986), Modern Spectrum Analysis, IEEE Press, New York.

STRAND O. N. (1977),"Multichannel Complex Maximum Entropy (AutoRegressive) Spectral Analysis", IEEE Trans. on Automatic Control, AC-Vol. 22(4), Pages 634-640, (1977).

VAN DEN BOS A. (1971) "Alternative Interpretation of Maximum Entropy Spectral Analysis", IEEE Trans. Infor. Tech., Vol. IT-17, Pages 92-93.

WELCH P. D. (1967), The Use of Fast Fourier Transform for the Estimation of Power Spectra: A Method Based on Time Averaging over Short Modified Periodograms , IEEE Trans. Audio and Electroacoustics, Vol. AU-15, Pages 70-73.

WILKINSON J. H. (1965), The Algebraic Eigenvalue Problem, Oxford University Press, Oxford.

9

Spectral Subtraction

9.1 Spectral Subtraction
9.2 Processing Distortions
9.3 Nonlinear Spectral Subtraction
9.4 Implementation of Spectral Subtraction

Spectral subtraction is a method for restoration of the power or the magnitude spectrum of a signal observed in additive noise, through subtraction of an estimate of the average noise spectrum from the noisy signal spectrum. The noise spectrum is estimated, and updated, from the periods when the signal is absent and only the noise is present. The assumption is that the noise is a stationary or a slowly varying process, and that the noise spectrum does not change significantly in-between the update periods. For restoration of time-domain signals, an estimate of the instantaneous magnitude spectrum is combined with the phase of the noisy signal, and then transformed via an inverse discrete Fourier transform to the time domain. In terms of computational complexity spectral subtraction is relatively inexpensive. However, due to random variations of noise, spectral subtraction can result in negative estimates of the short-time magnitude or power spectrum. The magnitude and power spectrum are non-negative variables, and any negative estimate of these variables should be mapped into a non-negative value. This nonlinear rectification process distorts the distribution of the restored signal. The processing distortion becomes more noticeable as the signal to noise ratio decreases. In this chapter we study spectral subtraction, and the different methods of reducing and removing the processing distortions.

9.1 Spectral Subtraction

In applications where, in addition to the noisy signal, the noise is accessible on a separate channel, it may be possible to retrieve the signal by subtracting an estimate of the noise from the noisy signal. For example, the adaptive noise canceller of Section 1.3.1 takes as the inputs the noise and the noisy signal, and outputs an estimate of the signal. However, in many applications, such as at the receiver of a noisy communication channel, the only signal that is available is the noisy signal. In these situations it is not possible to cancel out the random noise, but it may be possible to reduce the *average effects* of the noise on the signal spectrum. The effect of additive noise on the magnitude spectrum of a signal is to increase the mean and the variance of the spectrum as illustrated in Figure 9.1. The increase in the variance of the signal spectrum results from the random fluctuations of the noise, and can not be cancelled out. The increase in the mean of the signal spectrum can be removed by subtraction of an estimate of the mean of the noise spectrum from the noisy signal spectrum. The noisy signal model in the time domain is given by

$$y(m) = x(m) + n(m) \tag{9.1}$$

where $y(m)$, $x(m)$, and $n(m)$ are the signal, the additive noise, and the noisy signal respectively, and m is the discrete time index.

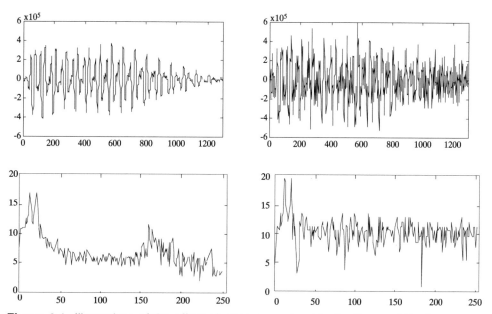

Figure 9.1 Illustrations of the effect of noise on a signal in the time and the frequency domains.

In the frequency domain, the noisy signal model is expressed as

$$Y(f) = X(f) + N(f) \tag{9.2}$$

where $Y(f)$, $X(f)$ and $N(f)$ are the Fourier transforms of the noisy signal, the original signal and the noise respectively, and f is the frequency variable. In spectral subtraction the incoming signal $x(m)$ is buffered and divided into segments of N samples length. Each segment is windowed, using a Hanning or a Hamming window, and then transformed via discrete Fourier transform (DFT) to N spectral samples. The windows alleviate the effects of the discontinuities at the endpoints of each segment. The windowed signal is given by

$$
\begin{aligned}
y_w(m) &= w(m)y(m) \\
&= w(m)[x(m) + n(m)] \\
&= x_w(m) + n_w(m)
\end{aligned}
\tag{9.3}
$$

The windowing operation can be expressed in the frequency domain as

$$
\begin{aligned}
Y_W(f) &= W(f) * Y(f) \\
&= X_W(f) + N_W(f)
\end{aligned}
\tag{9.4}
$$

where the operator * denotes the convolution. Throughout this chapter it is assumed that the signals are windowed, and hence for simplicity we drop the use of the subscript w for windowed signals.

Figure 9.2 illustrates a block diagram configuration of spectral subtraction. A more detailed implementation is described in Section 9.4. The equation describing spectral subtraction may be expressed as

$$\left|\hat{X}(f)\right|^b = |Y(f)|^b - \alpha \overline{|N(f)|^b} \tag{9.5}$$

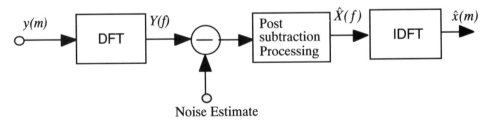

Figure 9.2 A block diagram illustration of spectral subtraction.

Where $\left|\hat{X}(f)\right|^b$ is an estimate of the original signal spectrum $|X(f)|^b$, and $\overline{|N(f)|^b}$ is the time-averaged noise spectra. It is assumed that the noise is a wide sense stationary random process. For magnitude spectral subtraction the exponent $b=1$, and for power spectral subtraction $b=2$. The parameter α , in Eq. (9.5), controls the amount of noise subtracted from the noisy signal. For full noise subtraction $\alpha=1$ and for over-subtraction $\alpha>1$. The time-averaged noise spectrum is obtained from the periods when the signal is absent and only the noise is present as

$$\overline{|N(f)|^b} = \frac{1}{K} \sum_{i=0}^{K-1} |N_i(f)|^b \tag{9.6}$$

In Eq. (9.6) $|N_i(f)|$ is the spectrum of the ith noise frame, and it is assumed that there are K frames in a noise-only period, where K is a variable. Alternatively the averaged noise spectrum can be obtained as the output of a first order digital lowpass filter as

$$\overline{|N_i(f)|^b} = \rho \, \overline{|N_{i-1}(f)|^b} + (1-\rho) \, |N_i(f)|^b \tag{9.7}$$

Where the low pass filter coefficient ρ is typically set between 0.8 and 0.95.

For restoration of a time domain signal the magnitude spectrum estimate $|\hat{X}(f)|$ is combined with the phase of the noisy signal, and then transformed into the time via the inverse discrete Fourier transform as

$$\hat{x}(m) = \sum_{k=0}^{N-1} \left|\hat{X}(k)\right| e^{j\theta_Y(k)} \, e^{-\frac{j2\pi}{N}km} \tag{9.8}$$

where $\theta_Y(k)$ is the phase of the noisy signal frequency $Y(k)$. The signal restoration Eq. (9.8) is based on the assumption that the audible noise is mainly due to the distortion of the magnitude spectrum, and that the phase distortion is largely inaudible. Evaluations of the perceptual effects of simulated phase distortions validate this assumption.

Due to the variations of the noise spectrum, spectral subtraction may produce negative estimates of the power or the magnitude spectrum. This outcome is more probable as signal to noise ratio (SNR) decreases. To avoid negative magnitude estimates the spectral subtraction output is post processed using a mapping function $T[.]$ of the form :

$$T\left[|\hat{X}(f)|\right] = \begin{cases} |\hat{X}(f)| & if \;\; |\hat{X}(f)| > \beta |Y(f)| \\ fn[|Y(f)|] & otherwise \end{cases} \tag{9.9}$$

For example, we may chose a rule such that if the estimate $|\hat{X}(f)| > 0.1|Y(f)|$ (for magnitude spectrum, 0.1 is equivalent to -20 dB), then $|\hat{X}(f)|$ should be set to some function of the noisy signal fn[$Y(f)$]. In its simplest form fn[$Y(f)$]=*noise floor*, where noise floor is a positive constant. An alternative choice is fn[$|Y(f)|$]=$\beta|Y(f)|$. In this case

$$T\left[|\hat{X}(f)|\right] = \begin{cases} |\hat{X}(f)| & \text{if } |\hat{X}(f)| > \beta \ |Y(f)| \\ \beta \ |Y(f)| & \text{otherwise} \end{cases} \tag{9.10}$$

Spectral subtraction may be implemented in the power or the magnitude spectral domains. The two methods are similar, although theoretically they result in somewhat different expected performance.

9.1.1 Power Spectrum Subtraction

The power spectrum subtraction, or squared-magnitude spectrum subtraction, is defined by the following equation

$$|\hat{X}(f)|^2 = |Y(f)|^2 - \overline{|N(f)|^2} \tag{9.11}$$

where it is assumed that α, the subtraction factor in Eq. (9.5), is unity. We refer to $\mathcal{E}\left[|Y(f)|^2\right]$ as the power spectrum, to its time-averaged version $\overline{|X(f)|^2}$, as the time-averaged power spectrum, and to $|X(f)|^2$ as the *instantaneous* power spectrum. By expanding the instantaneous power spectrum of the noisy signal, $|Y(f)|^2$ and grouping the appropriate terms, Eq. (9.11) may be rewritten as

$$|\hat{X}(f)|^2 = |X(f)|^2 + \underbrace{\left(|N(f)|^2 - \overline{|N(f)|^2}\right)}_{\textit{Noise variations}} + \underbrace{X^*(f)N(f) + X(f)N^*(f)}_{\textit{Cross products}} \tag{9.12}$$

Taking the expectations of both sides of Eq. (9.12) and assuming that the signal and the noise are uncorrelated ergodic processes we have

$$\mathcal{E}\left[|\hat{X}(f)|^2\right] = \mathcal{E}\left[|X(f)|^2\right] \tag{9.13}$$

From Eq. (9.13) the average of the estimate of the instantaneous power spectrum converges to the power spectrum of the noise-free signal. However, it must be noted that for nonstationary signals, such as speech, the objective is to recover the *instantaneous* or the short-time spectrum, and only a relatively small amount of averaging can be applied. Too much averaging will smear and obscure the temporal evolution of the spectral events.

9.1.2 Magnitude Spectrum Subtraction

The magnitude spectrum subtraction is defined as

$$\left|\hat{X}(f)\right| = |Y(f)| - \overline{|N(f)|} \tag{9.14}$$

where $\overline{|N(f)|}$ is the time-averaged magnitude spectrum of the noise. Taking the expectation of Eq. (9.14) we have

$$\begin{aligned}
\mathcal{E}\left[\left|\hat{X}(f)\right|\right] &= \mathcal{E}\left[|Y(f)|\right] - \mathcal{E}\left[\overline{|N(f)|}\right] \\
&= \mathcal{E}\left[|X(f) + N(f)|\right] - \mathcal{E}\left[\overline{|N(f)|}\right] \\
&\approx \mathcal{E}\left[|X(f)|\right]
\end{aligned} \tag{9.15}$$

For signal restoration the magnitude estimate is combined with the phase of the noisy signal and then transformed into the time using Eq. (9.8).

9.1.3 Spectral Subtraction Filter: Relation to Wiener Filters

The spectral subtraction equation can be expressed as the product of the noisy signal spectrum and the frequency response of a spectral subtraction filter as

$$\begin{aligned}
\left|\hat{X}(f)\right|^2 &= |Y(f)|^2 - \overline{|N(f)|^2} \\
&= H(f)|Y(f)|^2
\end{aligned} \tag{9.16}$$

where $H(f)$, the frequency response of the spectral subtraction filter, is defined as

$$\begin{aligned}
H(f) &= 1 - \frac{\overline{|N(f)|^2}}{|Y(f)|^2} \\
&= \frac{|Y(f)|^2 - \overline{|N(f)|^2}}{|Y(f)|^2}
\end{aligned} \tag{9.17}$$

The spectral subtraction filter $H(f)$ is a zero phase filter, with its magnitude response in the range $0 \leq H(f) \leq 1$. The filter acts as a SNR-dependent attenuator. The attenuation at each frequency increases with the decreasing SNR, and conversely decreases with the increasing SNR. The least mean squared error linear filter for noise removal is the Wiener filter. Implementation of a Wiener filter requires the power spectra (or equivalently the correlation functions) of the signal and the noise process, as discussed in Chapter 5. Spectral subtraction is used as a substitute for the Wiener filter when the signal power spectra is not available. In this section we discuss the close relation between the Wiener filter and the spectral subtraction. For restoration of a signal observed in additive noise, the equation describing the frequency response of the Wiener filter was derived in Chapter 5 as

$$W(f) = \frac{\mathcal{E}\left[|Y(f)|^2\right] - \mathcal{E}\left[|N(f)|^2\right]}{\mathcal{E}\left[|Y(f)|^2\right]} \tag{9.18}$$

A comparison of $W(f)$ and $H(f)$, from Eqs. (9.18) and (9.17), shows that the Wiener filter is based on the *ensemble-average* spectra of the signal and the noise, whereas the spectral subtraction filter uses the instantaneous spectra of the noisy signal and the *time-averaged* spectra of the noise. In Wiener filter theory the averaging operations are taken across the ensemble of different realisations of the signal and noise processes. In spectral subtraction we have access only to a single realisation of the process. However, assuming that the signal and noise are wide sense stationary ergodic processes, we may replace the instantaneous noisy signal spectrum $|Y(f)|^2$ in spectral subtraction Eq. (9.18) with the time-averaged spectrum $\overline{|Y(f)|^2}$, to obtain

$$H(f) = \frac{\overline{|Y(f)|^2} - \overline{|Y(f)|^2}}{\overline{|Y(f)|^2}} \tag{9.19}$$

For an ergodic process, as the time-averaged spectrum approaches the ensemble averaged spectrum, in the limit, the spectral subtraction filter approaches the Wiener filter . In practice many signals, such as speech and music, are highly nonstationary, and only a limited degree of beneficial time-averaging of the spectral parameters can be expected.

9.2 Processing Distortions

The main problem in spectral subtraction is the processing distortions caused by the random variations of the noise spectrum. From Eq. (9.11) and the constraint that the magnitude spectrum must be non-negative, we may identify three sources of distortions of the instantaneous estimate of the magnitude or power spectrum as :

(a) the variations of the instantaneous noise power spectrum about the mean,
(b) the cross product terms, and
(c) the nonlinear mapping of the spectral estimates that fall below a threshold.

The same sources of distortions appear in both the magnitude and the power spectrum subtraction. Of the three sources of distortions listed above, the dominant distortion is often due to the nonlinear mapping of the negative, or small valued, spectral estimates. This distortion produces a metallic sounding noise, known as "the musical noise" due to their narrow-band spectrum and the tin-like sound. The success of spectral subtraction depends on the ability of the algorithm to reduce the noise variations and to remove the processing distortions.

In its worst, and not uncommon, case the residual noise can have the following two forms: a) a sharp trough or peak in the signal spectra, and b) isolated narrow bands of frequencies. In the vicinity of a high amplitude signal frequency, the noise-induced trough or peak is often masked, and made inaudible, by the high signal energy. The main cause of audible degradations are the isolated frequency components also known as the musical noise illustrated in Figure 9.3. The musical noise is characterised as short lived narrow bands of frequencies surrounded by relatively.

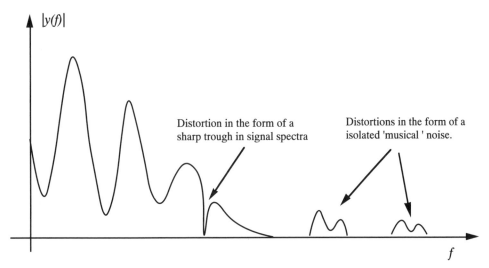

Figure 9.3 Illustration of distortions that may result from spectral subtraction.

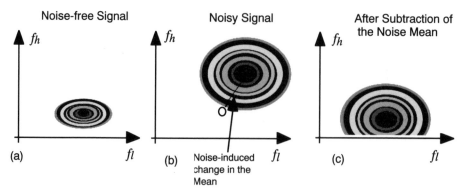

Figure 9.4 Illustration of the distorting effect of spectral subtraction on the space of the magnitude spectrum of a signal.

low level frequency components. In audio signal restoration the distortion caused by spectral subtraction can result in a significant deterioration of the signal quality. This is particularly true at low signal to noise ratios. The effects of a bad implementation of subtraction algorithm can result in a signal that is of a lower perceived quality, and lower information content, than the original noisy signal.

9.2.1 Effect of Spectral Subtraction on Signal Distribution

Figure 9.4 is an illustration of the distorting effect of spectral subtraction on the distribution of the magnitude spectrum of a signal. In this figure we have considered the simple case where the spectrum of a signal is divided into two parts; a low frequency band f_l and a high frequency band f_h. Each point in figure 9.4 is plot of the high frequency spectrum versus the low frequency spectrum, in a two-dimensional signal space. Figure 9.4(a) shows an assumed distribution of the spectral samples, of a signal, in the two-dimensional magnitude frequency space. The effect of the random noise, shown in Figure 9.4(b), is an increase in the mean and the variance of the spectrum, by an amount that depends on the mean and the variance of the magnitude spectrum of the noise. The increase in the variance constitutes an irrevocable distortion. The increase in the mean of the magnitude spectrum can be removed through spectral subtraction. Figure 9.4(c) illustrates the distorting effect of spectral subtraction on the distribution of the signal spectrum. As shown, due to the noise-induced increase in the variance of the signal spectrum, after subtraction of the average noise spectrum, a proportion of the signal population, particularly those with a low SNR, become negative and have to be mapped to non-negative values. This process distorts the distribution of the low SNR part of the signal spectrum.

9.2.2   Reducing the Noise Variance

The distortions that result from spectral subtraction are due to the variations of the noise spectrum. In Section 8.2 we considered the methods of reducing the variance of the estimate of a power spectrum. For a white noise process with variance σ_n^2, it can be shown that the variance of the DFT spectrum of the noise $N(f)$ is given by

$$Var\left[|N(f)|^2\right] \approx P_{NN}^2(f) = \sigma_n^4 \qquad (9.20)$$

And the variance of the running average of K independent spectral components is

$$Var\left[\frac{1}{K}\sum_{i=0}^{K-1}|N_i(f)|^2\right] \approx \frac{1}{K}P_{NN}^2(f) \approx \frac{1}{K}\sigma_n^4 \qquad (9.21)$$

Form Eq. (9.21), the noise variations can be reduced by averaging the noisy signal frequency components. The fundamental limitation is, that the averaging process, in addition to reducing the noise variance, also has the undesirable effect of smearing and obscuring the time variations of the signal spectrum. Therefor an averaging process should reflect a compromise between the conflicting requirements to reduce the noise variance, and to retain the time resolution of the nonstationary spectral events. This is important because time resolution plays a particularly important part in both the quality and the intelligibility of audio signals.
In spectral subtraction, the noisy signal $y(m)$ is segmented into blocks of N samples. Each signal block is then transformed via a DFT into a block of N spectral samples $Y(f)$. Successive blocks of spectral samples form a two-dimensional frequency-time matrix and can be denoted as $Y(f,t)=Y_t(f)$. The signal $Y(f,t)$ can be considered as a band-pass channel f which contains a time-varying signal $X(f,t)$, plus a random noise component $N(f,t)$. One method for reducing the noise variations is to lowpass filter the magnitude spectrum at each frequency. A simple recursive first order digital lowpass filter is given by

$$|Y_{LP}(f,t)| = \rho\,|Y_{LP}(f,t-1)| + (1-\rho)\,|Y(f,t)| \qquad (9.22)$$

Where the subscript LP denotes the output of the lowpass filter, and the smoothing coefficient, ρ, controls the bandwidth, and the time constant, of the low pass filter.

9.2.3   Filtering Out the Processing Distortions

Audio signals, such as speech and music, are composed of sequences of nonstationary acoustic events. The acoustic events are "born", have a varying life-

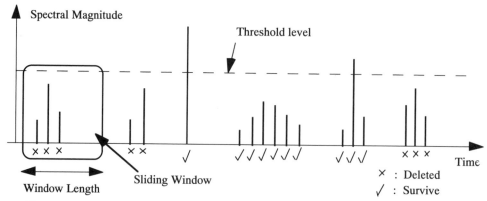

Figure 9.5 Illustration of a method for identification and filtering of "musical noise".

time, disappear, and then reappear with a different intensity and spectral composition. The time varying nature of audio signals plays an important role in conveying information, sensation and quality. The musical noise, introduced as an undesirable by-product of spectral subtraction, is also time-varying. However, there are significant differences between the characteristics of most audio signals and the so called musical noise. The differences may be used to identify and remove some of the more annoying distortions. Identification of musical noise may be achieved by examining the variations of the signal in the time and frequency domains. The main characteristics of the musical notes are that they tend to be, relatively short lived, random isolated bursts of narrow band signals, with relatively small amplitudes.

Using a DFT block size of 128 samples, at a sampling rate of 20 kHz, experiments indicate that the great majority of musical noise tend to last no more than three frames, whereas genuine signal frequencies have a considerably longer duration. This observation was used as the basis of an effective musical noise suppression system. Figure 9.5 demonstrates a method for identification of musical noise. Each DFT channel is examined to identify short-lived frequency events. If a frequency component has a duration shorter than a preselected time window, an amplitude smaller than a threshold, and is not masked by signal components in the adjacent frequency bins, then it is classified as distortion and deleted.

9.3 Non-linear Spectral Subtraction

The use of spectral subtraction in its basic form of Eq. (9.5) may cause a deterioration in the quality and the information content of a signal. For example, in audio signal restoration the musical tones can cause a degradation in the perceived quality of the signal, and in speech recognition the basic spectral subtraction can result in deterioration of the recognition accuracy. In literature, there are a number of variants of spectral subtraction that aim to provide consistent performance improvement across a range of SNRs. These methods basically differ; in their approach to estimation of the noise spectrum, in the method of averaging the noisy signal spectrum, and in the post processing method for the removal of processing distortions.

Non-linear spectral subtraction are heuristic methods that utilise estimates of the local signal to noise ratio, and the observation that at a low signal to noise ratio over-subtraction can produce improved results. For an explanation of the improvement that can result from over-subtraction, consider the following expression of the basic spectral subtraction equation :

$$\left|\hat{X}_t(f)\right| = |Y(f)| - \overline{|N(f)|}$$

$$\approx |X(f)| + |N(f)| - \overline{|N(f)|} \tag{9.23}$$

$$\approx |X(f)| + V_N(f)$$

where $V_N(f)$ is the zero mean random component of the noise spectrum. If $V_N(f)$ is well above the signal $X(f)$, then the signal may be considered as lost to the noise. In this case over-subtraction, followed by nonlinear processing of the negative estimates, results in a higher overall attenuation of the noise. This argument explains why subtracting more than the noise average can sometimes produce better results.

The nonlinear variants of spectral subtraction may be described by the following equation

$$\left|\hat{X}(f)\right| = |Y(f)| - \alpha(SNR(f)) \, \overline{|N(f)|}_{NL} \tag{9.24}$$

where $\alpha(SNR(f))$ is an SNR-dependent subtraction factor, and $\overline{|N(f)|}_{NL}$ is a nonlinear estimate of the noise spectrum. The spectral estimate is further processed to avoid negative estimates as

$$\hat{X}(f) = \begin{cases} \hat{X}(f) & \text{if } \hat{X}(f) > \beta \, Y(f) \\ \beta \, Y(f) & \text{otherwise} \end{cases} \tag{9.25}$$

One form of an SNR-dependent subtraction factor for Eq. (9.25) is given by

$$\alpha(SNR(f)) = \left(1 + \frac{sd(|N(f)|)}{|N(f)|}\right) \tag{9.26}$$

where the function $sd(|N(f)|)$ is the standard deviation of the noise at frequency f. For white noise $sd(|N(f)|) = \sigma_n$ where σ_n^2 is the noise variance. Substitution of Eq. (9.26) in Eq. (9.24) yields

$$|\hat{X}(f)| = |Y(f)| - \left(1 + \frac{sd(|N(f)|)}{|N(f)|}\right)\overline{|N(f)|} \qquad (9.27)$$

In Eq. (9.27) the subtraction factor depends on the mean and the variance of the noise. Note that the amount over-subtracted is the standard deviation of the noise. This formula is appealing because on one extreme for deterministic noise, such as a sine wave, $\alpha(SNR(f))=1$, and in the other extreme for white noise $\alpha(SNR(f))=2$. In application of spectral subtraction to speech recognition it is found the best subtraction factor is in between 1 and 2.

In the nonlinear spectral subtraction method of Lockwood and Boudy, the spectral subtraction filter is obtained from

$$H(f) = \frac{\overline{|Y(f)|^2} - \overline{|N(f)|^2}_{NL}}{\overline{|Y(f)|^2}} \qquad (9.28)$$

Lockwood and Boudy suggested the following function as a nonlinear estimator of the noise spectrum

$$\overline{|N(f)|^2}_{NL} = \Phi\left(\underset{over\,M\,frames}{Max}\left(|N(f)|^2\right), SNR(f), \overline{|N(f)|^2}\right) \qquad (9.29)$$

The estimate of the noise spectrum is a function of the maximum value of noise spectrum over M frames, and the signal to noise ratio. One form for the nonlinear function $\Phi(.)$ is

$$\Phi\left(\underset{over\,M\,frames}{Max}\left(|N(f)|^2\right), SNR(f)\right) = \frac{\underset{over\,M\,frames}{Max}\left(|N(f)|^2\right)}{1 + \gamma\,SNR(f)} \qquad (9.30)$$

Where γ is a design parameter. From Eq. (9.30) as the SNR decreases the output of the nonlinear estimator $\Phi(.)$ approaches $\max(|N(f)|^2)$, and as the SNR increases it approaches zero. For over subtraction, the noise estimate is forced to be an over-estimation by using the following limiting function

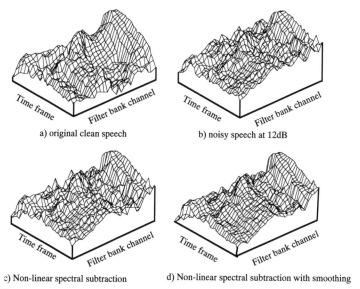

a) original clean speech   b) noisy speech at 12dB

c) Non-linear spectral subtraction   d) Non-linear spectral subtraction with smoothing

Figure 9.6 Illustration of the effects of non-linear spectral subtraction.

$$\overline{|N(f)|^2} \leq \Phi\left(\underset{over\ M\ frames}{Max} \left(|N(f)|^2 \right), SNR(f), \overline{|N(f)|^2} \right) \leq 3\,\overline{|N(f)|^2} \qquad (9.31)$$

The maximum attenuation of the spectral subtraction filter is limited to $H(f) \geq \beta$ where usually $\beta \geq 0.01$. Figure 9.6 illustrates the effects of nonlinear spectral subtraction and smoothing in restoration of the spectrum of a speech signal.

9.4 Implementation of Spectral Subtraction

Figure 9.7 is a block diagram illustration of a spectral subtraction system. It includes the following subsystems :

(a) a silence detector, for detection of the periods of signal inactivity. The noise spectra is updated during the periods of signal inactivity.
(b) a discrete Fourier transformer (DFT), for transforming the time domain signal to the frequency domain. The DFT is followed by a magnitude operator.
(c) a lowpass filter (LPF) for reducing the noise variance. The purpose of the LPF is to reduce the processing distortions due to noise variations.
(d) a post-processor for removing the processing distortions introduced by spectral subtraction.

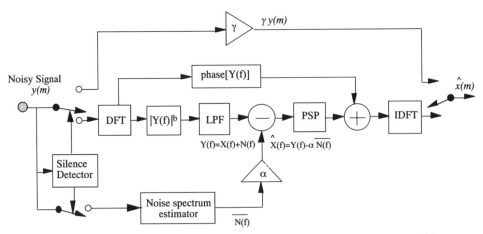

Figure 9.7 Block diagram configuration of a spectral subtraction system. PSP = post spectral subtraction processing.

(e) an inverse discrete Fourier transform (IDFT) for transforming the processed signal to the time domain.

(f) an attenuator, γ, for attenuation of the noise during silent periods.

The DFT based spectral subtraction is a block processing algorithm. The incoming audio signal is buffered and divided into overlapping blocks of N samples as shown in Figure 9.7. Each block is Hanning (or Hamming) windowed, and then transformed via a DFT to the frequency domain. After spectral subtraction, the magnitude spectrum is combined with the phase of the noisy signal, and transformed back to the time domain. Each signal block is then overlapped and added to the preceding and succeeding blocks to form the final output.

The choice of the block length, for spectral analysis, is a compromise between the conflicting requirements of the time resolution and the spectral resolution. Typically a block length of 5 to 50 milliseconds is used. At a sampling rate of say 20 kHz this translates to a value for N in the range of 100 to 1000 samples. The frequency resolution of the spectrum is directly proportional to the number of samples N. A larger value of N produces a better estimate of the spectrum. This is particularly true for the lower part of the frequency spectrum, as low frequency components vary slowly with the time, and require a larger window for a stable estimate. The conflicting requirement is that due to the non-stationary nature of audio signals the window length should not be too large, so that short duration events are not obscured.

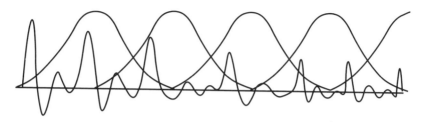

Figure 9.8 Illustration of the window and overlap process in spectral subtraction

The main function of the window and the overlap operations is to alleviate discontinuities at the end-points of each output block. Although, there are a number of useful windows with different frequency/time characteristics, in most implementations of the spectral subtraction a Hanning window is used. In removing distortions introduced by spectral subtraction, the post processor algorithm makes use of such information as the correlation of each frequency channel from one block to the next, and the durations of the signal events and the distortions. The correlation of the signal spectral components, along the time dimension, can be partially controlled by the choice of the window length and the overlap. The correlation of spectral components along the time domain increases with decreasing window length and increasing overlap. However, increasing the overlap can also increase the correlation of noise frequencies along the time dimension.

9.4.1 Application to Speech Restoration and Recognition

In speech restoration the objective is to estimate the instantaneous spectrum $X(f)$. The restored magnitude spectrum is combined with the phase of the noisy signal to form the restored speech signal. In contrast, speech recognition systems are more concerned with restoration of the envelope of the short-time spectrum than the detailed structure of the spectrum. Averaged values, such as the envelope of a spectrum, can often be estimated with more accuracy than the instantaneous values. However, in speech recognition, as in signal restoration, the processing distortion due to the negative spectral estimates cause deterioration in performance. A careful implementation of spectral subtraction can result in a significant improvement in the recognition performance.

Figure 9.9 Illustrates the effects of spectral subtraction in restoring a signal contaminated with white noise. Figure 9.10 illustrates the improvement that can be obtained from application of spectral subtraction to recognition of noise speech. The figures were obtained for a hidden Markov model base spoken digit recognition in the presence of Helicopter type noise.

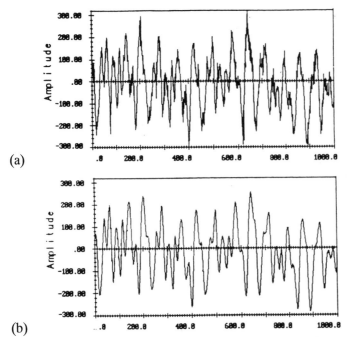

(a)

(b)

Figure 9.9 (a) a noisy signal, (b) restored signal after spectral subtraction.

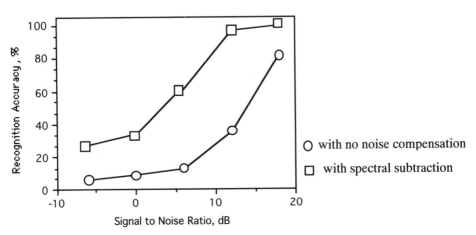

Figure 9.10 The effect of spectral subtraction in improving speech recognition (for a spoken digit data base) in the presence of helicopter noise.

Summary

This chapter began with an introduction to spectral subtraction and its relation to Wiener filter. The main attraction of spectral subtraction is its relative simplicity, in that it only requires an estimate of the noise power spectrum. However this can also be viewed as a fundamental limitation in that spectral subtraction does not utilise the statistics and the distributions of the signal process. The main problem in spectral subtraction is the processing distortions caused by the random variations of the noise. The estimates of the magnitude and power spectral variables, that due to noise variations are negative, have to be mapped into non-negative values. In Section 9.2 we considered the processing distortions, and illustrated the effects of rectification of negative estimates on the distribution of the signal spectrum. In Section 9.3 a number of nonlinear variants of spectral subtraction method were considered. In signal restoration and in applications of spectral subtraction to speech recognition it is found that over-subtraction, that is subtracting more than the average noise value, can lead to improved results; if a frequency component is immersed in noise then over-subtraction can cause further attenuation of the noise. A formula is proposed in which the over-subtraction factor is made dependent on the noise variance. As mentioned earlier, the fundamental problem with spectral subtraction is that it employs relatively too little prior information, and for this reason it is outperformed by Wiener filter and the Bayesian statistical restoration methods.

Bibliography

BOLL S.F (1979), Suppression of Acoustic Noise in Speech Using Spectral Subtraction IEEE Transactions, Vol. ASSP-27, No 2, Pages 113-20, April.

BROUTI M., SCHWARTZ R., MAKHOUL J. (1979), Enhancement of Speech Corrupted by Acoustic Noise, Proc. IEEE, Int. Conf. on Acoustics, Speech and Signal Processing, ICASSP-79, Pages 208-11.

CROZIER P.M.*et al* (1993), The Use of Linear Prediction and Spectral Scaling For Improving Speech Enhancement, EUROSPEECH-93, Pages 231-2343.

EPHRAIM Y., (1992), Statistical Model Based Speech Enhancement systems, Proc. IEEE, Vol. 80, No. 10, Pages 1526-55.

EPHRAIM Y. and VAN TREES H.L. (1993), A Signal Subspace Approach for Speech Enhancement, ICASSP-93, Pages 355-58.

JUANG B. H., RABINER L. R. (1987), Signal Restoration by Spectral Mapping, Proc. IEEE, Int. Conf. on Acoustics, Speech and Signal Processing, ICASSP-87 Texas.

KOBAYASHI T. *et al* (1993), Speech recognition under the nonstationary noise based on the noise hidden Markov model and spectral subtraction, EUROSPEECH-93, Pages 833-837.

LIM J. S. (1978), Evaluations of Correlation Subtraction Method for Enhancing Speech Degraded by Additive White Noise, IEEE Trans. Acoustics, Speech and Signal Processing, Vol. ASSP-26, No. 5, Pages 471-472.

LOCKWOOD P., BOUDY J. (1992),Experiments with a Nonlinear Spectral Subtractor (NSS), Hidden Markov Models and the Projection, for Robust Speech Recognition in Car, Speech Communications, Elsevier, Pages 215-228.

LOCKWOOD P. *et al* (1992), Non-Linear Spectral Subtraction and Hidden Markov Models for Robust Speech Recognition in Car Noise Environments, ICASSP-92, Pages 265-268.

MILNER B. P. (1995), Speech Recognition in Adverse Environments, PhD Thesis, University of East Anglia, UK.

MCAULAY R.J., MALPASS M.L.(1980), Speech enhancement using a soft-decision noise suppresion filter, IEEE Trans. ASSP, Vol. 28, no. 2, Pages 137-145, April.

NOLAZCO-FLORES JA, YOUNG SJ. (1994),Adapting a HMM-based Recogniser for Noisy Speech Enhanced by Spectral Subtraction, Proc. IEEE, Int. Conf. on Acoustics, Speech and Signal Processing, ICASSP-94 Adelaide.

PORTER J.E., BOLL S.F. (1984), Optimal Estimators for Spectral Restoration of Noisy Speech, ICASSP-84, Pages 18A.2.1-18A.2.4.

O'SHAUGHNESSY D. (1989), Enhancing Speech Degraded by Additive Noise or Interfering Speakers, IEEE Commun. Mag. Pages 46-52.

POLLAK P. *et al* (1993), Noise suppression system for a car, EUROSPEECH-93, Pages 1073-1076.

SORENSON H.B(1993), Robust Speaker Independent Speech Recognition Using Non-Linear Spectral Subtraction Based IMELDA, EUROSPEECH-93, Pages 235-238.

SONDHI M. M., SCHMIDT C. E., RABINER R. (1981),Improving the Quality of a Noisy Speech Signal, Bell Syst. Tech. J., Vol. 60, No. 8, Pages 1847-1859.

VAN COMPERNOLLE D. (1989), Noise Adaptation in a Hidden Markov Model Speech Recognition System, Computer Speech and Language, Vol. 3, Pages 151-167.

VASEGHI S.V., FRAYLING-CORCK R. (1993), Restoration of Archived Gramophone Records, Journal of Audio Engineering Society.

XIE F.(1993), Speech Enhancement by Non-Linear Spectral Estimation a Unifying Approach, EUROSPEECH-93, Pages 617-620.

10

Interpolation

10.1 Introduction
10.2 Polynomial Interpolation
10.3 Statistical Interpolation

Interpolation is the estimation of the unknown, or the lost, samples of a signal using a weighted average of a number of known samples at the neighbourhood points. Interpolators are used in various forms in most signal processing and decision making systems. Applications of interpolators include conversion of a discrete-time signal to a continuous-time signal, sampling rate conversion in multi-rate systems, low bit rate speech coding, upsampling of a signal for improved graphical representation, and restoration of a sequence of samples irrevocably distorted by transmission errors, impulsive noise, drop outs etc.

This chapter begins with a study of the ideal interpolation of a band limited signal, a simple model for the effects of a number of missing samples, and the factors that affect interpolation. The classical approach to interpolation is to construct a polynomial that passes through the known samples. In Section 10.2 a general form of polynomial interpolation, and its special forms Lagrange, Newton, Hermite, and cubic spline interpolators are considered. Optimal interpolators utilise predictive and statistical models of the signal process. In Section 10.3 a number of model-based interpolation methods are considered. These methods include maximum a posterior interpolation and least squared error interpolation based on an autoregressive model. Finally we consider time-frequency interpolation, and interpolation through search of an adaptive codebook for the best signal.

10.1 Introduction

The objective in interpolation is to obtain a high fidelity reconstruction of the unknown or the missing samples of a signal. The emphasis in this chapter is on the interpolation of a *sequence* of lost samples. However, first in this section, the theory of ideal interpolation of a band-limited signal is introduced, and its applications in conversion of a discrete-time signal to a continuous-time signal, and in conversion of the sampling rate of a digital signal are considered. Then, a simple distortion model is used to gain some insights, on the effects of a sequence of missing samples, and the methods for recovery of the lost samples. The factors that affect interpolation error are also considered in this section.

10.1.1 Interpolation of a Sampled Signal

A common application of interpolation, is the reconstruction of a continuous-time signal $x(t)$ from a discrete-time signal $x(m)$. The condition for the recovery of a continuos time signal from its samples is stated by the Nyquist sampling theorem. The Nyquist theorem states that a band-limited signal, with a highest frequency content of f_c Hz, can be reconstructed from its samples *if* the sampling speed is greater than $2f_c$ samples per second. Consider a bandlimited continuous-time signal

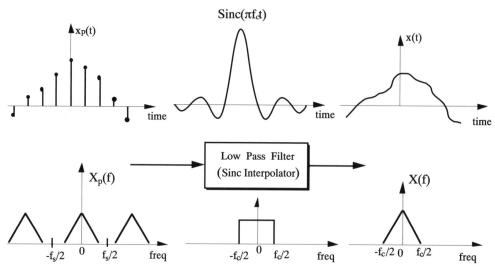

Figure 10.1 Reconstruction of a continuous-time signal from its samples. In frequency domain interpolation is equivalent to filtering out the frequencies introduced by sampling.

$x(t)$, sampled at a rate of f_s samples per second. The discrete-time signal $x(m)$ may be expressed as the following product

$$x(m) = x(t)\,p(t) = \sum_{m=-\infty}^{\infty} x(t)\,\delta(t - mT_s) \qquad (10.1)$$

where $p(t)=\sum\delta(t-mT_s)$ is the sampling function, and $T_s=1/f_s$ is the sampling interval. Taking the Fourier transform of Eq. (10.1) it can be shown that the spectrum of the sampled signal is given by

$$X_s(f) = X(f) * P(f) = \sum_{k=-\infty}^{\infty} X(f + kf_s) \qquad (10.2)$$

where $X(f)$ and $P(f)$ are the spectra of the signal $x(t)$ and the sampling function $p(t)$ respectively, and the operator * denotes the convolution. Eq. (10.2), illustrated in Figure 10.1, states that; the spectrum of a sampled signal is composed of the original baseband spectrum $X(f)$, and the repetitions, or images, of $X(f)$ spaced uniformly at frequency intervals of $f_s=1/T_s$. When the sampling frequency is above the Nyquist rate, the based-band spectrum $X(f)$ is not overlapped by its images $X(f \pm kf_s)$, and the original signal can be recovered by a lowpass filter as shown in Figure 10.1. Hence the ideal interpolator of a band-limited sampled signal is an ideal lowpass filter with a sinc impulse response. The recovery of a continuous-time signal through sinc interpolation can be expressed as

$$x(t) = \sum_{m=-\infty}^{\infty} x(m)\,T_s f_c \,\mathrm{sinc}\big(\pi f_c(t - mT_s)\big) \qquad (10.3)$$

In practice the sampling rate should be sufficiently greater than $2f_c$, say $2.5f_c$, in order to accommodate the transition bandwidth of the interpolating lowpass filter.

10.1.2   Digital Interpolation by a Factor of *I*

Applications of digital interpolators include sampling rate conversion in multi-rate systems, and upsampling for improved graphical representation. To change a sampling rate by a factor of $V=I/D$ (where I and D are integers) the signal is first interpolated by a factor of I, and then the interpolated signal is decimated by a factor of D.

Consider a band limited discrete-time signal $x(m)$ with a based band spectrum $X(f)$ as shown in Figure 10.2. The sampling rate can be increased by a factor of I through

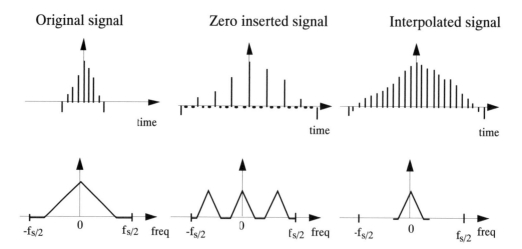

Figure 10.2 Illustration of upsampling by a factor of 3 using a two stage process of zero-insertion and digital lowpass filtering.

interpolation of I-1 samples in between every two samples of $x(m)$. In the followings it is shown that digital interpolation by a factor of I can be achieved through a two stage process of : (a) insertion of I-1 zeros in between every two samples, and (b) lowpass filtering of the zero-inserted signal by a filter with a cutoff frequency of $f_s/2I$. Consider the signal $x_z(m)$ obtained by inserting I-1 zeros in between every two samples of $x(m)$ and expressed as

$$x_z(m) = \begin{cases} x\left(\dfrac{m}{I}\right) & m = 0, \pm I, \pm 2I,... \\ 0 & otherwise \end{cases} \tag{10.4}$$

The spectrum of the zero-inserted signal is related to the spectrum of the original discrete time signal by

$$X_z(f) = \sum_{m=-\infty}^{\infty} x_z(m)e^{-j2\pi fm}$$

$$= \sum_{m=-\infty}^{\infty} x(m)e^{-j2\pi fmI} \tag{10.5}$$

$$= X(I.f)$$

Equation (10.5) states that the spectrum of the zero-inserted signal $X_z(f)$ is a frequency-scaled version of the spectrum of the original signal $X(f)$. Figure 10.2 shows that the baseband spectrum of the zero-inserted signal is composed of I repetitions of the based band spectrum of the original signal. The interpolation of the zero-inserted signal is therefore equivalent to filtering out the repetitions of $X(f)$ in the base band of $X_z(f)$, as illustrated in Figure 10.2. Note that to maintain the real-time duration of the signal the sampling rate of the interpolated signal needs to be increased by a factor of I.

10.1.3   Interpolation of a Sequence of Lost Samples

In this section we introduce the problem of interpolation of a sequence of M missing samples of a signal given a number of samples on both side of the gap as illustrated in Figure 10.3. Perfect interpolation is only possible if the missing samples are redundant, in the sense that they carry no more information than that conveyed by the known neighbouring samples. This would be the case if the signal is a completely predictable signal such as a sine wave, or in the case of a random signal if the sampling rate is greater than M times the Nyquist rate. However in many practical cases the signal is a realisation of a random process, and the sampling rate is only marginally above the Nyquist rate. In such cases, the lost samples can not be perfectly recovered and some interpolation error is inevitable.
A simple distortion model for a signal $y(m)$ with M missing samples, illustrated in Figure 10.3, is given by

$$\begin{aligned} y(m) &= x(m)d(m) \\ &= x(m)[1 - r(m)] \end{aligned} \tag{10.6}$$

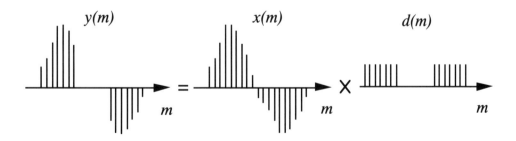

Figure 10.3 Illustration of a distortion model form a signal with a sequence of missing samples.

where the distortion operator $d(m)$ is defined as

$$d(m) = 1 - r(m) \tag{10.7}$$

and $r(m)$ is a rectangular pulse of duration M samples starting at the sampling time k

$$r(m) = \begin{cases} 1 & k \le m \le k + M - 1 \\ 0 & otherwise \end{cases} \tag{10.8}$$

In the frequency domain Eq. (10.6) becomes

$$\begin{aligned} Y(f) &= X(f) * D(f) \\ &= X(f) * [\delta(f) - R(f)] \\ &= X(f) - X(f) * R(f) \end{aligned} \tag{10.9}$$

where $D(f)$ is the spectrum of the distortion $d(m)$, $\delta(f)$ is the Kronecker delta function, and $R(f)$, the frequency spectrum of the rectangular pulse $r(m)$, is given by

$$R(f) = e^{-j2\pi f(k+(M-1)/2)} \frac{\sin(\pi f M)}{\sin(\pi f)} \tag{10.10}$$

In general the distortion $d(m)$ is a non-invertible, many-to-one transformation, and perfect interpolation with zero error is not possible. However, as discussed in Section 10.3, the interpolation error can be minimised by optimal utilisation of the signal models and the information contained in the neighbouring samples.

Example 10.1 Interpolation of missing samples of a sinusoidal signal.
Consider a cosine waveform of amplitude A and frequency f_0 with M missing samples modelled as

$$\begin{aligned} y(m) &= x(m)\, d(m) \\ &= A\left(\cos 2\pi f_0 m\right)(1 - r(m)) \end{aligned} \tag{10.11}$$

where $r(m)$ is the rectangular pulse defined in Eq. (10.7). In the frequency domain, the distorted signal can be expressed as

$$\begin{aligned} Y(f) &= \frac{A}{2}(\delta(f - f_0) + \delta(f + f_0)) * (\delta(f) - R(f)) \\ &= \frac{A}{2}(\delta(f - f_0) + \delta(f + f_0) - R(f - f_0) - R(f + f_0)) \end{aligned} \tag{10.12}$$

where $R(f)$ is the spectrum of the rectangular pulse $r(m)$ as in Eq. (10.9).
From Eq. (10.12) it is evident that, for a cosine signal of frequency f_0, the distortion in the frequency domain due to the missing samples is manifested in the appearance of sinc functions centred at $\pm f_0$. The distortion can be removed by filtering the signal with a very narrow band-pass filter. Note that for a cosine signal perfect restoration is possible only because the signal has infinitely narrow bandwidth, or equivalently because the signal is completely predictable. In fact, for this example the distortion can also be removed using a linear prediction model which, for a cosine signal, can be regarded as a data adaptive narrow bandpass filter.

10.1.4 Factors that Affect Interpolation

Interpolation is affected by a number of factors, the most important of which are as follows :
(a) The predictability, or correlation structure of the signal. As the correlation of successive samples increases, the predictability of a sample from the neighbouring samples increases. In general, interpolation improves with the increasing correlation structure, or equivalently the decreasing bandwidth, of a signal.
(b) The sampling rate. As the sampling rate increases, adjacent samples become more correlated, the redundant information increases, and interpolation improves.
(c) Nonstationary characteristics of the signal. For time varying signals the available samples, some distance in time away from the missing samples, may not be relevant because the signal characteristics may have completely changed. This is particularly important in interpolation of a large sequence of samples.
(d) The length of the missing samples. In general interpolation quality decreases with increasing length of the missing samples.
(e) Finally interpolation depends on the optimal use of the data and the efficiency of the interpolator.
The classical approach to interpolation is to construct a polynomial that passes through the known samples. We continue this chapter with a study of the general form of polynomial interpolation, and consider Lagrange, Newton, Hermite and cubic spline interpolators. Polynomial interpolators are *not* optimal or well suited to make efficient use of a relatively large number of known samples, or to interpolate a relatively large segment of missing samples.
In Section 10.3 we study several statistical digital signal processing methods for interpolation of a sequence of missing samples. These include model-based methods which are well suited for interpolation of small to medium size gaps of missing samples. We also consider frequency-time interpolation methods, and interpolation through waveform substitution, which have the ability to replace relatively large gaps of missing samples.

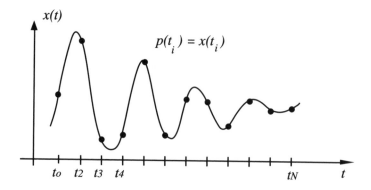

Figure 10.4 Illustration of an Interpolation curve through a number of samples.

10.2 Polynomial Interpolation

The classical approach to interpolation, is to construct a polynomial that passes through the known samples. Polynomial interpolators may be formulated in various forms such as power series, Lagrange interpolation, and Newton interpolation. These various forms are mathematically equivalent and can be transformed from one into another. Suppose the data consists of $N+1$ samples $\{x(t_0), x(t_1), \ldots x(t_N)\}$, where $x(t_n)$ denotes the amplitude of the signal $x(t)$ at time t_n. The polynomial of order N that passes through the $N+1$ known samples is unique, and may be written in a power series form as

$$\hat{x}(t) = P_N(t) = a_0 + a_1 t + a_2 t^2 + a_3 t^3 + \cdots + a_N t^N \tag{10.13}$$

where $P_N(t)$ is a polynomial of order N, and a_k's are the polynomial coefficients. From Eq. (10.13), and a set of $N+1$ known samples, a system of $N+1$ linear equations with N unknown coefficients can be formulated as

$$
\begin{aligned}
x(t_0) &= a_0 + a_1 t_0 + a_2 t_0^2 + a_3 t_0^3 + \cdots + a_N t_0^N \\
x(t_1) &= a_0 + a_1 t_1 + a_2 t_1^2 + a_3 t_1^3 + \cdots + a_N t_1^N \\
&\vdots \qquad\qquad \vdots \qquad\qquad\qquad \ddots \\
x(t_N) &= a_0 + a_1 t_N + a_2 t_N^2 + a_3 t_N^3 + \cdots + a_N t_N^N
\end{aligned}
\tag{10.14}
$$

From Eq. (10.14) the polynomial coefficients are given by

$$\begin{pmatrix} a_0 \\ a_1 \\ a_2 \\ \vdots \\ a_N \end{pmatrix} = \begin{pmatrix} 1 & t_0 & t_0^2 & t_0^3 & \cdots & t_0^N \\ 1 & t_1 & t_1^2 & t_1^3 & \cdots & t_1^N \\ 1 & t_2 & t_2^2 & t_2^3 & \cdots & t_2^N \\ \vdots & \vdots & \vdots & \vdots & \ddots & \vdots \\ 1 & t_N & t_N^2 & t_N^3 & \cdots & t_N^N \end{pmatrix}^{-1} \begin{pmatrix} x(t_0) \\ x(t_1) \\ x(t_2) \\ \vdots \\ x(t_N) \end{pmatrix} \tag{10.15}$$

The matrix in Eq. (10.15) is called a Vandermonde matrix. For a large number of samples, N, the Vandermonde matrix becomes large and ill-conditioned. An ill-conditioned matrix is sensitive to small computational errors, such as quantisation errors, and can easily produce inaccurate results.

There are alternative methods of implementation of the polynomial interpolator which are simpler to program and/or better structured, such as Lagrange and Newton methods. However, it must be noted that these variants of the polynomial interpolation also become ill-conditioned for a large number of samples N.

10.2.1   Lagrange Polynomial Interpolation

To introduce the Lagrange interpolation, consider a line interpolator passing through two points $x(t_0)$ and $x(t_1)$

$$\hat{x}(t) = p_1(t) = x(t_0) + \underbrace{\frac{x(t_1) - x(t_0)}{t_1 - t_0}}_{slope\ of\ line} (t - t_0) \tag{10.16}$$

The line Equation (10.16) may be rearranged and expressed as

$$p_1(t) = \frac{t - t_1}{t_0 - t_1} x(t_0) + \frac{t - t_0}{t_1 - t_0} x(t_1) \tag{10.17}$$

Equation (10.17) is in the form of a Lagrange polynomial. Note that the Lagrange form of a line-interpolator is composed of the weighted combination of two lines as illustrated in Figure 10.2. In general, the Lagrange polynomial, of degree N, passing through $N+1$ samples $\{x(t_0), x(t_1), \ldots x(t_N)\}$ is given by the polynomial equation

$$P_N(t) = L_0(t)x(t_0) + L_1(t)x(t_1) + \cdots + L_N(t)x(t_N) \tag{10.18}$$

where each Lagrange coefficient $L_N(t)$ is itself a polynomial of degree N given by

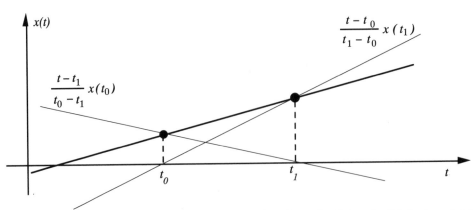

Figure 10.5 The Lagrange line interpolator passing through $x(t_0)$ and $x(t_1)$, is described in terms of the combination of two lines: one passing through $(x(t_0), t_1)$ and the other through $(x(t_1), t_0)$.

$$L_i(t) = \frac{(t - t_0) \cdots (t - t_{i-1})(t - t_{i+1}) \cdots (t - t_N)}{(t_i - t_0) \cdots (t_i - t_{i-1})(t_i - t_{i+1}) \cdots (t_i - t_N)} = \prod_{\substack{n=0 \\ n \neq i}}^{N} \frac{t - t_n}{t_i - t_n} \quad (10.19)$$

Note that the ith Lagrange polynomial coefficient $L_i(t)$ becomes unity at the ith known sample point $(L_i(t_i)=1)$, and zero at every other known sample $(L_i(t_j)=0 \ i \neq j)$. Therefore $P_N(t_i)=L_i(t_i)x(t_i)=x(t_i)$, and the polynomial passes through the known data points as required.

The main drawbacks of Lagrange interpolation are as follows: (a) the computational complexity is large, (b) the coefficients of a polynomial of order N can not be used in the calculations of the coefficients of a higher order polynomial, and (c) the evaluation of the interpolation error is difficult. The Newton polynomial, introduced in the next section, overcomes some of these difficulties.

10.2.2 Newton Interpolation Polynomial

Newton polynomials have a recursive structure, such that a polynomial of order N can be constructed by extension of a polynomial of order $N-1$ as follows

$$p_0(t) = a_0 \qquad\qquad\qquad\qquad\qquad\qquad\qquad \text{d.c}$$

$$\begin{aligned} p_1(t) &= a_0 \ + \ a_1 \cdot (t - t_0) \\ &= p_0(t) \ + \ a_1 \cdot (t - t_0) \end{aligned} \qquad\qquad \text{ramp}$$

$$p_2(t) = \underbrace{a_0 + a_1 \cdot (t - t_0)}_{} + a_2 \cdot (t - t_0)(t - t_1)$$

$$= \underbrace{p_1(t)}_{} + a_2 \cdot (t - t_0)(t - t_1) \qquad \text{quadratic}$$

$$p_3(t) = \underbrace{a_0 + a_1 \cdot (t - t_0) + a_2 \cdot (t - t_0)(t - t_1)}_{} + a_3 \cdot (t - t_0)(t - t_1)(t - t_2)$$

$$= \underbrace{p_2(t)}_{} + a_3 \cdot (t - t_0)(t - t_1)(t - t_2) \qquad \text{cubic}$$

$$(10.20)$$

and in general the recursive, *order update,* form of a Newton polynomial can be formulated as

$$p_N(t) = p_{N-1}(t) + a_N \cdot (t - t_0)(t - t_1) \cdots (t - t_{N-1}) \tag{10.21}$$

For a sequence of $N+1$ samples $\{x(t_0), x(t_1), \ldots x(t_N)\}$, the polynomial coefficients are obtained using the constraint $p_N(t_i) = x(t_i)$ as follows :
To solve for the coefficient a_0 equate the polynomial Eq. (10.21) at $t=t_0$ to $x(t_0)$ as

$$p_N(t_0) = p_0(t_0) = x(t_0) = a_0 \tag{10.22}$$

To solve for the coefficient a_1, the first-order polynomial $p_1(t)$ is evaluated at $t=t_1$

$$p_1(t_1) = x(t_1) = a_0 + a_1(t_1 - t_0) = x(t_0) + a_1(t_1 - t_0) \tag{10.23}$$

from which

$$a_1 = \frac{x(t_1) - x(t_0)}{t_1 - t_0} \tag{10.24}$$

Note that the coefficient a_1 is the slope of the line passing through the points $[x(t_0), x(t_1)]$. To solve for the coefficient a_2 the second-order polynomial $p_2(t)$ is evaluated at $t=t_2$

$$p_2(t_2) = x(t_2) = a_0 + a_1(t_2 - t_0) + a_2(t_2 - t_0)(t_2 - t_1) \tag{10.25}$$

Substituting a_0 and a_1 from Eqs. (10.22) and (10.24) in (10.25) we obtain

$$a_2 = \left(\frac{x(t_2) - x(t_1)}{t_2 - t_1} - \frac{x(t_1) - x(t_0)}{t_1 - t_0} \right) \bigg/ (t_2 - t_0) \tag{10.26}$$

Each term in the bracket of Eq. (10.26) is a slope term, and the coefficient a_2 is the slope of slope. To formulate a solution for the higher order coefficients, we need to introduce the concept of divided differences. Each of the two ratios in the numerator bracket of Eq. (10.26) is a so called "divided difference". The divided difference between two points t_i and t_{i-1} is defined as

$$d_1(t_{i-1}, t_i) = \frac{x(t_i) - x(t_{i-1})}{t_i - t_{i-1}} \qquad (10.27)$$

The divided difference between two points may be interpreted as the average difference or the slope of line passing through the two points. The second order divided difference (i.e. the divided difference of divided difference) over three points t_{i-2}, t_{i-1} and t_i is given by

$$d_2(t_{i-2}, t_i) = \frac{d_1(t_{i-1}, t_i) - d_1(t_{i-2}, t_{i-1})}{t_i - t_{i-2}} \qquad (10.28)$$

and the third order divided difference is

$$d_3(t_{i-3}, t_i) = \frac{d_2(t_{i-2}, t_i) - d_2(t_{i-3}, t_{i-1})}{t_i - t_{i-3}} \qquad (10.29)$$

and so on. In general the jth order divided difference can be formulated in terms of the divided differences of order j-1, in an order-update equation given as

$$d_j(t_{i-j}, t_i) = \frac{d_{j-1}(t_{i-j+1}, t_i) - d_{j-1}(t_{i-j}, t_{i-1})}{t_i - t_{i-j}} \qquad (10.30)$$

Note that $a_1 = d_1(t_0, t_1)$, $a_2 = d_2(t_0, t_2)$, $a_3 = d_3(t_0, t_3)$, and in general the Newton polynomial coefficients are obtained from the divided differences using the relation

$$a_i = d_i(t_0, t_i) \qquad (10.31)$$

A main advantage of the Newton polynomial is its computational efficiency, in that a polynomial of order N-1 can be easily extended to a higher order polynomial of order N. This is useful in the selection of the best polynomial order, for a given set of data.

10.2.3 Hermite Interpolation Polynomials

Hermite polynomials are formulated to fit not only to the signal samples, but also to the derivatives of the signal as well. Suppose the data consists of $N+1$ samples and assume that all the derivatives up to the Mth order derivative are available. Let the data set, the signal samples and the derivatives, be denoted as $\left[x(t_i), x'(t_i), x''(t_i), \cdots, x^{(M)}(t_i), \; i = 0, \cdots, N\right]$. There are altogether $K=(N+1)(M+1)$ data points and a polynomial of order $K-1$ can be fitted to the data as

$$p(t) = a_0 + a_1 t + a_2 t^2 + a_3 t^3 + \cdots + a_{K-1} t^{K-1} \tag{10.32}$$

To obtain the polynomial coefficients we substitute the given samples in the polynomial and its M derivatives as

$$
\begin{aligned}
p(t_i) &= x(t_i) \\
p'(t_i) &= x'(t_i) \\
p''(t_i) &= x''(t_i) \\
\vdots &= \vdots \\
p^{(M)}(t_i) &= x^{(M)}(t_i) \qquad i = 0, 1, \dots, N
\end{aligned}
\tag{10.33}
$$

In all, there are $K=(M+1)(N+1)$ equations in (10.33), and these can be used to calculate the coefficients of the polynomial Eq. (10.32). In theory, the constraint that the polynomial must also fit the derivatives should result in a better interpolating polynomial that passes through the sampled points and is also consistent with the known underlying dynamics (the derivatives) of the curve. However, even for moderate values of N and M the size of Eq. (10.33) becomes too large for most practical purposes.

10.2.4 Cubic Spline Interpolation

A polynomial interpolator of order N is constrained to pass through $N+1$ known samples, and can have $N-1$ maxima and minima. In general, the interpolation error increases rapidly with increasing polynomial order, as the interpolating curve has to wiggle through the $N+1$ samples. When a large number of samples are to be fitted with a smooth curve, it may be better to divide the signal into a number of smaller intervals, and to fit a low order interpolating polynomial to each small interval. Care must be taken that the polynomial curves are continuous at the end-points of each interval. In cubic spline interpolation, a cubic polynomial is fitted to each interval between two samples. A cubic polynomial has the form:

$$p(t) = a_0 + a_1 t + a_2 t^2 + a_3 t^3 \qquad (10.34)$$

A cubic polynomial has four coefficients, and needs four conditions for the determination of a unique set of coefficients. For each interval, two conditions are set by the samples at the end-points of the interval. Two further conditions are met by the constraints that the first derivatives of the polynomial should be continuous at the two end-points. Consider an interval $t_i \le t \le t_{i+1}$ of length $T_i = t_{i+1} - t_i$ as shown in the Figure 10.6. Using a local co-ordinate $\tau = t - t_i$, the cubic polynomial becomes

$$p(\tau) = a_0 + a_1 \tau + a_2 \tau^2 + a_3 \tau^3 \qquad (10.35)$$

At $\tau = 0$ we obtain the first coefficient a_0 as

$$a_0 = p(\tau = 0) = x(t_i) \qquad (10.36)$$

The second derivative of $p(\tau)$ is given by

$$p''(\tau) = 2a_2 + 6a_3 \tau \qquad (10.37)$$

Evaluation of the second derivative at $\tau = 0$ (i.e. $t = t_i$)gives the coefficient a_2

$$a_2 = \frac{p_i''(\tau = 0)}{2} = \frac{p_i''}{2} \qquad (10.38)$$

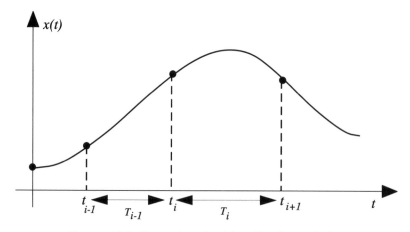

Figure 10.6 Illustration of cubic spline interpolation.

Similarly, evaluating the second derivative at point t_{i+1} (i.e. $\tau = T_i$) yields the fourth coefficient

$$a_3 = \frac{p''_{i+1} - p''_i}{6T_i} \tag{10.39}$$

Now to obtain the coefficient a_1, we evaluate $p(\tau)$ at $\tau = T_i$

$$p(\tau = T_i) = a_0 + a_1 T_i + a_2 T_i^2 + a_3 T_i^3 = x(t_{i+1}) \tag{10.40}$$

and substitute a_0, a_2 and a_3 from equations (10.36), (10.38) and (10.39) in (10.40)

$$a_1 = \frac{x(t_{i+1}) - x(t_i)}{T_i} - \frac{p''_{i+1} + 2p''_i}{6} T_i \tag{10.41}$$

The cubic polynomial can now be written as

$$p(\tau) = x(t_i) + \left(\frac{x(t_{i+1}) - x(t_i)}{T_i} - \frac{p''_{i+1} + 2\,p''_i}{6} T_i \right) \tau + \frac{p''_i}{2} \tau^2 + \frac{p''_{i+1} - p''_i}{6T_i} \tau^3 \tag{10.42}$$

To determine the coefficients of the polynomial in Eq. (10.42), we need the second derivatives p''_i and p''_{i+1}. These are obtained from the constraint that the first derivatives of the curves at the endpoints of each interval must be continuous. From Eq. (10.42) the first derivatives of $p(\tau)$ evaluated at the endpoints t_i and t_{i+1} are

$$p'_i = p'(\tau = 0) = -\frac{T_i}{6} \left[p''_{i+1} + 2p''_i \right] + \frac{1}{T_i} \left[x(t_{i+1}) - x(t_i) \right] \tag{10.43}$$

$$p'_{i+1} = p'(\tau = T_i) = \frac{T_i}{6} \left[2p''_{i+1} + p''_i \right] + \frac{1}{T_i} \left[x(t_{i+1}) - x(t_i) \right] \tag{10.44}$$

Similarly, for the preceding interval, $t_{i-1} < t < t_i$, the first derivative of the cubic spline curve evaluated at $\tau = t_i$ is

$$p'_i = p'(\tau = t_i) = \frac{T_{i-1}}{6} \left[2p''_i + p''_{i-1} \right] + \frac{1}{T_{i-1}} \left[x(t_i) - x(t_{i-1}) \right] \tag{10.45}$$

For continuity of the first derivative at t_i, p'_i at the end of the interval (t_{i-1}, t_i) must be equal to the p'_i at the start of the interval (t_i, t_{i+1}). Equating the right hand sides of Eqs. (10.43) and (10.45), and repeating this exercise yields

$$T_{i-1} p''_{i-1} + 2 (T_{i-1} + T_i) p''_i + T_i p''_{i+1} = 6 \left(\frac{1}{T_{i-1}} x(t_{i-1}) - \left(\frac{1}{T_{i-1}} + \frac{1}{T_i} \right) x(t_i) + \frac{1}{T_i} x(t_{i+1}) \right)$$

$$i = 1, 2, \ldots, N\text{-}1 \qquad (10.46)$$

In Eq. (10.46) there are N-1 equations in N+1 unknowns p''_i. For a unique solution we need to specify the second derivatives at the points t_0 and t_N. This can be done in two ways : (a) set the second derivatives at the end points, t_0 and t_N, (i.e. p''_0 and p''_N), to zero, or (b) extrapolate the derivatives from the inside data.

10.3 Statistical Interpolation

The statistical signal processing approach to interpolation of a sequence of lost samples is based on the utilisation of a predictive or a probabilistic model of the signal. In this section we study the maximum a posterior interpolation, an autoregressive model based interpolation, a frequency-time interpolation method, and interpolation through searching a signal record for the best replacement.
Figure 10.7 illustrates the problem of interpolation of a sequence of lost samples. It is assumed that we have a signal record of N samples, and that within this record a segment of M samples, starting at time k, $x_{Uk} = \{x(k), x(k+1), \ldots, x(k+M-1)\}$ are missing. The objective is to make an optimal estimate of the missing segment x_{Uk}, using the remaining N-k samples x_{Kn} and a model of the signal process.

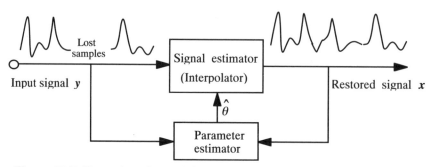

Figure 10.7 Illustration of a model-based iterative signal interpolation system.

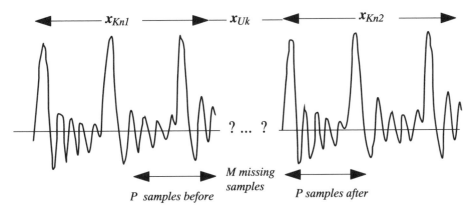

Figure 10.8 A signal with M missing samples and N–M known samples. On each side of the missing segment, P samples are used to interpolate the unknown samples.

An N-sample signal vector x, composed of M unknown samples and N-M known samples, can be written as

$$x = \begin{pmatrix} x_{Kn_1} \\ x_U \\ x_{Kn_2} \end{pmatrix} = \begin{pmatrix} x_{Kn_1} \\ 0 \\ x_{Kn_2} \end{pmatrix} + \begin{pmatrix} 0 \\ x_{Uk} \\ 0 \end{pmatrix} = K x_{Kn} + U x_{Uk} \qquad (10.47)$$

where the vector $x_{Kn}=[x_{Kn1}\, x_{Kn2}]^T$ is composed of the known samples, and the vector x_{Uk} is composed of the unknown samples as illustrated in Figure 10.8. The matrices K and U in Eq. (10.47) are rearrangement matrices that assemble the vector x from x_{Kn} and x_{Uk}.

10.3.1 Maximum a Posterior Interpolation

The posterior pdf of an unknown signal segment x_{Uk}, given a number of neighbouring samples x_{Kn}, can be expressed using the Bayes rule as

$$\begin{aligned} f_X(x_{Uk}|x_{Kn}) &= \frac{f_X(x_{Kn}, x_{Uk})}{f_X(x_{Kn})} \\ &= \frac{f_X(x = K x_{Kn} + U x_{Uk})}{f_X(x_{Kn})} \end{aligned} \qquad (10.48)$$

In Eq. (10.48), for a given sequence of samples x_{Kn}, $f_X(x_{Kn})$ is a constant. Therefore the estimate that maximises the posterior pdf, the MAP estimate, is given by

$$\hat{x}_{Uk}^{MAP} = \underset{x_{Uk}}{\arg\max} \; f_X(K x_{Kn} + U x_{Uk}) \tag{10.49}$$

Example 10.2 MAP Interpolation of a Gaussian Signal

Assume that an observation signal $x=Kx_{Kn}+Ux_{Uk}$, from a zero-mean Gaussian process, is composed of a sequence of M missing samples x_{Uk} and $N-M$ known neighbouring samples as in Eq. (10.47). The pdf of the signal x is given by

$$f_X(x) = \frac{1}{(2\pi)^{N/2}|\Sigma_{xx}|^{1/2}} \exp(-\frac{1}{2}x^T \Sigma_{xx}^{-1} x) \tag{10.50}$$

where Σ_{xx}. is the covariance matrix of the Gaussian vector process x. Substitution of Eq. (10.50) in Eq. (10.48) yields the conditional pdf of the unknown signal x_{Uk} given a number of samples x_{Kn} as

$$f_X(x_{Uk}|x_{Kn}) = \frac{1}{f_X(x_{Kn})} \frac{1}{(2\pi)^{N/2}|\Sigma_{xx}|^{1/2}} \times$$
$$\exp\left(-\frac{1}{2}(K x_{Kn} + U x_{Uk})^T \Sigma_{xx}^{-1}(K x_{Kn} + U x_{Uk})\right) \tag{10.51}$$

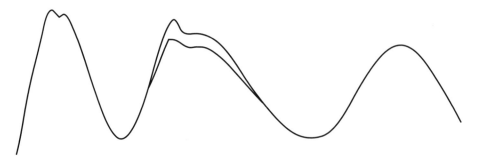

Figure 10.9 Illustration of MAP interpolation of a segment of 20 samples.

The MAP signal estimate, obtained by setting the derivative of the log-likelihood function $\ln f_X(x|x_{Kn})$ of Eq. (10.51) with respect to x_{Uk} to zero is given by

$$x_{Uk} = -\left(U^T \, \Sigma_{xx}^{-1} \, U\right)^{-1} U^T \, \Sigma_{xx}^{-1} \, K \, x_{Kn} \tag{10.52}$$

10.3.2   Least Squared Error Autoregressive Interpolation

In this Section we describe interpolation based on an Autoregressive (AR) model of the signal process. The term "autoregressive model" is an alternative terminology for the linear predictive models considered in Chapter 7. In this section, the terms "linear predictive model" and "autoregressive model" are used inter-changeably. The autoregressive interpolation algorithm is a two stage process : in the first stage the AR model coefficients are estimated from the incomplete signal, and in the second stage the estimates of the model coefficients are used to interpolate the missing samples. For high quality interpolation, the estimation algorithm should utilise all the correlation structures of the signal process including periodic or pitch period structures. In Section 10.3.4 the AR interpolation method is extended to include the pitch-period correlations.

10.3.3   Interpolation Based on a Short-term Prediction Model

An autoregressive (AR), or linear predictive, signal $x(m)$ is described as

$$x(m) = \sum_{k=1}^{P} a_k \, x(m-k) + e(m) \tag{10.53}$$

where $x(m)$ is the AR signal, a_k are the model coefficients and $e(m)$ is a zero mean excitation signal. The excitation may be a random signal, a quasi-periodic impulse train, or a mixture of the two. The AR coefficients, a_k, model the correlation structure or equivalently the spectral patterns of the signal.

Assume that we have a signal record of N samples and that within this record a segment of M samples, starting from the sample k, $x_{Uk} = \{x(k), ..., x(k+M-1)\}$ are missing. The objective is to estimate the missing samples x_{Uk}, using the remaining $N-k$ samples and an AR model of the signal. Figure 10.8 illustrates the interpolation problem. For this signal record of N samples, the AR Eq. (10.53) can be expanded to form the following matrix equation :

$$
\begin{pmatrix}
e(P) \\
e(P+1) \\
\vdots \\
\hline
e(k-1) \\
\hline
e(k) \\
e(k+1) \\
e(k+2) \\
\vdots \\
e(k+M+P-2) \\
e(k+M+P-1) \\
\hline
e(k+M+P) \\
e(k+M+P+1) \\
\vdots \\
e(N-1)
\end{pmatrix}
=
\begin{pmatrix}
x(P) \\
x(P+1) \\
\vdots \\
\hline
x(k-1) \\
\hline
x_{Uk}(k) \\
x_{Uk}(k+1) \\
x_{Uk}(k+2) \\
\vdots \\
x(k+M+P-2) \\
x(k+M+P-1) \\
\hline
x(k+M+P) \\
x(k+M+P+1) \\
\vdots \\
x(N-1)
\end{pmatrix}
-
\begin{pmatrix}
x(P-1) & x(P-2) & \cdots & x(0) \\
x(P) & x(P-1) & \cdots & x(1) \\
\vdots & \vdots & \ddots & \vdots \\
x(k-2) & x(k-3) & \cdots & x(k-P-1) \\
x(k-1) & x(k-2) & \cdots & x(k-P) \\
x_{Uk}(k) & x(k-1) & \cdots & x(k-P+1) \\
x_{Uk}(k+1) & x_{Uk}(k) & \cdots & x(k-P+2) \\
\vdots & \vdots & \ddots & \vdots \\
x(k+M+P-3) & x(k+M+P-2) & \cdots & x_{Uk}(k+M-2) \\
x(k+M+P-2) & x(k+M+P-1) & \cdots & x_{Uk}(k+M-1) \\
x(k+M+P-1) & x(k+M+P) & \cdots & x(k+M) \\
x(k+M+P) & x(k+M+P+1) & \cdots & x(k+M+1) \\
\cdots & \cdots & \ddots & \cdots \\
x(N-2) & x(N-3) & \cdots & x(N-P-1)
\end{pmatrix}
\begin{pmatrix}
a_1 \\
a_2 \\
a_3 \\
\vdots \\
a_P
\end{pmatrix}
$$

$$(10.54)$$

Where the subscript Uk denotes the unknown samples. Equation (10.54) can be rewritten in a compact vector notation form as

$$e(x_{UK}, a) = x - Xa \qquad (10.55)$$

where the error vector, $e(x_{Uk}, a)$ is expressed as a function of the unknown samples and the unknown model coefficient vector. In this section, the optimality criterion, for the estimation of the model coefficient vector a and the missing samples x_{Uk}, is the minimum mean squared error given by the inner vector product

$$e^T e(x_{UK}, a) = x^T x + a^T X^T X a - 2a^T X^T x \qquad (10.56)$$

The squared error function in Eq. (10.56) involves nonlinear unknown terms of fourth order as $a^T X^T X a$, and cubic order as $a^T X^T x$. The least squared error formulation, obtained by differentiating $e^T.e(x_{Uk}, a)$, w.r.t the vectors a or x_{Uk}, results in a set of nonlinear equations of cubic order whose solution is non-trivial. A sub-optimal, but practical and mathematically tractable, approach is to solve for the missing samples and the unknown model coefficients in two separate stages. This is an instance of the general estimate and maximise (EM) algorithm, and is similar to the linear-predictive model-based restoration considered in Section 6.7. In the first stage of the solution, Eq. (10.54) is *linearised* by either assuming that the missing samples have zero values, or discarding the set of equations in matrix Eq. (10.54), between the two dashed lines, which involve the unknown signal samples. The linearised equations are used to solve for the AR model coefficient vector a by forming the equation

$$\hat{a} = \left(X_{Kn}^T X_{Kn} \right)^{-1} \left(X_{Kn}^T x_{Kn} \right) \qquad (10.57)$$

Where the vector $\hat{a}$ is an estimate of the model coefficients, obtained from the available signal samples.

The second stage of the solution involves the estimation of the unknown signal samples x_{Uk}. For an AR model of order P, and an unknown signal segment of length M, there are $2M+P$ nonlinear equations in (10.54) which involve the unknown samples, these are :

$$
\begin{pmatrix}
e(k) \\
e(k+1) \\
e(k+2) \\
\vdots \\
e(k+M+P-2) \\
e(k+M+P-1)
\end{pmatrix}
=
\begin{pmatrix}
x_{Uk}(k) \\
x_{Uk}(k+1) \\
x_{Uk}(k+2) \\
\vdots \\
x(k+M+P-2) \\
x(k+M+P-1)
\end{pmatrix}
-
\begin{pmatrix}
x(k-1) & x(k-2) & \cdots & x(k-p) \\
x_{Uk}(k) & x(k-1) & \cdots & x(k-p+1) \\
x_{Uk}(k+1) & x_{Uk}(k) & \cdots & x(k-p+2) \\
\vdots & & \ddots & \vdots \\
x_{Uk}(k+M+P-3) & x_{Uk}(k+M+P-4) & \cdots & x_{Uk}(k+M-2) \\
x_{Uk}(k+M+P-2) & x_{Uk}(k+M+P-3) & \cdots & x_{Uk}(k+M-1)
\end{pmatrix}
\begin{pmatrix}
a_1 \\
a_2 \\
a_3 \\
\vdots \\
a_{P-1} \\
a_P
\end{pmatrix}
$$

$$(10.58)$$

The estimate of the predictor coefficient vector $\hat{a}$ obtained from the first stage of the solution, is substituted in Eq. (10.58), so the only remaining unknowns in (10.58) are the missing signal samples. Equation (10.58) may be partitioned and rearranged in vector notation in the following form :

$$
\begin{pmatrix}
e(k) \\
e(k+1) \\
e(k+2) \\
e(k+3) \\
e(k+4) \\
\vdots \\
e(k+P-1) \\
e(k+P) \\
e(k+P+1) \\
\vdots \\
e(k+m+P-2) \\
e(k+m+P-1)
\end{pmatrix}
=
\begin{pmatrix}
1 & 0 & 0 & 0 & \cdots & 0 \\
-a_1 & 1 & 0 & 0 & \cdots & 0 \\
-a_2 & -a_1 & 1 & 0 & \cdots & 0 \\
-a_3 & -a_2 & -a_1 & 1 & \cdots & 0 \\
-a_4 & -a_3 & -a_2 & -a_1 & \cdots & 0 \\
\vdots & \vdots & \vdots & \vdots & \ddots & \vdots \\
-a_P & -a_{P-1} & -a_{P-2} & -a_{P-3} & \cdots & 0 \\
0 & -a_P & -a_{P-1} & -a_{P-2} & \cdots & 0 \\
0 & 0 & -a_P & -a_{P-1} & \cdots & 0 \\
\vdots & \vdots & \vdots & \vdots & \ddots & \vdots \\
0 & 0 & 0 & 0 & \cdots & -a_{P-1} \\
0 & 0 & 0 & 0 & \cdots & -a_P
\end{pmatrix}
\begin{pmatrix}
x_{Uk}(k) \\
x_{Uk}(k+1) \\
x_{Uk}(k+2) \\
x_{Uk}(k+3) \\
\vdots \\
x_{Uk}(k+M-1)
\end{pmatrix}
+
$$

$$
\begin{pmatrix}
-a_P & -a_{P-1} & -a_{P-2} & \cdots & -a_1 & 0 & \cdots & 0 & 0 & 0 & \cdots & 0 \\
0 & -a_P & -a_{P-1} & \cdots & -a_2 & 0 & \cdots & 0 & 0 & 0 & \cdots & 0 \\
0 & 0 & -a_P & \cdots & -a_3 & 0 & \cdots & 0 & 0 & 0 & \cdots & 0 \\
\vdots & \vdots & \vdots & \ddots & \vdots & \vdots & \ddots & \vdots & \vdots & \vdots & \ddots & \vdots \\
0 & 0 & 0 & \cdots & -a_P & 0 & \cdots & 0 & 0 & 0 & \cdots & 0 \\
0 & 0 & 0 & \cdots & 0 & 0 & \cdots & 0 & 0 & 0 & \cdots & 0 \\
0 & 0 & 0 & \cdots & 0 & 0 & \cdots & 1 & 0 & 0 & \cdots & 0 \\
0 & 0 & 0 & \cdots & 0 & 0 & \cdots & -a_1 & 1 & 0 & \cdots & 0 \\
0 & 0 & 0 & \cdots & 0 & 0 & \cdots & -a_2 & -a_1 & 1 & \cdots & 0 \\
0 & 0 & 0 & \cdots & 0 & 0 & \cdots & -a_3 & -a_2 & -a_1 & \cdots & 0 \\
\vdots & \vdots & \vdots & \ddots & \vdots & \vdots & \ddots & \vdots & \vdots & \vdots & \ddots & 0 \\
0 & 0 & 0 & \cdots & 0 & 0 & \cdots & -a_{P-1} & -a_{P-2} & -a_{P-3} & \cdots & -a_1
\end{pmatrix}
\begin{pmatrix}
x(k-P) \\
x(k-P+1) \\
x(k-P+2)) \\
\vdots \\
x(k-1) \\
0 \\
\vdots \\
x(k+M) \\
x(k+M+1) \\
x(k+M+2) \\
\vdots \\
x(k+M+P-1)
\end{pmatrix}
\quad (10.59)
$$

In Eq. (10.59) the unknown and the known samples are rearranged and grouped into two separate vectors. In a compact vector-matrix notation Eq. (10.58) can be written in the form

$$e = A_1 x_{Uk} + A_2 x_{Kn} \qquad (10.60)$$

where e is the error vector, A_1 is the first coefficient matrix, x_{Uk} is the unknown signal vector being estimated, A_2 is the second coefficient matrix and the vector x_{Kn} consists of the *known* samples in the signal matrix and vectors of Eq. (10.58). The total squared error is given by

$$e^T e = \left(A_1 x_{Uk} + A_2 x_{Kn}\right)^T \left(A_1 x_{Uk} + A_2 x_{Kn}\right) \qquad (10.61)$$

The least squared AR (LSAR) interpolation is obtained by minimisation of the squared error function w.r.t. the unknown signal samples x_{Uk} as

$$\frac{\partial e^T e}{\partial x_{Uk}} = 2 A_1^T A_1 x_{Kn} + 2 A_1^T A_2 x_{Kn} = 0 \qquad (10.62)$$

from Eq. (10.62) we have

$$\hat{x}_{Uk}^{LSAR} = -\left(A_1^T A_1\right)^{-1}\left(A_1^T A_2\right) x_{Kn} \qquad (10.63)$$

The solution in Eq. (10.62) gives the vector $\hat{x}_{Uk}^{LSAR}$ which is the least squared error estimate of the unknown data vector.

10.3.4 Interpolation Based on Long-term and Short-term Correlations

For the best results a model-based interpolation algorithm should utilise all the correlation structures of the signal process including any periodic structures. For example, the main correlations in a voiced speech signal are the short-term correlation due to the resonance of the vocal tract, and the long-term correlation due to the quasi-periodic excitation pulses of the glottal cords. For voiced speech, interpolation based on the short-term correlation does not perform well if the missing samples coincide with an underlying excitation pulse. In this section the AR interpolation is extended to include both the long-term and the short-term correlations. For most audio signals the short-term correlation of each sample with the immediately preceding samples decays exponentially with time, and can be modelled with an AR model of order 10 to 20. In order to include the pitch periodicities in the AR model of Eq. (10.53), the model order must be greater than the pitch period. For speech, the pitch period is normally in the range 4 to 20 milliseconds, equivalent to 40 to 200 samples at a sampling rate of 10 kHz.

Implementation of an AR model of this order is not practical due to stability problems and computational complexity.

A more practical AR model that includes the effects of the long term correlations, is illustrated in Figure 10.10. This modified AR model may be expressed by the following equation

$$x(m) = \sum_{k=1}^{P} a_k x(m-k) + \sum_{k=-Q}^{Q} p_k x(m-T-k) + e(m) \qquad (10.64)$$

The AR model of Eq. (10.64) is composed of a *short term predictor* $\sum a_k x(m-k)$ that models the contribution of the P immediate past samples, and a *long term predictor* $\sum p_k x(m-T-k)$ that models the contribution of $2Q+1$ samples a pitch period away. The parameter T is the pitch period, it can be estimated from the autocorrelation function as the time difference between the peak of the autocorrelation, which is at the correlation lag zero, and the second largest peak which should happen a pitch period away from the lag zero.

The AR model of Eq. (10.64) is specified by the parameter vector $c=[a_1, ..., a_P, p_{-Q}, ..., p_Q]$ and the pitch period T. Note that in Figure 10.9 the sample marked '?' coincides with the onset of an excitation pulse. This sample is not well predictable from the P past samples because they do not include a pulse event. The sample is more predictable from the $2Q+1$ samples a pitch period away as they include the effects of a similar excitation pulse. The predictor coefficients are estimated (see Chapter 7) using the so called normal equations as

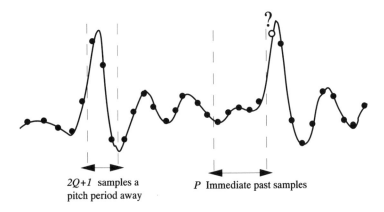

2Q+1 samples a   P Immediate past samples
pitch period away

Figure 10.10 A quasi-periodic waveform. The sample marked " ? " is predicted using P immediate past samples and 2Q+1 samples a pitch period away.

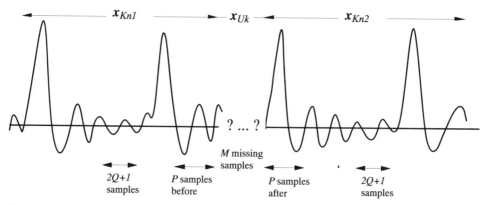

Figure 10.11 A signal with M missing samples. On each side of the missing sequence P immediate samples and Q samples a pitch period away are used for interpolation.

$$c = R_{xx}^{-1} r_{xx} \tag{10.65}$$

where R_{xx} is the autocorrelation matrix of signal x and r_{xx} is the correlation vector. In an expanded form Eq. (10.65) can be written as

$$
\begin{pmatrix} a_1 \\ a_2 \\ a_3 \\ \vdots \\ a_P \\ P_{-Q} \\ P_{-Q+1} \\ \vdots \\ P_{+Q} \end{pmatrix}
=
\begin{pmatrix}
r(0) & r(1) & \cdots & r(P-1) & r(T+Q-1) & r(T+Q) & \cdots & r(T-Q-1) \\
r(1) & r(0) & \cdots & r(P-2) & r(T+Q-2) & r(T+Q-1) & \cdots & r(T+Q-2) \\
r(2) & r(1) & \cdots & r(P-3) & r(T+Q-3) & r(T+Q-2) & \cdots & r(T+Q-3) \\
\vdots & \vdots & \ddots & \vdots & \vdots & \vdots & \ddots & \vdots \\
r(P-1) & r(P-2) & \cdots & r(0) & r(T+Q-P) & r(T+Q-P+1) & \cdots & r(T+Q-P) \\
r(T+Q-1) & r(T+Q-2) & \cdots & r(T+Q-P) & r(0) & r(1) & \cdots & r(2Q) \\
r(T+Q) & r(T+Q-1) & \cdots & r(T+Q-P+1) & r(1) & r(0) & \cdots & r(2Q-1) \\
\vdots & \vdots & \ddots & \vdots & \vdots & \vdots & \ddots & \vdots \\
r(T-Q-1) & r(T-Q-2) & \cdots & r(T-Q-P) & r(2Q) & r(2Q-1) & \cdots & r(0)
\end{pmatrix}^{-1}
\begin{pmatrix} r(1) \\ r(2) \\ r(3) \\ \vdots \\ r(P) \\ r(T+Q) \\ r(T+Q-1) \\ \vdots \\ r(T-Q) \end{pmatrix}
\tag{10.66}
$$

The modified AR model can be used for interpolation in the same way as the conventional AR model described in the previous Section. Again, it is assumed that within a data window of N speech samples, a segment of M samples commencing from the sample point k, $x_{Uk}=\{x(k), x(k+1) \ldots,x(k+M-1)\}$ is missing. Figure 10.11 illustrates the interpolation problem. The missing samples are estimated using P samples in the immediate vicinity and Q samples a pitch period away on both side of the missing signal. For the signal record of N samples, the modified AR Eq. (10.64) can be written in a matrix form as

$$
\begin{pmatrix}
e(T+Q) \\
e(T+Q+1) \\
\vdots \\
e(k-1) \\
e(k) \\
e(k+1) \\
e(k+2) \\
\vdots \\
e(k+M+P-2) \\
e(k+M+P-1) \\
e(k+M+P) \\
e(k+M+P+1) \\
\vdots \\
e(N-1)
\end{pmatrix}
=
\begin{pmatrix}
x(T+Q) \\
x(T+Q+1) \\
\vdots \\
x(k-1) \\
x_{UK}(k) \\
x_{UK}(k+1) \\
x_{UK}(k+2) \\
\vdots \\
x(k+M+P-2) \\
x(k+M+P-1) \\
x(k+M+P) \\
x(k+M+P+1) \\
\vdots \\
x(N-1)
\end{pmatrix}
-
\begin{pmatrix}
x(T+Q-1) & \cdots & x(T+Q-P) & x(2Q) & \cdots & x(0) \\
x(T+Q) & \cdots & x(T+Q-P+1) & x(2Q+1) & \cdots & x(1) \\
\vdots & \ddots & \vdots & \vdots & \ddots & \vdots \\
x(k-2) & \cdots & x(k-P-1) & x(k-T+Q-1) & \cdots & x(k-T-Q-1) \\
x(k-1) & \cdots & x(k-P) & x(k-T+Q) & \cdots & x(k-T-Q) \\
x_{UK}(k) & \cdots & x(k-P+1) & x(k-T+Q+1) & \cdots & x(k-T-Q+1) \\
x_{UK}(k+1) & \cdots & x(k-P+2) & x(k-T+Q+2) & \cdots & x(k-T-Q+2) \\
\vdots & \ddots & \vdots & \vdots & \ddots & \vdots \\
x(k+M+P-3) & \cdots & x_{UK}(k+M-2) & x(k+M+P-T+Q-2) & \cdots & x(k+M+P-T-Q-2) \\
x(k+M+P-2) & \cdots & x_{UK}(k+M-1) & x(k+M+P-T+Q-1) & \cdots & x(k+M+P-T-Q-1) \\
x(k+M+P-1) & \cdots & x(k+M) & x(k+M+P-T+Q) & \cdots & x(k+M+P-T-Q) \\
x(k+M+P) & \cdots & x(k+M+1) & x(k+M+P-T+Q+1) & \cdots & x(k+M+P-T-Q+1) \\
\vdots & \ddots & \vdots & \vdots & \ddots & \vdots \\
x(N-2) & \cdots & x(N-P-1) & x(N-T+Q-1) & \cdots & x(N-T-Q-1)
\end{pmatrix}
\begin{pmatrix}
a_1 \\
a_2 \\
a_3 \\
\vdots \\
a_P \\
P_{-Q} \\
\vdots \\
P_{+Q}
\end{pmatrix}
$$

$$(10.67)$$

Where the subscript *Uk* denotes the unknown samples. In compact matrix notation this set of equation can be written in the form

$$
e(x_{Uk}, c) = x + Xc \tag{10.68}
$$

As in Section 10.3.2, the interpolation problem is solved in two stages :
(a) In the first stage the known samples on both side of the missing signal are used to estimate the AR coefficient vector *c*.
(b) In the second stage the AR coefficient estimates are substituted in Eq. (10.68) so that the only unknowns are the data samples.
The solution follows the same steps as those described in Section 10.3.2.

10.3.5 LSAR Interpolation Error

In this section we discuss the effects of the signal characteristics, the model parameters and the number of unknown samples on the interpolation error. The interpolation error, $v(m)$, may be defined as the difference between an original sample, $x(m)$, and the interpolated sample $\hat{x}(m)$ given by

$$
v(m) = \hat{x}(m) - x(m) \tag{10.69}
$$

A common measure of signal distortion is the mean squared error distance defined as

$$
D(c, M) = \frac{1}{M} \mathcal{E}\left(\sum_{m=0}^{M-1} (x(k+m) - \hat{x}(k+m))^2 \right) \tag{10.70}
$$

Where E [.] is the expectation operator. In Eq. (10.70) the average distortion D is expressed as a function of the number of the unknown samples M, and also the model coefficient vector c. In general the quality of interpolation depends on the following factors :

(a) The signal correlation structure. For deterministic signals such as sine waves the theoretical interpolation error is zero. However information bearing signals have a degree of randomness that makes perfect interpolation impossible.

(b) The length of the missing segment. The amount of information lost, and hence the interpolation error, increase with the number of missing samples. Within a sequence of missing samples the error is usually largest for the samples in the middle of the gap. The interpolation Eq. (10.63), becomes increasingly ill-conditioned as the length of the missing samples increases.

(c) The nature of the excitation underlying the missing samples. The LSAR interpolation can not account for any random excitation underlying the missing samples. In particular, the interpolation quality suffers, when the missing samples coincide with the onset of an excitation pulse. In general the MMSE criterion causes the interpolator to under-estimate the energy of the underlying excitation signal. The inclusion of long term prediction and the use of quasi-periodic structure of signals improves the ability of the interpolator to restore the missing samples.

(d) AR model order, and the method used for estimation of the AR coefficients. The interpolation error depends on the AR model order. Usually a model order of 2 to 3 times the length of missing data sequence achieves good result. The interpolation error also depends on how well the AR parameters can be estimated from the incomplete data. In Eq. (10.54), in the first stage of the solution, where the AR coefficients are estimated, two different approaches may be employed to linearise the system of equations. In the first approach all equations, between the dashed lines, which involve nonlinear terms are discarded. This approach has the advantage that no assumption is made about the missing samples. In fact, from a signal ensemble point of view, the effect of discarding some equations is equivalent to that of having a smaller signal record. In the second method starting from an initial estimate of the unknown vector (such as $x_{Uk}=0$) Eq. (10.54) is solved to obtain the AR parameters. The AR coefficients are then used in the second stage of the algorithm to estimate the unknown samples. These estimates may be improved in further iterations of the algorithm. The algorithm usually converges after one or two iterations.

Figures 10.12 and 10.13 show the results of application of the least squared error AR interpolation method to speech signals. The interpolated speech segments were chosen to coincide with the onset of an excitation pulse. In these experimental cases the original signals are available for comparison. Each signal was interpolated by the AR model of Eq. (10.53) and also by the extended AR model of Eq. (10.64). The

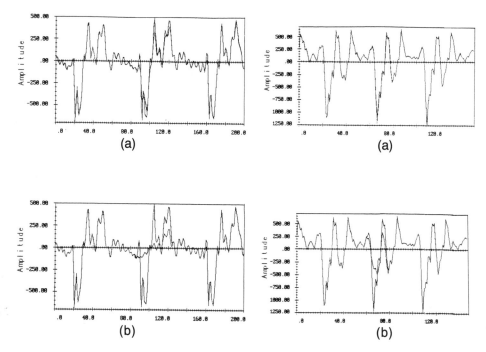

Figure 10.12 (a) A section of voiced speech showing interpolation, of a segment of 40 samples starting from the sample point 80, (b) Interpolation when only short-term correlation is used.

Figure 10.13 (a) A section of voiced speech showing the interpolation, of a segment of 20 samples starting from the sample point 60, (b) Interpolation when only short-term correlation is used.

length of the conventional linear predictor model was set to 20. The modified linear AR model of Eq. (10.64) is of order (20,7), that is the short-term predictor has 20 coefficients, and the long-term predictor has 7 coefficients. The figures clearly demonstrate that the modified AR model which include the long as well as the short term correlation structures outperform the conventional AR model.

10.3.6   Interpolation in Frequency-Time Domain

Time-domain, AR model based, interpolation methods are effective for the interpolation of a relatively short length of samples (say less than 100 samples at a 20 kHz sampling rate), but suffer severe performance degradations when used for interpolation of large sequence of samples. This is partly due to the numerical problems associated with the inversion of a large matrix, involved in the time domain interpolation of a large number of samples, Eq. (10.58).

Spectral-time representation provides a useful form for the interpolation of a large gap of missing samples. For example, through spectral-time representation, the problem of interpolation of a gap of N samples in the time domain can be converted into the problem of interpolation of a gap of one sample, along the time, in each of N orthogonal frequency channels.

Spectral-Time Representation with STFT

A relatively simple and practical method for spectral-time representation of a signal is the short time Fourier transform (STFT) method. To construct a two-dimensional, STFT, from a one dimensional function of time $x(m)$, the input signal

is segmented into overlapping blocks of N samples, as illustrated in Figure 10.14. Each block is windowed, prior to Fourier transform, to reduce the spectral leakage due to the effects of discontinuities at the edges of the block.

The frequency spectrum of the mth block is given by the discrete Fourier transform

$$X(k,m) = \sum_{i=0}^{N-1} w(i)\, x(m(N-D)+i)\, e^{-j\frac{2\pi}{N}ik} \quad k= 0, ..., N\text{--}1 \quad (10.71)$$

where $X(k,m)$ is a spectral-time representation with the time index m, and the frequency index k, N is the number of samples in each block and D is the block overlap. In STFT it is assumed that the signal frequency composition is time-invariant within the duration of each block, but it may vary across the blocks.

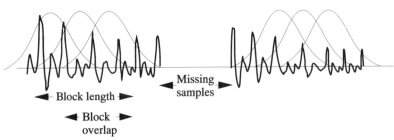

Figure 10.14 Illustration of segmentation of a signal (with a missing gap) for spectral-time representation.

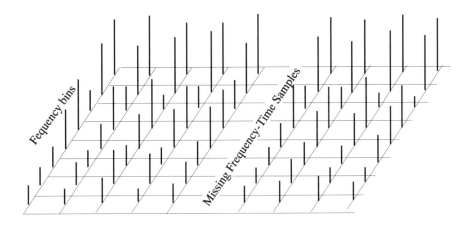

Time (Blocks)

Figure 10.15 Spectral-time representation of a signal with a missing gap.

In general, the kth spectral component of a signal has a time-varying character, it is "born", it evolves for some time, disappears, and then reappears with a different intensity and a different characteristics. Figure 10.15 illustrates a spectral-time signal with a missing block of samples. The aim of interpolation is to fill in the signal gap such that, at the beginning and at the end of the gap, the continuity of both the magnitude and the phase of each frequency component is maintained. For most time-varying signals (such as speech), a low order polynomial interpolator of magnitude and phase, making use of the few adjacent blocks on either side of the gap, would produce satisfactory results.

10.3.7 Interpolation using Adaptive Code Books

In LSAR interpolation method, described in Section 10.3.2, the signals are modelled as the output of an AR model excited by a random input. Given enough samples, the AR coefficients can be estimated with reasonable accuracy. However, the instantaneous values of the random excitation during the periods when the signal is missing can not be recovered. This leads to a consistent under-estimation of the amplitude and the energy of the interpolated samples. One solution to this problem is to use a zero-input signal model. Zero-input models are feedback oscillator systems that produce an output signal without requiring an input.

The general form of the equation describing a digital nonlinear oscillator can be expressed as

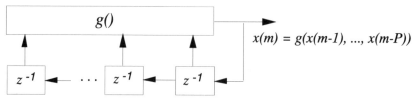

Figure 10.16 Configuration of a digital oscillator.

$$x(m) = g_f\big(x(m-1), x(m-2), \cdots, x(m-P)\big) \qquad (10.72)$$

The mapping function $g_f()$ may be a parametric, or a non-parametric, mapping. The model in Eq. (10.72) can be considered as a nonlinear predictor and the subscripts f denotes forward prediction based on the past samples.

A parametric model of a nonlinear oscillator can be formulated using a Volterra filter model. However, in this section we considered a nonparametric method for its ease of formulation and stable characteristics. KUBIN and KLEIJIN described a non-parametric oscillator based on a codebook model of the signal process.

In this method each entry in the codebook has $P+1$ samples where the $(P+1)^{th}$ sample is intended as an output. Given P input samples $x=[x(m-1), ..., x(m-P)]$, the

codebook output is the $(P+1)^{th}$ sample of the vector in the codebook whose first P samples have a minimum distance from the input signal x. For a signal record of length N samples, a codebook of size $N-P$ vectors can be constructed by dividing the signal into overlapping segments of $P+1$ samples with the successive segments having an overlap of P samples. Similarly a backward oscillator can be expressed as

$$x_b(m) = g_b\big(x(m+1), x(m+2), \cdots, x(m+P)\big) \qquad (10.73)$$

As in the case of a forward oscillator, the backward oscillator can be designed using a non-parametric method based on an adaptive codebook of the signal process. In this case each entry in the codebook has $P+1$ samples where the 1st sample is intended as an output sample. Given P input samples $x=[x(m), ..., x(m+P-1)]$ the codebook output is the first sample of the codebook vector whose next P samples have a minimum, distance from the input signal x.

For interpolation of M missing samples, the forward and backward nonlinear oscillators are can be combined as

$$\hat{x}(k+m) = \left(\frac{M-1-m}{M-1}\right)\hat{x}_f(k+m) + \left(\frac{m}{M-1}\right)\hat{x}_b(k+m) \qquad (10.74)$$

where it is assumed that the missing samples start at k.

10.3.8 Interpolation Through Signal Substitution

Audio signals often have a time-varying but quasi-periodic repetitive structure. Therefore most acoustic events in a signal record *reoccur* with some variations. This observation forms the basis for interpolation through pattern matching, where a missing segment of a signal is substituted by the best match from a signal record. Consider a relatively long signal record of N samples, with a gap of M missing samples at its centre. A section of the signal with the gap in the middle can be used to search for the best match segment in the record. The missing samples are then substituted by the corresponding section of the best match signal. This interpolation method is particularly useful when the length of the missing signal segment is very large. For a given class of signals we may be able to construct a library of patterns for use in waveform substitution, BOGNER (1989).

Summary

Interpolators, in their various forms, are used in most signal processing applications. The obvious example is the estimation of a sequence of missing samples. However the use of interpolator covers a much wider range of applications, from low bit rate speech coding to the pattern recognition and decision making systems. We started this Chapter with a study of the ideal interpolation of a band limited signal, and its applications in digital to analog conversion and in multi-rate signal processing. In this Chapter, various interpolation methods are categorised and studied in two different Sections; one on polynomial interpolation which is the more traditional numerical computing approach, and the other on statistical interpolation which is the digital signal processing approach.

The general form of the polynomial interpolator was formulated and its special forms Lagrange, Newton, Hermite and cubic spline interpolators were considered. The polynomial methods are not equipped to make optimal use of the predictive and statistical structures of the signal, and are impractical for interpolation of a relatively large number of samples. A number of useful statistical interpolators were studied. These include the maximum a posterior interpolation, the least squared error AR interpolation, frequency-time interpolation, and an adaptive code book interpolator. The model-based interpolation method based on an autoregressive model is satisfactory for most audio applications so long as the length of the missing samples is not to large. For interpolation of a relatively large number of samples the time-frequency interpolation method and the adaptive code book method are more suitable.

Bibliography

BOGNER R.E., LI T. (1989), Pattern Search Prediction of Speech, Proc. IEEE Int. Conf. on Acoustics, Speech and Signal Processing, ICASSP-89, Pages 180-183, Glasgow.

COHEN L. (1989), Time-Frequency Distributions- A review, Proceedings of the IEEE, Vol. 77(7), Pages 941-81.

CROCHIERE R.E., RABINER L.R. (1981), Interpolation and Decimation of Digital Signals-A Tutorial review, Proc. IEEE, Vol. 69, Pages. 300-331, March.

GODSILL S.J. (1993), The Restoration of Degraded Audio Signals, Ph.D. Thesis, Cambridge University, Cambridge.

GODSILL S.J. and P.J.W. Rayner (1993), Frequency domain interpolation of sampled signals , IEEE Int. Conf., Speech and Signal Processing, ICASSP-93, Minneapolis.

JANSSEN A.J., VELDHUIS R., VRIES L.B (1984),Adaptive Interpolation of Discrete-Time Signals th/at can be Modelled as Autoregressive Processes, Proc. IEEE Trans. Acoustics, Speech and Signal Processing, Vol. ASSP-34, No. 2, Pages 317-330 June.

JANSSEN A.J., VRIES L.B (1984), Interpolation of Band-Limited Discrete-Time Signals by Minimising Out-of Band Energy, Proc. IEEE Int. Conf. on Acoustics, Speech and Signal Processing, ICASSP-84.

KAY S.M. (1983), Some Results in Linear Interpolation theory, IEEE Trans. Acoustics Speech and Signal Processing, Vol. ASSP-31, Pages 746-49, June.

KAY S. M. (1988) Modern Spectral Estimation : Theory and Application. Prentice-Hall, Englewood Cliffs, N. J.

KOLMOGROV A.N. (1939),Sur l' Interpolation et Extrapolation des Suites Stationaires, Comptes Rendus de l'Academie des Sciences, Vol. 208, Pages 2043-45.

KUBIN G., KLEIJIN W.B. (1994) Time-Scale Modification of Speech Based on a Nonlinear Oscillator Model, Proc. IEEE Int. Conf., Speech and Signal Processing, ICASSP-94, Pages I453-56, Adelaide.

LOCHART G.B., GOODMAN D.J. (1986), Reconstruction of missing speech packets by waveform substitution Signal processing 3: Theories and Applications, Pages 357-360.

MARKS R.J. (1983), Restoring Lost Samples from an over-sampled band-limited signal, IEEE Trans. Acoustics, Speech and Signal Processing, Vol. ASSP-31, No 2. Pages .752-55, June.

MARKS R.J. (1991), Introduction to Shannon Sampling and Interpolation Theory, Springer Verlag.

MATHEWS J.H. (1992), Numerical Methods for Mathematics, Science and Engineering, Prentice-Hall, Englewood Cliffs, N. J.

MUSICUS B. R. (1982), Iterative Algorithms for Optimal Signal Reconstruction and Parameter Identification Given Noisy and Incomplete Data, Ph.D. Thesis, MIT August .

PLATTE H.J., ROWEDDA V. (1985), A Burst Error Concealment Method for Digital Audio Tape Application. AES preprint, 2201:1-16.

PRESS W.H., FLANNERY B.P., TEUKOLSKY S.A., VETTERELING W.T. (1992), Numerical Recepies in C, Second edition, Cambridge University Press, Cambridge.

NAKAMURA S. (1991), Applied Numerical Methods with Software, Prentice-Hall, Englewood Cliffs, N. J.

SCHAFER, R.W., RABINER, L.R. (1973), A Digital Signal Processing Approach to Interpolation, Proc. IEEE, Vol.. 61, Pages 692-702, June.

STEELE R., JAYANT N. S. (1980), Statistical Block Coding for DPCM-AQF Speech, IEEE Trans On Communications, Vol. COM-28, No. 11, Pages 1899-1907, Nov.

TONG H.(1990), Nonlinear Time Series A Dynamical System Approach, Oxford University Press.

VASEGHI S.V.(1988), Algorithms for Restoration of Gramophone Records, Ph.D. Thesis, Cambridge University, Cambridge.

VELDHUIS R. (1990), Restoration of Lost samples in Digital Signals. Prentice-Hall.

VERHELST W., ROELANDS M. (1993), An Overlap-Add Technique Based on Waveform Similarity (Wsola) for High Quality Time-Scale Modification of Speech, Proc. IEEE Int. Conf. on Acoustics, Speech and Signal Processing, ICASSP-93, Pages II-554-II-557, Adelaide.

WIENER N. (1949), Extrapolation, Interpolation and Smoothing of Stationary Time Series With Engineering Applications, MIT Press, Cambridge, Mass.

11

Impulsive Noise

11.1   Impulsive Noise
11.2   Stochastic Models for Impulsive Noise
11.3   Median Filters
11.4   Impulsive Noise Removal Using Linear Prediction Models
11.5   Robust Parameter Estimation
11.6   Restoration of Archived Records

Impulsive noise are relatively short duration "on/off" pulses, caused by switching noise or adverse channel environments in a communication system, drop outs or surface degradation of audio recordings, clicks from computer keyboards etc. An impulsive noise filter can be used for enhancing the quality and intelligibility of noisy signals, and for achieving robustness in pattern recognition and adaptive control systems. This chapter begins with a study of the characteristics of an impulsive noise, and then proceeds to consider several methods for statistical modelling of an impulsive noise process. The classical method for removal of impulsive noise is the median filter. However, the median filter often results in some signal degradation. For optimal performance, an impulsive noise removal system should utilise (a) the distinct features of the noise and the signal, (b) the statistics of the signal and the noise, and (c) a model of the physiology of the signal and the noise generation. We describe a model-based system that detects each impulsive noise, and then proceeds to replace the samples obliterated by the impulse. We also consider the methods for introducing robustness to impulsive noise in parameter estimation.

11.1 Impulsive Noise

In this section, first the mathematical concept of a digital impulse is introduced, and then the various forms of actual impulsive noise in communication systems are considered. A digital impulse, $\delta(m)$, is defined as a signal with an "on" duration of one sample, and is expressed as :

$$\delta(m) = \begin{cases} 1 & m = 0 \\ 0 & m \neq 0 \end{cases} \tag{11.1}$$

where the variable m designates the discrete-time index. Using the Fourier transform relation, the frequency spectrum of an impulse is given by

$$\Delta(f) = \sum_{m=-\infty}^{\infty} \delta(m)\, e^{-j2\pi f m} = 1.0 \quad -\infty < f < \infty \tag{11.2}$$

where f is the frequency variable. From Eq. (11.2), it is evident that the energy of an impulse is spread equally among all frequencies, as shown in Figure 11.1.a. In communication systems, real impulsive-type noise have a duration which is normally more than one sample long. For example, in the context of audio signals, short duration, sharp pulses, of up to 3 milliseconds (60 samples at a 20 kHz sampling rate) may be considered as impulsive type noise. Figures 11.1(b) and 11.1(c) illustrate two examples of short duration pulses and their respective spectra.

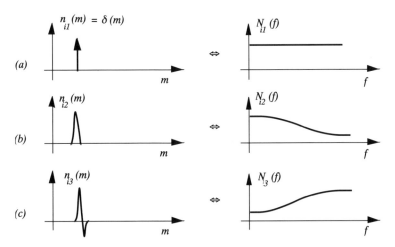

Figure 11.1 Time and frequency sketches of a) an ideal impulse, b) and c) short duration pulses.

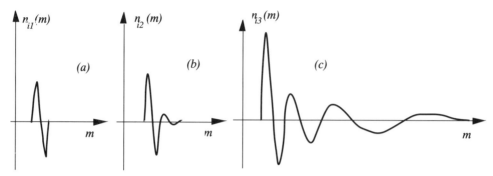

Figure 11.2 Illustration of variations of the impulse response of a non-linear system with increasing amplitude of the impulse.

In a communication system, an impulsive noise originates at some point in time and space, and then propagates through the channel to the receiver. The received noise is shaped by the channel, and can be considered as the channel impulse response. In general, the characteristics of a communication channel may be linear or non-linear, stationary or time varying. Furthermore, many communication systems, in response to a large-amplitude impulse, exhibit a nonlinear characteristic.

Figure 11.2 illustrates some examples of impulsive noise, typical of that observed on an old gramophone recording. In this case the communication channel is the playback system, and may be assumed time-invariant. The figure also shows some variations of the channel characteristics with the amplitude of impulsive noise. These variations may be attributed to the non-linear characteristics of the playback mechanism.

An important consideration in the development of a noise processing system is the choice of an appropriate domain, the time or the frequency, for signal representation.

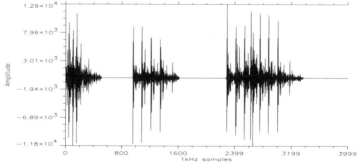

Figure 11.3 A machine gun noise, the oscillatory decay can be considered as the response of the gun and the acoustic environment.

The choice should depend on the specific objective of the system. In signal restoration the objective is to separate the noise from the signal, and the representation domain must be the one that emphasises the distinguishing features of the signal and the noise. Impulsive noise is more distinct and detectable in time than in frequency, and it is appropriate to use time domain signal processing for noise detection and removal. In signal classification and parameter estimation, the objective may be to compensate for the average effects of the noise over a number of samples, and in some cases it may be more appropriate to process the impulsive noise in frequency domain where the effect of noise is a change in the mean of the power spectrum of the signal.

11.1.1 Autocorrelation and Power Spectrum of Impulsive Noise

Impulsive noise is a nonstationary binary-state sequence of impulses with random amplitudes and random positions of occurrence. The nonstationary nature of impulsive noise can be seen by considering the power spectrum of a noise process with a few impulses per second: when the noise is absent the process has zero power, and when an impulse is present the noise power is the power of the impulse. Therefore the power spectrum and hence the autocorrelation of an impulsive noise is a binary state, time-varying process. An impulsive noise sequence can be modelled as an amplitude-modulated binary-state sequence, and expressed as

$$n_i(m) = n(m)\ b(m) \tag{11.3}$$

where $b(m)$ is a binary-state random sequence of one's and zero's, and $n(m)$ is a random noise process. Assuming that impulsive noise is an uncorrelated random process, the autocorrelation of impulsive noise may be defined as a binary-state process:

$$r_{nn}(k,m) = \mathcal{E}[n_i(m)n_i(m+k)] = \sigma_n^2\ \delta(k)b(m) \tag{11.4}$$

where $\delta(k)$ is the Kronecker delta function. Since it is assumed that the noise is an uncorrelated process, the autocorrelation is zero for $k{\neq}0$, therefore Eq. (11.4) may be written as

$$r_{nn}(0,m) = \sigma_n^2\ b(m) \tag{11.5}$$

Note that for a zero mean noise process, $r_{nn}(0,m)$ is the time-varying noise power. The power spectrum of an impulsive noise sequence is obtained, by taking the Fourier transform of the autocorrelation function Eq. (11.4), as

$$P_{N_I N_I}(f, m) = \sigma_n^2\, b(m) \tag{11.6}$$

In Eq. (11.5) and (11.6) the autocorrelation and power spectrum are expressed as binary state functions that depend on the "on/off" state of impulsive noise at time m.

11.2 Stochastic Models for Impulsive Noise

In this section we study a number of statistical models for the characterisation of an impulsive noise process. An impulsive noise sequence, $n_i(m)$, consists of short duration pulses of a random amplitude, duration, and time of occurrence, and may be modelled as the output of a filter excited by an amplitude-modulated random binary sequence as

$$n_i(m) = \sum_{k=0}^{P-1} h_k n(m - k) b(m - k) \tag{11.7}$$

Figure 11.4 illustrates the impulsive noise model of Eq. (11.7). In Eq. (11.7) $b(m)$ is a binary-valued random sequence model of the time of occurrence of impulsive noise, $n(m)$ is a continuous-valued random process model of impulse amplitude, and $h(m)$ is the impulse response of a filter that models the duration and shape of each impulse. Two important statistical processes for modelling impulsive noise as an amplitude-modulated binary sequence are the Bernoulli-Gaussian process and the Poisson-Gaussian process, these are discussed next.

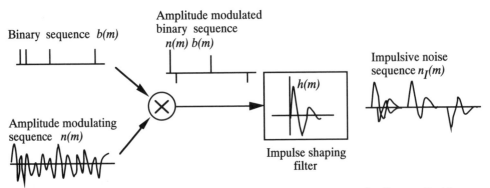

Figure 11.4 Illustration of an impulsive noise model as the output of a filter excited by an amplitude-modulated binary sequence.

11.2.1   Bernoulli-Gaussian Model of Impulsive Noise

In a Bernoulli-Gaussian model of an impulsive noise process, the random time of occurrence of impulses is modelled by a binary Bernoulli process $b(m)$, and the amplitude of the impulses is modelled by a Gaussian process $n(m)$. A Bernoulli process, $b(m)$, is a binary-valued process that takes a value of "1" with a probability of α, and a value of "0" with a probability of $1-\alpha$. The probability mass function of a Bernoulli process is given by

$$P_B(b(m)) = \begin{cases} \alpha & for \quad b(m)=1 \\ 1-\alpha & for \quad b(m)=0. \end{cases} \tag{11.8}$$

A Bernoulli process has a mean of $\mu_b=\alpha$ and a variance of $\sigma_b^2 = \alpha(1-\alpha)$. A zero mean Gaussian pdf model of the random amplitudes of impulsive noise is given by

$$f_N(n(m)) = \frac{1}{\sqrt{2\pi}\,\sigma_n} \exp\left(-\frac{n^2(m)}{2\sigma_n^2}\right) \tag{11.9}$$

where σ_n^2 is the variance of noise amplitude. In a Bernoulli-Gaussian model the probability density function of an impulsive noise $n_i(m)$ is given by

$$f_N^{BG}(n_i(m)) = (1-\alpha)\,\delta(n_i(m)) + \alpha\,f_N(n_i(m)) \tag{11.10}$$

where $\delta(n_i(m))$ is the Kronecker delta function. Note that the function $f_N^{BG}(n_i(m))$ is a mixture of discrete probability mass function $\delta(n_i(m))$, and a continuous probability density function $f_N(n_i(m))$.

An alternative model for impulsive noise is a binary-state Gaussian process (Section 2.5.4) with a low-variance state modelling the absence of impulses, and a relatively high-variance state modelling the amplitude of impulsive noise.

11.2.2   Poisson-Gaussian Model of Impulsive Noise

In a Poisson-Gaussian model the probability of occurrence of an impulsive noise event is modelled by a Poisson process, and the distribution of the random amplitude of impulsive noise is modelled by a Gaussian process. The Poisson process, described in Chapter 2, is a random event-counting process. In a Poisson model the probability of occurrence of k impulsive noise in a time interval of T is given by

$$P(k,T) = \frac{(\lambda T)^k}{k!} e^{-\lambda T} \qquad (11.11)$$

Where λ is a rate function with the following properties:

$$Prob(1 \; Impulse \; in \; a \; small \; time \; interval \; \Delta t) \; = \; \lambda \Delta t$$
$$Prob(0 \; Impulse \; in \; a \; small \; time \; interval \; \Delta t) \; = \; 1 - \lambda \Delta t \qquad (11.12)$$

It is assumed that no more than one impulsive noise can occur in a time interval of Δt. In a Poisson-Gaussian model, the pdf of an impulsive noise $n_i(m)$ in a small time interval of Δt. is given by

$$f_{N_I}^{PG}(n_i(m)) = (1 - \lambda \Delta t)\delta(n_i(m)) + \lambda \Delta t \; f_N(n_i(m)) \qquad (11.13)$$

where $f_N(n_i(m))$ is the Gaussian pdf of Eq. (11.9).

11.2.3   A Binary State Model of Impulsive Noise

An impulsive noise process may also be modelled by a binary-state model as shown in Figure 11.4. In this binary model, the state S_0 corresponds to the "*off*" condition when impulsive noise is absent, in this state the model emits zero-valued samples. The state S_1 corresponds to the "on" condition and in this state the model emits short duration pulses of a random amplitude and duration. The probability of a transition from state S_i to state S_j is denoted by a_{ij}. In its simplest form, as shown in Figure 11.5, the model is memoryless, and the probability of a transition to state S_i is independent of the current state of the model. In this case, the probability that at time $t+1$ the model is in the state S_0, is independent of the state at time t and is given by

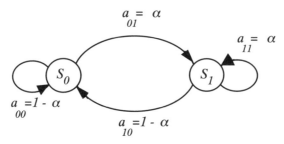

Figure 11.5 A binary-state model of an impulsive noise generator.

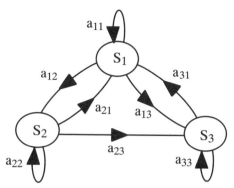

Figure 11.6 A 3 state model of impulsive noise and the decaying oscillations that often follow impulses.

$$P\big(s(t+1) = S_0 \big| s(t) = S_0\big) \; = \; P\big(s(t+1) = S_0 \big| s(t) = S_1\big) \; = \; 1 - \alpha \qquad (11.14)$$

where s_t denotes the state at time t. Likewise, the probability that at time $t+1$ the model is in state S_1 is given by

$$P\big(s(t+1) = S_1 \big| s(t) = S_0\big) \; = \; P\big(s(t+1) = S_1 \big| s(t) = S_1\big) \; = \; \alpha \qquad (11.15)$$

In a more general form of the binary-state model, a Markovian state-transition can model the dependencies in the noise process. The model then becomes a 2-state hidden Markov model considered in Chapter 4.

In one of its simplest forms, the state S_1 emits samples from a zero mean Gaussian random process. The impulsive noise model in state S_1 can be configured to accommodate a variety of impulsive noise of different shapes, durations and pdfs. A practical method for modelling a variety of impulsive noise is to use a codebook of M prototype impulsive noise, and their associated probabilities [(n_{i1}, p_{i1}), (n_{i2}, p_{i2}), ..., (n_{iM}, p_{iM})], where p_j denotes the probability of impulsive noise of the type n_j. The impulsive noise code book may be designed by classification of a large number of "training" impulsive noise into a relatively small number of clusters. For each cluster, the average impulsive noise is chosen as the representative of the cluster. The number of impulses in the cluster of type j, divided by the total number of impulses in all clusters gives p_j, the probability of an impulse of type j.

Figure 11.6 shows a three-state model of the impulsive noise and the decaying oscillations that might follow the noise. In this model the state S_0 corresponds to the absence of noise, the state S_1 models the impulsive noise and the state S_2 models any oscillations that can result from the impulse.

11.2.4   Signal to Impulsive Noise Ratio

For impulsive noise the average signal to impulsive noise ratio, averaged over an entire noise sequence including the time instances when the impulses are absent, depends on two parameters: (a) the average power of each impulsive noise, and (b) the rate of occurrence of impulsive noise. Let $P_{impulse}$ denote the power of each impulse, and P_{Signal} the signal power. We may define a "local" time-varying signal to impulsive noise ratio as

$$SINR(m) = \frac{P_{\text{Signal}}(m)}{b(m)P_{\text{Impulse}}(m)} \tag{11.16}$$

The average signal to noise ratio, assuming that the parameter α is the fraction of signal samples contaminated by impulsive noise, can be defined as

$$SINR = \frac{P_{\text{Signal}}}{\alpha \cdot P_{\text{Impulse}}} \tag{11.17}$$

Note that from Eq. (11.16), for a given signal power, there are many values of α and P_{Signal} that can yield the same average SINR.

11.3   Median Filters

The classical approach to removal of impulsive noise is the median filter. The median of a set of samples $\{x(m)\}$ is a member of the set $x_{Med}(m)$ such that; half the population of the set are larger than $x_{Med}(m)$ and half are smaller than $x_{Med}(m)$. Hence the median of a set of samples is obtained by sorting the samples in ascending or descending order, and selecting the mid-value. In median filtering, a window of predetermined length slides sequentially over the signal and the mid-sample within the window is replaced by the median of all the samples that are inside the window, as illustrated in Figure 11.7.

The output, $\hat{x}(m)$, of a median filter with input $y(m)$, and a median window of length $2K+1$ samples is given by

$$
\begin{aligned}
\hat{x}(m) &= y_{Med}(m) \\
&= Median\left[y(m-K), \ldots, y(m), \ldots, y(m+K)\right]
\end{aligned}
\tag{11.18}
$$

The median of a set of numbers is a non-linear statistics of the set, with the useful property that it is insensitive to the presence of a sample with an unusually large

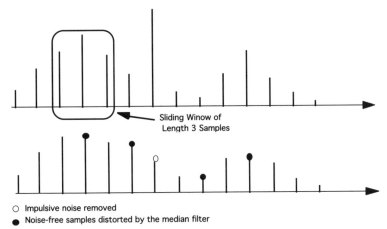

Figure 11.7 Input and output of a median filter. Note that in addition to suppressing the impulsive outlier, the filter also distorts some genuine signal components.

value, a so called outlier, in the set. In contrast, the mean, and in particular the variance, of a set of numbers are sensitive to the presence of impulsive type noise. An important property of median filters, particularly useful in image processing, is that they preserves edges or stepwise discontinuities in the signal. Median filters can be used for removing impulses in an image without smearing the edge information; this is of significant importance in image processing. However, experiments with median filters, for removal of impulsive noise from audio signals, demonstrate that median filters are unable to produce high quality audio restoration. The median filters can not deal with "real" impulsive noise, which are often more than one or two samples long. Furthermore, median filters introduce a great deal of processing distortion by modifying genuine signal samples that are mistaken for impulsive noise. The performance of median filters may be improved by employing an adaptive threshold, so that a sample is replaced by the median, only if the difference between the sample and the median is above the threshold :

$$\hat{x}(m) = \begin{cases} y(m) & if \ \ |y(m) - y_{Med}(m)| < k \cdot \theta(m) \\ y_{Med}(m) & otherwise \end{cases} \tag{11.19}$$

where $\theta(m)$ is an adaptive threshold that may be related to a robust estimate of the average of $|y(m) - y_{Med}(m)|$, and k is a tuning parameter. Median filters are not optimal, because they do not make efficient use of prior knowledge of the physiology of signal generation, or a model of the signal and noise statistical distributions. In the following section we describe a autoregressive model-based impulsive removal system, capable of producing high quality audio restoration.

11.4   Impulsive Noise Removal Using Linear Prediction Models

In this section we study a model-based impulsive noise removal system. Impulsive disturbances usually contaminate a relatively small fraction, α, of the total samples. Since a large fraction, $1-\alpha$, of samples remain unaffected by impulsive noise, it is advantageous to locate individual noise pulses, and correct *only* those samples that are distorted. This strategy avoids the unnecessary processing. and compromise in the quality, of the relatively large fraction of samples that are not disturbed by impulsive noise. The impulsive noise removal system, shown in Figure 11.8, consists of two subsystems: a detector and an interpolator. The detector locates the position of each noise pulse, and the interpolator replaces the distorted samples using the samples on both sides of the impulsive noise. The detector is composed of a linear prediction analysis system, a matched filter and a threshold detector. The output of the detector is a binary switch and controls the interpolator; a detector output of "0" signals the absence of impulsive noise and the interpolator is bypassed. A detector output of "1" signals the presence of impulsive noise and the interpolator is activated to replace the samples obliterated by noise.

11.4.1   Impulsive Noise Detection

A simple method for detection of impulsive noise is to employ an amplitude threshold, and classify the samples with an amplitude above the threshold as noise. This method works fairly well for relatively large-amplitude impulses, but fails when

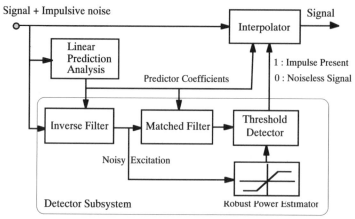

Figure 11.8   Configuration of an impulsive noise removal system incorporating a detection and an interpolation subsystem.

the noise amplitude falls below the signal. Detection can be improved by utilising the characteristic differences between the impulsive noise and the signal. An impulsive noise, or a short duration pulse, introduces uncharacteristic discontinuity in a correlated signal. The discontinuity becomes more detectable when the signal is differentiated. The differentiation, or for digital signals the differencing, operation is equivalent to decorrelation or spectral whitening. In this section we describe a model based decorrelation method for improving impulsive noise detectability. The correlation structure of the signal is modelled by a linear predictor and the process of decorrelation is achieved by inverse filtering. Linear prediction, and inverse filtering are covered in Chapter 7. Figure 11.9 shows a model for noisy signal. The noise-free signal, $x(m)$, is described by a linear prediction model as

$$x(m) = \sum_{k=1}^{P} a_k\, x(m-k) + e(m) \tag{11.20}$$

where $a=[a_1, a_2, ..., a_P]^T$ is the coefficients of a linear predictor of order P, and $e(m)$ the excitation is either a noise-like signal, or a mixture of a random noise and a quasi-periodic train of pulses as illustrated in Figure 11.9. The impulsive noise detector is based on the observation that linear predictors are a good model of the correlated signals but not the uncorrelated binary-state impulsive-type noise. Transforming the noisy signal, $y(m)$, to the excitation signal of the predictor has the following effects:
  (a) The scale of signal amplitude is reduced to almost that of the original excitation signal, whereas the scale of noise amplitude remains unchanged or increases,
  (b) The signal is decorrelated, whereas the impulsive noise is smeared and transformed to a scaled version of the impulse response of the inverse filter.
Both effects improve noise delectability. Speech, or music, are composed of random excitations spectrally shaped and amplified by the resonances of vocal tract or the musical instruments. The excitation is more random than the speech, and often has a

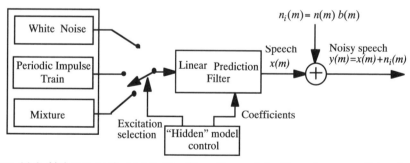

Figure 11.9 Noisy speech model. The Signal is modelled by a linear predictor. Impulsive noise is modelled as an amplitude modulated binary-state process.

much smaller amplitude range. The improvement in noise pulse detectability, obtained by inverse filtering, can be substantial and depends on the time-varying correlation structure of the signal. Note that this method effectively reduces the impulsive noise detection to the problem of separation of outliers from a random noise excitation signal using some optimal thresholding device.

11.4.2 Analysis of Improvement in Noise Detectability

In the followings the improvement in noise detectability that results from inverse filtering is analysed. From Eq. (11.20) we can rewrite the noisy signal model as

$$y(m) = x(m) + n_i(m)$$

$$= \sum_{k=1}^{P} a_k x(m-k) + e(m) + n_i(m) \qquad (11.21)$$

Where $y(m)$, $x(m)$ and $n_i(m)$ are the noisy signal, the signal and the noise respectively. Using an estimate $\hat{a}$ of the predictor coefficient vector a, the noisy signal $y(m)$ can be inverse filtered and transformed to the noisy excitation signal $v(m)$

$$v(m) = y(m) - \sum_{k=1}^{P} \hat{a}_k y(m-k)$$

$$= x(m) + n_i(m) - \sum_{k=1}^{P} (a_k - \tilde{a}_k)[x(m-k) + n_i(m-k)] \qquad (11.22)$$

Where $\tilde{a}_k$ is the error in the estimate of the predictor coefficient. Using Eq. (11.20) Eq. (11.22) can be rewritten in the following form

$$v(m) = e(m) + n_i(m) + \sum_{k=1}^{P} \tilde{a}_k x(m-k) - \sum_{k=1}^{P} \hat{a}_k n_i(m-k) \qquad (11.23)$$

From Eq. (11.23) there are essentially three terms that contribute to the noise in the excitation sequence. These are :
(a) the impulsive disturbance, $n_i(m)$, this is usually the dominant term,
(b) the effect of the past P noise samples, smeared to the present time by the action of the inverse filtering, $\sum \hat{a}_k n_i(m-k)$, and
(c) the increase in the variance of the excitation signal, caused by the error in the parameter vector estimate, and expressed by the term $\sum \tilde{a}_k x(m-k)$.

The improvement resulting from the inverse filter can be formulated as follows. The impulsive noise to signal ratio for the noisy signal is given by

$$\frac{Impulsive\,noise\,power}{Signal\,power} = \frac{\mathcal{E}[n_i^2(m)]}{\mathcal{E}[x^2(m)]} \qquad (11.24)$$

Where $\mathcal{E}[.]$ is the expectation operator. Note that in impulsive noise detection, the signal of interest is the impulsive noise to be detected from the accompanying signal. Assuming that the dominant noise term in the noisy excitation signal $v(m)$, is the impulse $n_i(m)$, then the impulsive noise to excitation signal ratio is given by

$$\frac{Impulsive\,noise\,power}{Excitation\,power} = \frac{\mathcal{E}[n_i^2(m)]}{\mathcal{E}[e^2(m)]} \qquad (11.25)$$

The overall gain in impulsive noise to signal ratio is obtained, by dividing Eqs. (11.24) and (11.25), as

$$\frac{\mathcal{E}[x^2(m)]}{\mathcal{E}[e^2(m)]} = Gain \qquad (11.26)$$

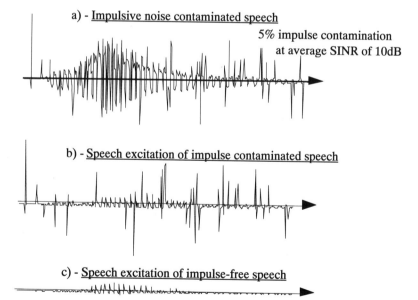

a) - Impulsive noise contaminated speech

5% impulse contamination at average SINR of 10dB

b) - Speech excitation of impulse contaminated speech

c) - Speech excitation of impulse-free speech

Figure 11.10 Illustration of the effects of inverse filtering on detectability of Impulsive noise.

This simple analysis demonstrates that the improvement in impulsive noise detectability depends on the power amplification characteristics, due to resonances, of the linear predictor model. For speech signals the scale of the amplitude of the noise-less speech excitation is in the order of 10^{-1} to 10^{-4} of that of the speech itself, therefore substantial improvement in impulsive noise detectability can be expected through inverse filtering of the noisy speech signals.

Figure 11.8 illustrates the effect of inverse filtering in improving the detectability of impulsive noise. The inverse filtering has the effect that the signal, $x(m)$, is transformed to an uncorrelated excitation signal, $e(m)$, whereas the impulsive noise is smeared to a scaled version of the inverse filter impulse response $[1, -a_1, ..., -a_P]$, as indicated by the term $\sum \hat{a}_k n_i(m-k)$ in Eq. (11.23). Assuming that the excitation is a white noise Gaussian signal, a filter matched to the inverse filter coefficients may enhance the delectability of the smeared impulsive noise from the excitation signal.

11.4.3 Two-sided Predictor for Impulsive Noise Detection

In the previous section it is shown that impulsive noise detectability can be improved by decorrelating the speech signal. The process of decorrelation can be taken further by the use of a two-sided linear prediction model. The two-sided linear prediction of a sample $x(m)$ is based on the P past samples and the P future samples, and is defined by the equation

$$x(m) = \sum_{k=1}^{P} a_k x(m-k) + \sum_{k=1}^{P} a_{k+P} x(m+k) + e(m) \qquad (11.27)$$

Where a_k are the two-sided predictor coefficients, and $\varepsilon(m)$ is the excitation signal. All the analysis for the case of one-sided linear predictor equally apply to the two-sided model. However, the variance of the excitation of two-sided model is less than that of the one sided model because in Eq. (11.27) the correlation of each sample with the future, as well as the past, samples are removed. Although Eq. (11.27) is a noncausal filter, its inverse, required in the detection subsystem, is causal. The use of a two sided predictor can result in further improvement in noise detectability.

11.4.4 Interpolation of Discarded Samples

Samples irrevocably distorted by an impulsive noise are discarded and the gap so left is interpolated. For interpolation imperfections to remain inaudible a high fidelity interpolator is required. A number of interpolators for replacement of a sequence of missing samples are introduced in Chapter 10. The least squared autoregressive (LSAR) interpolation algorithm of Section 10.3.2 produces high quality results for relatively small number of missing samples left by an impulsive noise. The LSAR interpolation method is a two stage process. In the first stage, the available samples on both sides of the noise pulse are used to estimate the parameters of a linear prediction model of the signal. In the second stage the estimated model parameters, and the samples on both sides of the gap, are used to interpolate the missing samples. The use of this interpolator, in replacement of audio signals distorted by impulsive noise, has produced high quality results.

11.5 Robust Parameter Estimation

In Figure 11.8 the threshold, used for detection of impulsive noise from the excitation signal, is derived from a nonlinear robust estimate of the excitation power. In this section we consider robust estimation of a parameter, such as power, in the presence of impulsive noise.

A *robust* estimator is one that is not over sensitive to deviations of the input signal from the assumed distribution. In a robust estimator, an input sample with an unusually large amplitude has only a limited effect on the estimation results. Most signal processing algorithms developed for adaptive filtering, speech recognition, speech coding *etc.* are based on the assumption that the signal and the noise are Gaussian distributed, and employ a mean squared distance measure as the optimality criterion. The mean squared error criterion is sensitive to non-Gaussian events such as impulsive noise. A large impulsive noise in a signal can substantially overshadow the influence of noise-free samples.

Figure 11 illustrates the variations of several cost of error function with a parameter θ. Figure 11.11(a) shows a least squared error cost function, and its influence function. The influence function is the derivative of the cost function and, as the name implies, it has a direct influence on the estimation results. It can be seen from the influence function of Figure 11.11(a) that an unbounded sample has an unbounded influence on the estimation results.

A method for introducing robustness is to use a non-linear function and limit the influence of any one sample on the overall estimation results. The ABSolute value of error is a robust cost function, as shown by the influence function in Figure 11.11(b). One disadvantage of this function is that it is not continuous at the origin. A further drawback is that it does not allow for the fact that in practice a large proportion of the samples are not contaminated with impulsive noise, and may well be modelled with Gaussian densities.

Many processes may be regarded as Gaussian for the sample values that cluster about the mean. For such processes it is desirable to have an influence function that limits the influence of outliers and at the same time is linear and optimal for the large

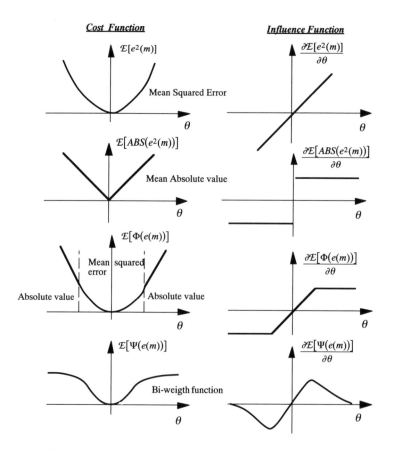

Figure 11.11 Illustration of a number of cost of error functions and the corresponding influence functions.

number of relatively small-amplitude samples that may be regarded as Gaussian distributed. One such function is Huber's function defined as

$$\psi[e(m)] = \begin{cases} e^2(m), & if \ |e(m)| \le k; \\ k|e(m)|, & otherwise. \end{cases} \tag{11.28}$$

Huber's function, shown in Figure 11.11(c) is a hybrid of the least mean square and the absolute value of error functions. Tukeys bi-weight function which is a redescending robust objective function, is defined as

$$\psi[e(m)] = \begin{cases} [1-(1-e^2(m))^3]/6, & if \ |e(m)| \le 1; \\ 1/6 & otherwise. \end{cases} \tag{11.29}$$

As shown in Figure 11.11(d) the influence function is linear for small signal values but introduces attenuation as the signal value exceeds some threshold. The threshold may be obtained from a robust median estimate of the signal power.

11.6 Restoration of Archived Gramophone Records

This Section describes the application of the impulsive noise removal system of Figure 11.8 to restoration of archived audio records. As the bandwidth of archived recordings is limited to 7 to 8 kHz, a lowpass, anti-aliasing, filter with a cutoff frequency of 8 kHz is used to remove the out of band noise. Playedback signals were sampled at a rate of 20 kHz, and digitised to 16 bits. Figure 11.12(a) shows a 50 ms segment of noisy music and song from an old 78-rpm gramophone record. The impulsive interferences are due to faults in the record stamping process, granularity's of the record material, or physical damage. This signal is modelled by a predictor of order 20. The excitation signal obtained from the inverse filter, and the matched filter

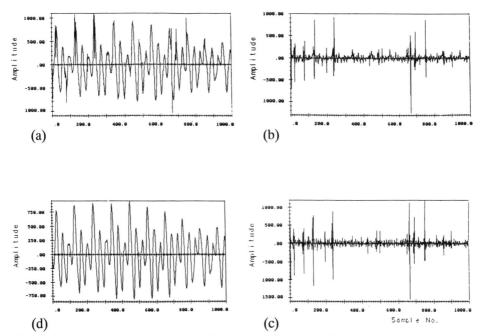

Figure 11.12 (a) A noisy audio signal from a 78 rpm record, (b) noisy excitation signal, (c) matched filter output, (d) restored signal.

output, are shown in Figures 11.12(b) and 11.12(c) respectively. A close examination of these figures show that some of the ambiguities between the noise pulses and the genuine signal excitation pulses are resolved after matched filtering.

The amplitude threshold for detection of impulsive noise from the excitation signal, is adapted on a block basis, and is set to $k.\,\sigma_e^2$, where σ_e^2 is a robust estimate of the excitation power. The robust estimate is obtained by passing the noisy excitation signal through a soft nonlinearity that rejects outliers. The scalar k is a tuning parameter, the choice of k reflects a trade off between the hit rate and the false alarm rate of the detector. As k decreases, smaller noise pulses are detected but the false detection rate also increases. When an impulse is detected, a few samples are discarded and replaced by the LSAR interpolation algorithm described in Chapter 10. Figure 11.12.(d) shows the signal with the impulses removed. The impulsive noise removal system of Figure 11.8 was successfully applied to restoration of numerous examples of archived gramophone records. The system is also effective in suppressing impulsive noise in examples of noisy telephone conversations.

Summary

The classic linear time invariant theory on which much of the signal processing methods are based is not equipped to deal with the nonstationary impulsive noise problem. In this chapter we considered the impulsive noise as a random on/off process and studied several stochastic models for impulsive noise including the Bernoulli-Gaussian model, the Poisson-Gaussian and the hidden Markov model (HMM). The HMM provides a particularly interesting frame work because the theory of HMM studied in Chapter 4 is well developed, and also because the state sequence of an HMM of noise can be used to provide an estimate of the presence or the absence of the noise pluses. By definition an impulsive noise is a short and sharp event uncharacteristic of the signal that it contaminates. In general differencing operation enhance the detectibility of impulsive noise. Based on this observation, in Section 11.4 we considered an algorithm based on a linear prediction model of the signal for detection of impulsive noise.

In the next Chapter we expand the materials we considered in this chapter for the modelling, detection, and removal of transient noise pulses.

Bibliography

DEMPSTER A. P , LAIRD N.M and RUBIN D.B. (1971), Maximum likelihood from Incomplete Data via the EM Algorithm, Journal of the Royal Statistical Society, Ser. Vol. 39, Pages 1-38.

GODSIL S. (1993), Restoration of Degraded Audio Signals, Cambridge University Press, Cambridge.

GALLAGHER N.C., WISE G.L. (1981), A Theoretical Analysis of the Properties of Median Filters, IEEE Transactions on Acoustics, Speech and Signal Processing, Vol. ASSP-29, Pages 1136-41

JAYNAT N.S. (1976), Average and Median Based Smoothing for Improving Digital Speech Quality in the Presence of Transmission Errors, IEEE Trans. Commun. pages 1043-45, Sept issue.

KELMA V.C and LAUB A.J (1980), The Singular Value Decomposition : Its Computation and Some Applications, IEEE Trans. Automatic Control, Vol. AC-25, Pages 164-76.

KUNDA A., MITRA S., VAIDYANATHAN P. (1984), Applications of Two Dimensional Generalised Mean Filtering for Removal of Impulsive Noise from Images, IEEE Trans, ASSP, Vol. 32, No. 3, Pages 600-609, June.

MILNER B. P. (1995), Speech Recognition in Adverse Environments, PhD Thesis, University of East Anglia, UK.

NIEMINEN, HEINONEN P, NEUVO Y (1987), Suppression and Detection of Impulsive Type Interference using Adaptive Median Hybrid Filters, IEEE. Proc. ICASSP-87 Pages 117-20.

TUKEY J.W. (1971), Exploratory Data Analysis, Addison Wesley, Reading, Mass.

RABINER L.R., SAMBUR M.R. and SCHMIDT C.E. (1984), Applications of a Nonlinear Smoothing Algorithm to Speech Processing, IEEE Trans, ASSP, Vol. 32, No.3 June.

VASEGHI S.V., RAYNER P.J.W (1990), Detection and Suppression of Impulsive Noise in Speech Communication Systems, IEE Proc-I Communications Speech and Vision, Pages 38-46, February.

VASEGHI S. V., MILNER B. P. (1995), Speech Recognition in Impulsive Noise, IAcoustics, Speech and Signal Processing, CASSP-95, Pages 437-40.

12

Transient Noise

12.1   Transient Noise Waveforms
12.2   Transient Noise Pulse Models
12.3   Detection of Noise Pulses
12.4   Removal of Noise Pulse Distortions

Transient noise pulses differ from short duration impulsive noise, in that they have a longer duration, a relatively higher proportion of low-frequency energy content, and usually occur less frequently than impulsive noise. The sources of transient noise pulses are varied and may be electronic, acoustic, or due to physical defects in the recording medium. Examples of transient noise pulses include switching noise in telephony, noise pulses due to adverse radio transmission environments, scratches and defects on damaged records, click sounds from a computer keyboard etc. The noise pulse removal methods considered in this chapter are based on the observation that transient noise pulses can be regarded as the response of the communication channel, or the playback system, to an impulse. In this chapter we study the characteristics of transient noise pulses and consider a template-based method, a linear predictive model, and a hidden Markov model for the modelling and removal of transient noise pulses. The subject of this chapter closely follows that of Chapter 11 on impulsive noise.

12.1 Transient Noise Waveforms

Transient noise pulses are often composed of a relatively short and sharp initial pulse followed by decaying low frequency oscillations. The oscillations are due to resonances of the communication channel excited by the initial pulse. In a telecommunication system, a noise pulse originates at some point in time and space, and propagates through the channel to the receiver. The noise is shaped by the channel, and may be considered as the channel impulse response. Thus we expect to be able to characterise the noise pulses with the same degree of success as we have in characterising the channels through which the pulses propagate.

As an illustration of a transient noise pulse, consider scratch pulses from a damaged gramophone record. Scratch pulses are acoustic manifestations of the response of the stylus and the associated electro-mechanical playback system to a sharp physical discontinuity in the recording medium. Since scratches are essentially the impulse response of the playback mechanism, it is expected that for a given system, various scratch pulses exhibit a similar characteristics. As shown in Figure 12.1, a typical scratch waveform often exhibits two distinct regions : (a) the initial response of the playback system to the physical discontinuity on the record, followed by (b) decaying oscillations that causes additive distortion. The initial pulse is relatively short and has a duration in the order of 1 to 5 ms, whereas the oscillatory tail has a longer duration and may last up to 50 ms. Note in Figure 12.1 that the frequency of the decaying oscillations decreases with time. This behaviour may be attributed to the nonlinear modes of response of the electro-mechanical playback system excited by the physical scratch discontinuity.

Observations of many scratch waveforms from damaged gramophone records reveals that they have a well defined profile, and can be characterised by a relatively small number of typical templates. A similar argument can be applied to describe the transient noise pulses in other systems as the response of the system to an impulsive type noise.

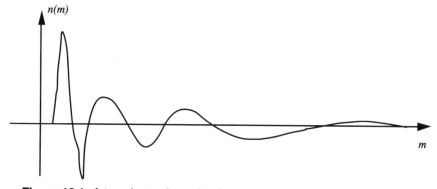

Figure 12.1 A transient noise pulse from a scratched gramophone record.

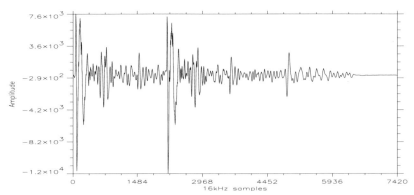

Figure 12.2 Example of transient noise pulses.

The observation that transient noise pulses exhibit certain distinct and definable characteristics can be used for the modelling, detection and the removal of these disturbances. Figure 12.2 shows examples of a machine gun noise. Note the noise can be considered as the impulse response of the gun and the acoustic environment.

12.2 Transient Noise Pulse Models

To a first approximation, a transient noise pulse $n(m)$ can be modelled as the impulse response of a linear time-invariant filter model of the channel as

$$n(m) = \sum_k h_k A\delta(m - k) = A h_m \qquad (12.1)$$

where A is the amplitude of the driving impulse and h_k is the channel impulse response. A burst of overlapping, or closely spaced, noise pulses can be modelled as the response of a channel to a sequence of impulses as

$$n(m) = \sum_k h_k \sum_j A_j\delta\big((m - T_j) - k\big) = \sum_j A_j h_{m-T_j} \qquad (12.2)$$

where it is assumed that the jth transient pulse is due to an impulse of amplitude A_j at time T_j. In practice, a noise model should be able to deal with the statistical variations of a variety of noise and channels. In this section we consider three methods for modelling the temporal, spectral and durational characteristics of a transient noise pulse process, these are: (a) a template-based model, (b) a linear-predictive model, and (c) a hidden Markov model.

12.2.1 Noise Pulse Templates

A widely used method for modelling the space of a random process is to design a codebook of templates containing a number of centroids of the signal space. The centroids can be obtained through a clustering process. The signal space is partitioned into a number of regions or clusters, and the "centre" of the space within each cluster is taken as a centroid of the signal process. Similarly, a codebook of transient noise pulses can be designed by collecting a large number of training examples of the noise, and then using a clustering technique to group, or partition, the noise data base into a number of clusters of noise pulses. The centre of each cluster is taken as a centroid of the noise space. Clustering techniques can be used to obtain a number of prototype templates for the characterisation of a set of transient noise pulses. The clustering of a signal process are based on a set of signal features that best characterise the signal. Signal features derived from the magnitude-spectrum are commonly used for the characterisation of many random processes. For transient noise pulses, the most important features are the pulse shape, the temporal-spectral characteristics of the pulse, the pulse duration, and the pulse energy. The design of a codebook of signal templates is described in Section 3.6.

12.2.2 Autoregressive Model of Transient Noise Pulses

Model-based methods have the advantage, over the template-based methods, that overlapped noise pulses can be modelled as the response of the model to a number of closely spaced impulsive inputs. In this section we consider an autoregressive (AR) model of transient noise pulses. The AR model for a single noise pulse $n(m)$ can be described as

$$n(m) = \sum_{k=1}^{P} c_k n(m-k) + A\delta(m) \tag{12.3}$$

where c_k are the AR model coefficients, and the excitation is an impulse function $\delta(m)$ of amplitude A. A number of closely spaced and overlapping transient noise pulses can be modelled as the response of the predictive model to a sequence of impulses as

$$n(m) = \sum_{k=1}^{P} c_k n(m-k) + \sum_{j}^{M} A_j \delta(m-T_j) \tag{12.4}$$

where it is assumed that T_k is the start of the kth pulse in a burst of M excitation pulses. An AR model for transient noise, proposed by Godsill, is driven by a binary-state excitation : in state S_0 the excitation is a zero-mean Gaussian process of a small variance σ_0^2, and in state S_1 the excitation is a zero-mean Gaussian process of variance $\sigma_1^2 >> \sigma_0^2$. In state S_1 a short duration, and relatively large amplitude, excitation generates a linear model of the transient noise pulse. In state S_0 the model generates a low amplitude excitation that partially models the inaccuracies of approximating a transient noise pulse by a linear predictive model. The binary state excitation signal can be expressed as

$$e_n(m) = \left[\sigma_1 b(m) + \sigma_0 \bar{b}(m)\right] u(m) \tag{12.5}$$

where $u(m)$ is an uncorrelated zero mean unit variance Gaussian process, and $b(m)$ indicates the state of the excitation signal : $b(m)=1$ indicates that the excitation has a variance of σ_1^2, and $b(m)=0$ (or its binary complement $\bar{b}(m)=1$) indicates the excitation has a smaller variance of σ_0^2. The time-varying variance of $e_n(m)$ can be expressed as

$$\sigma_{e_n}^2(m) = \sigma_1^2 b(m) + \sigma_0^2 \bar{b}(m) \tag{12.6}$$

Assuming that the excitation pattern $b(m)$ is given, and that the excitation amplitude is Gaussian, the pdf of an N sample long noise pulse $\boldsymbol{n}$ is given by

$$f_N(\boldsymbol{n}) = \frac{1}{(2\pi)^{N/2}\left|\Lambda_{e_n e_n}\right|^{1/2}} \exp\left(-\frac{1}{2}\boldsymbol{n}^T C^T \Lambda_{e_n e_n}^{-1} C \boldsymbol{n}\right) \tag{12.7}$$

where C is a matrix of coefficients of the AR model of the noise (as described in Section 7.4), and $\Lambda_{e_n e_n}$ is the diagonal covariance matrix of the input to the noise model. The diagonal elements of $\Lambda_{e_n e_n}$ are given by Eq. (12.6).

12.2.3 Hidden Markov Model of a Noise Pulse Process

A hidden Markov model (HMM), described in Chapter 4, is a finite state statistical model for nonstationary random processes such as speech or transient noise pulses. In general, we may identify three distinct states for a transient noise pulse process. These are (a) the periods during which noise pulses are absent, (b) the initial, and often short and sharp, pulse of a transient noise, and (c) the decaying oscillatory tail of a transient pulse. Figure 12.3 illustrates a 3-state HMM of transient noise pulses. The state S_0 models the periods when the noise pulses are absent. In this state the

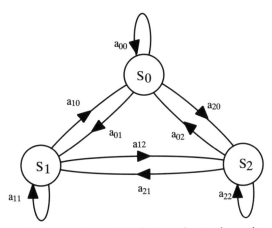

Figure 12.3 A three-state model of a transient noise pulse process.

noise process may be zero-valued. This state can also be used to model a different noise process such as a white noise process. The state S_1 models the relatively sharp pulse which forms the initial part of many transient noise pulses. The state S_2 models the decaying oscillatory part of a noise pulse that usually follows the initial pulse of a transient noise. A code book of waveforms in states S_1 and S_2 can model a variety of different noise pulses. Note that in the HMM model of Figure 12.3, the self loop transition provides a mechanism for the modelling of the variations in the duration of each noise segment. The skip-state transitions provide a mechanism for the modelling of those noise pulses which do not exhibit either the initial nonlinear pulse or the decaying oscillatory part.

A hidden Markov model of noise can be employed for both the detection and the removal of transient noise pulses. As described in Section 12.3.3, the maximum likelihood state-sequence, of the noise HMM, provides an estimate of the state of the noise at each time instant. The estimates of the states of the signal and the noise can be used for the implementation of an optimal state-dependent signal restoration algorithm.

12.3 Detection of Noise Pulses

For detection of a pulse process $n(m)$ observed in an additive signal $x(m)$, the signal and the pulse can be modelled as

$$y(m) = b(m)n(m) + x(m) \tag{12.8}$$

where $b(m)$ is a binary process that signals the presence or absence of a noise pulse. Using the model of Eq. (12.8), the detection of a noise pulse process can be considered as the estimation of the underlying binary state process $b(m)$. In this section we consider three different methods for detection of transient noise pulses using the noise template model, the linear predictive model of noise, and the hidden Markov model described in Section 12.2.

12.3.1 Matched Filter for Noise Pulse Detection

The inner product of two signal vectors provides a measure of similarity of the signals. Since filtering is basically an inner product operation, it follows that the output of a filter should provide a measure of similarity of the filter input and the filter impulse response. The classical method for detection of a signal is to use a filter whose impulse response is *matched* to the shape of the signal to be detected. The derivation of a matched filter for the detection of a pulse $n(m)$, is based on maximisation of the amplitude of the filter output when the input contains the pulse $n(m)$. The matched filter for detection of a pulse $n(m)$ observed in a "background" signal $x(m)$, is defined as

$$H(f) = K \frac{N^*(f)}{P_{XX}(f)} \qquad (12.9)$$

where $P_{XX}(f)$ is the power spectrum of $x(m)$ and $N^*(f)$ is the complex conjugate of the spectrum of the noise pulse. When the "background" signal process $x(m)$ is a zero mean uncorrelated signal with variance σ_x^2, the matched-filter for detection of the transient noise pulse $n(m)$ becomes

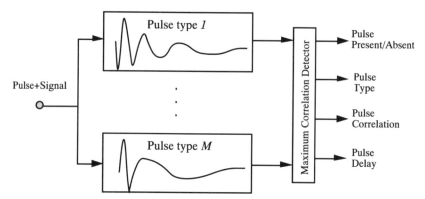

Figure 12.4 A bank of matched filters for detection of transient noise pulses.

$$H(f) = \frac{K}{\sigma_x^2} N^*(f) \tag{12.10}$$

The impulse response of the matched filter corresponding to Eq. (12.10) is given by

$$h(t) = C n(-t) \tag{12.11}$$

where the scaling factor C is given by $C = K/\sigma_x^2$. Let $z(m)$ denote the output of the matched filter. The output $z(m)$ is passed through a nonlinearity and a decision is made on the presence or the absence of the pulse as

$$\hat{b}(m) = \begin{cases} 1 & \text{if } |z(m)| \geq \text{Threshold} \\ 0 & \text{otherwise} \end{cases} \tag{12.12}$$

In Eq. (12.12) when the matched filter output exceeds a threshold the detector flags the presence of the signal at the input. Figure 12.4 shows a noise pulse detector composed of a bank of M different matched filters. The detector signals the presence or the absence of a noise pulse. If a pulse is present then additional information provide the type of the pulse, the maximum cross correlation of the input and the noise pulse template, and a time delay which can be used to align the input noise and the template. These information can be used for subtraction of the noise pulse as described in Section 12.4.1.

12.3.2   Noise Detection Based on Inverse Filtering

The initial part of a transient noise pulse is often a relatively short and sharp impulsive-type event, and can be used as a distinct feature for the detection of the noise pulses. The detectibility of a sharp noise pulse $n(m)$, observed in a correlated "background" signal $y(m)$, can often be improved by using a differencing operation which has the effect of enhancing the relative amplitude of the impulsive type noise. The differencing operation can be accomplished by an inverse linear predictor model of the back ground signal $y(m)$. An alternative interpretation is that the inverse filtering is equivalent to a spectral whitening operation : it effects the energy of the signal spectrum whereas the theoretically flat spectrum of the impulsive noise is largely unaffected. The use of an inverse linear predictor for the detection of an impulsive type event was considered in detail in Section 11.4. Note that the inverse filtering operation reduces the detection problem to that of detecting a pulse in additive white noise.

12.3.3 Noise Detection Based on HMM

In the 3-state hidden Markov model of a transient noise pulse process, described in Section 12.2.3, the states S_0, S_1 and S_2 correspond to the noise-absent state, the initial noise pulse state, and the decaying oscillatory noise state respectively. As described in Chapter 4, an HMM denoted by the symbol M, is defined by a set of Markovian state transition probabilities and Gaussian state observation pdfs. The statistical parameters of the HMM of a noise pulse process, can be obtained from a sufficiently large number of training examples of the process.

Given an observation vector $y = [y(0), y(1), ..., y(N-1)]$, the maximum likelihood state sequence $s = [s(0), s(1), ..., s(N-1)]$, of the HMM M is obtained as

$$s_{ML} = \arg\max_{s} f_{Y|S}(y|s,M) \tag{12.13}$$

where, for a hidden Markov model, the likelihood of an observation sequence $f_{Y|S}(y|s,\lambda)$ can be expressed as

$$f_{Y|S}(y(0),y(1),...,y(N-1)|s(0),s(1),...,s(N-1))$$
$$= \pi_{s(0)}f_{s(0)}(y(0)) \cdot a_{s(0),s(1)}f_{s(1)}(y(1)) \cdots a_{s(N-2),s(N-1)}f_{s(N-1)}(y(N-1)) \tag{12.14}$$

where $\pi_{s(i)}$ is the initial state probability, $a_{s(i),s(j)}$ denotes the probability of a transition from state $s(i)$ to state $s(j)$, and $f_{s(i)}(y(i))$ is the state observation pdf for the state $s(i)$. The maximum likelihood state sequence, derived using the Viterbi algorithm, is an estimate of the underlying states of the noise pulse process and can be used as a detector of the presence or absence of a noise pulse.

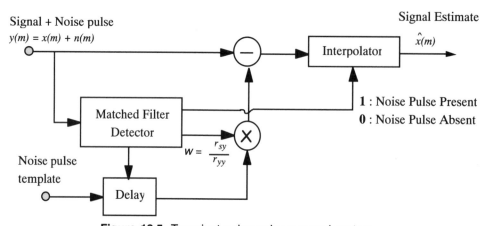

Figure 12.5 Transient noise pulse removal system.

12.4 Removal of Noise Pulse Distortions

In this section we consider two methods for the removal of transient noise pulses. These are : (a) an adaptive noise subtraction method, and (b) an autoregressive (AR) model-based restoration method. The noise removal methods assume that a detector signals the presence or the absence of a noise pulse, and provides additional information on the timing and the underlying the states of the noise pulse

12.4.1 Adaptive Subtraction of Noise pulses

The transient noise removal system, shown in Figure 12.5, is composed of a matched filter for detection of noise pulses, a linear adaptive noise subtractor for cancellation of the linear transitory part of a noise pulse, and an interpolator for the replacement of samples irrevocably distorted by the initial part of each pulse. Let $x(m)$, $n(m)$ and $y(m)$ denote the signal, the noise pulse and the noisy signal respectively, the noisy signal model is

$$y(m) = x(m) + b(m)n(m) \tag{12.15}$$

where $b(m)$ indicates the presence or the absence of a pulse. Assume that each noise pulse $n(m)$ can be modelled as the amplitude-scaled and time-shifted version of the noise pulse template $\bar{n}(m)$ so that

$$n(m) \approx w\, \bar{n}(m - D) \tag{12.16}$$

where w is an amplitude scalar and the integer D denotes the relative delay (time shift) between the noise pulse template and the detected noise. From Eq. (12.15) and (12.16) the noisy signal can be modelled as

$$y(m) \approx x(m) + w\bar{n}(m - D) \tag{12.17}$$

From Eq. (12.17) an estimate of the signal $x(m)$ can be obtained by subtracting an estimate of the noise pulse from that of the noisy signal as

$$\hat{x}(m) = y(m) - w\bar{n}(m - D) \tag{12.18}$$

where the time delay D required for time-alignment of the noisy signal $y(m)$ and the noise template $\bar{n}(m)$ is obtained from the cross correlation function CCF as

$$D = \arg\max_{k}\left[CCF(y(m),\ \bar{n}(m - k))\right] \tag{12.19}$$

When a noise pulse is detected, the time lag corresponding to the maximum of the cross-correlation function is used to delay and time-align the pulse template with the individual noise pulse. The template energy is adaptively matched to that of the noise pulse by an adaptive scaling coefficient w. The scaled and time-aligned noise template is subtracted from the noisy signal to remove linear additive distortions. The adaptation coefficient, w, is estimated as follows. The correlation of the noisy signal $y(m)$ with the delayed noise pulse template $\bar{n}(m - D)$ gives

$$
\begin{aligned}
\sum_{m=0}^{N-1} y(m)\bar{n}(m - D) &= \sum_{m=0}^{N-1} [x(m) + w\bar{n}(m - D)]\bar{n}(m - D) \\
&= \sum_{m=0}^{N-1} x(m)\bar{n}(m - D) + w \sum_{m=0}^{N-1} \bar{n}(m - D)\bar{n}(m - D)
\end{aligned}
\tag{12.20}
$$

Where N is the pulse template length. Since the signal $x(m)$ and the noise $n(m)$ are uncorrelated, the term $\Sigma x(m)\bar{n}(m - D)$ in the right hand side of Eq. (12.20) vanishes and we have

$$
w = \frac{\sum_{m} x(m)\bar{n}(m - D)}{\sum_{m} \bar{n}^2(m - D)}
\tag{12.21}
$$

Note that if a false detection of a noise pulse occurs then the cross correlation term and hence the adaptation coefficient w could be small. This will keep the signal distortion resulting from false detections to a minimum.

Samples that are irrevocably distorted by the initial scratch pulse are discarded and replaced by one of the interpolators introduced in Chapter 10. When there is no noise pulse, the coefficient w is zero, the interpolator is bypassed and the input signal is passed through unmodified. Figure 12.6(b) shows the result of processing the noisy signal of Figure 12.6(a) The linear oscillatory noise is completely removed by the adaptive subtraction method. For this signal 80 samples irrevocably distorted by the initial scratch pulse were discarded and interpolated.

12.4.2   AR-based Restoration of Signals Distorted by Noise Pulses

A model-based approach provides a more compact method for characterisation of transient noise pulses, and has the advantage that closely spaced pulses can be modelled as the response of the model to a number of closely spaced impulses. The

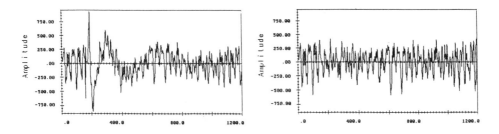

Figure 12.6 (a) A signal from an old gramophone record with a scratch noise pulse, (b) the restored signal.

signal $x(m)$ is modelled as the output of an AR model of order P_1 as

$$x(m) = \sum_{k=1}^{P_1} a_k x(m-k) + e(m) \tag{12.22}$$

Assuming that $e(m)$ is a zero-mean uncorrelated Gaussian process with variance σ_e^2, the pdf of a vector x of N successive signal samples is given by

$$f_X(x) = \frac{1}{(2\pi\sigma_e^2)^{N/2}} \exp\left(\frac{1}{2\sigma_e^2} x^T A^T A x\right) \tag{12.23}$$

where the elements of the matrix A are composed of the coefficients a_k of the linear predictor model as described in Section 7.4. In Eq. (12.23) it is assumed that the P_1 initial samples are known. The AR model for a single noise pulse waveform $n(m)$ can be written as

$$n(m) = \sum_{k=1}^{P_2} c_k n(m-k) + A\delta(m) \tag{12.24}$$

where c_k are the model coefficient, P_2 is the model order, and the excitation is a assumed to be an impulse of amplitude A. A number of closely spaced and overlapping noise pulses can be modelled as

$$n(m) = \sum_{k=1}^{P_2} a_k n(m-k) + \sum_{j}^{M} A_j \delta(m-T_j) \tag{12.25}$$

where it is assumed that T_k is the start of the kth excitation pulse in a burst of M pulses. A linear predictor model proposed by Godsill is driven by a binary-state excitation. The excitation waveform has two states : in state "0" the excitation is a zero-mean Gaussian process of variance σ_0^2, in state "1" the excitation is a zero-mean Gaussian process of variance $\sigma_1^2 >> \sigma_0^2$. In state "1" the model generates a short duration large amplitude excitation that largely models the transient pulse. In state "0", the model generates a low excitation that partially models the inaccuracies of approximating a nonlinear system by an AR model. The composite excitation signal can be written as

$$e_n(m) = \left[b(m)\sigma_1 + \bar{b}(m)\sigma_0\right]u(m) \tag{12.26}$$

where $u(m)$ is an uncorrelated zero mean Gaussian process of unit variance, $b(m)$ is a binary sequence that indicates the state of the excitation, and $\bar{b}(m)$ is the binary complement of $b(m)$. When $b(m)=1$ the excitation variance is σ_1^2 and when $b(m)=0$ the excitation variance is σ_0^2. The binary-state variance of $e_n(m)$ can be expressed as

$$\sigma_{e_n}^2(m) = b(m)\sigma_1^2 + \bar{b}(m)\sigma_0^2 \tag{12.27}$$

Assuming that the excitation pattern $b=[b(m)]$ is given, the pdf of an N sample noise pulse x is

$$f_N(n|b) = \frac{1}{(2\pi)^{N/2}\left|\Lambda_{e_ne_n}\right|^{1/2}} \exp\left(-\frac{1}{2}n^TC^T\Lambda_{e_ne_n}^{-1}Cn\right) \tag{12.28}$$

where the elements of the matrix C are composed of the coefficients c_k of the linear predictor model as described in Section 7.4. The posterior pdf of the signal x given the noisy observation y, $f_{X|Y}(x|y)$, can be expressed, using the Bayes rule, as

$$f_{X|Y}(x|y) = \frac{1}{f_Y(y)} f_{Y|X}(y|x)f_X(x)$$

$$= \frac{1}{f_Y(y)} f_N(y-x)f_X(x) \tag{12.29}$$

For a given observation $f_Y(y)$ is a constant. Substitution of Eq. (12.28) and (12.23) in Eq. (12.29) yields

$$f_{X|Y}(x|y) = \frac{1}{f_Y(y)} \frac{1}{(2\pi\sigma_e)^N |A_{e_n e_n}|^{1/2}} \times$$

$$\exp\left(-\frac{1}{2}(y-x)^T CT \Lambda_{e_n e_n}^{-1} C(y-x) + \frac{1}{2\sigma_e^2} x^T A^T A x\right)$$

(12.30)

The MAP solution obtained by maximisation of the log posterior function with respect to the undistorted signal x is

$$\hat{x}^{MAP} = \left(A^T A / \sigma_e^2 + C^T \Lambda_{e_n e_n}^{-1} C\right)^{-1} C^T \Lambda_{e_n e_n}^{-1} C \ y$$

(12.31)

Summary

In this chapter we considered the modelling, detection, and removal of transient noise pulses. Transient noise pulses are nonstationary events similar to impulsive noise but usually occur less frequently and have a longer duration than impulsive noise. An important observation in the modelling of transient noise is that the noise can be regarded as the impulse response of a communication channel, and hence may be modelled by one of a number of statistical methods used in the of modelling communication channels. In section we considered a template-based method, an AR model-based method and a hidden Markov model for removal of transient noise pulses.

Bibliography

GODSILL S. J. (1993), The Restoration of Degraded Audio Signals, Ph.D. Thesis, Cambridge University, Cambridge.

VASEGHI S. V. (1987), Algorithm for Restoration of Archived Gramophone Recordings, Ph.D. Thesis, Cambridge University, Cambridge.

13

Echo Cancellation

13.1 Telephone Line Echoes
13.2 Line Echo Cancellation
13.3 Acoustic Feedback Coupling Between Loudspeaker and Microphone
13.4 Sub-band Acoustic Echo Cancellation

Echo is the repetition of a wave due to reflection from points where the characteristics of the material through which the wave propagates changes. Acoustic echoes are due to reflection of the sound waves from walls, floors, ceilings, windows and other objects. Telephone line echoes result from impedance mismatch at the telephone exchange hybrids where the subscriber's two-wire line is connected to a four-wire line. Echoes can also result from a feedback path set up between the speaker and the microphone in a teleconference or hearing aid system. The perceptual effects of an echo depends on the time delay between the incident and the reflected waves, the strength of the reflected waves and the number of paths through which the waves are reflected. Acoustic echo is usually reflected from a multitude of different surfaces and travels through different paths. If the time delay is not too long then the acoustic echo may be perceived as a soft reverberation, and it may even add to the artistic quality of the sound. Concert halls and church halls with desirable reverberation characteristics can enhance the quality of a musical performance. Telephone line echoes, and acoustic feedback echoes in teleconference and hearing aid systems, are undesirable and annoying and can be quite disruptive. In this chapter we study the methods of removing line echoes from telephone and data telecommunication systems, and the acoustic feedback echoes from microphone-loudspeaker systems.

13.1 Telephone Line Echoes

Echoes on telephone lines are due to reflection of signals at the points of impedance mismatch on the connecting circuits. Every telephone in a given geographical area is connected to an exchange by a two-wire twisted line, called the subscriber's loop, which serves to receive and transmit signals. A local call is set up by establishing a direct connection, at the telephone exchange, between two subscriber's loops. For a local call, there is no noticeable echo either because there is not a significant impedance mismatch on the connecting lines, or because the distances are relatively small and hence echoes are perceived as a slight amplification effect. For long-distance communication between two exchanges it is necessary to use repeaters to amplify the speech signals, therefore a separate two-wire line is required for each direction of transmission. To establish a long distance call, at each end a two-wire local line from the subscriber's loop must be connected to a four-wire line at the exchange, as illustrated in Figure 13.1. The device that connects the subscriber's loop to the four-wire line is called a hybrid and is shown in Figure 13.2. As shown the hybrid is basically a three-port bridge circuit. If the hybrid bridge is perfectly balanced then there would be no reflection or echo. However, each hybrid serves a number of subscriber's loops. The subscriber's loops do not all have the same impedance characteristics, therefore it is not possible to achieve perfect balance for all subscribers at the hybrids. When the bridge is not perfectly balanced, some of the signal energy on the receiving pair of wires gets coupled back to the transmitting pair of wires and produces an echo. Telephone line echoes are undesirable and become annoying when the echo amplitude is relatively high and the echo delay is long. For example when a long-distance call is made via a satellite the round-trip echo delay can be as long as 600 ms, and echoes can become disruptive.

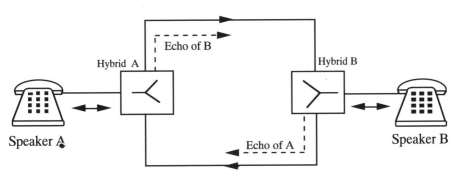

Figure 13.1 Illustration of a telephone call setup by connection of two-wire subscriber's via hybrids to four-wire lines at the exchange.

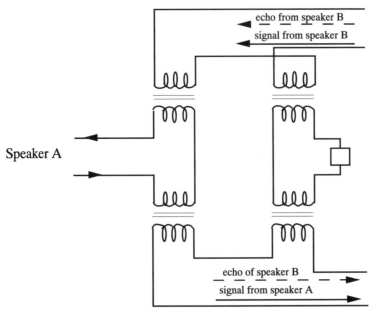

Figure 13.2 Two-wire to four-wire hybrid.

13.1.1 Telephone Line Echo Suppression

An echo suppresser attenuates the echo during the speech inactive periods. A line echo suppresser is controlled by an echo detection device. The echo detector monitors the signal levels on the incoming and the outgoing lines, and decides if the signal on a line from, say, speaker B to speaker A, is the speech from the speaker B to the speaker A, or the echo of speaker A. If the echo detector decides that the signal is an echo, then the signal is heavily attenuated. There is a similar echo suppression system from speaker A to speaker B. The performance of an echo suppresser depends on the accuracy of the echo/speech classification. Echo of speech often has a smaller amplitude level than the speech, but otherwise it has mainly the same spectral characteristics and statistics as those of the speech. Therefore the only basis to discriminate the speech from the echo is the signal level. As a result the speech/echo classifier may wrongly classify and let through high level echoes as speech, or attenuate low level speech as echo. For terrestrial circuits echo suppressers have been well designed, with an acceptable level of false decisions and a good performance. The performance of an echo suppresser depends on the time delay of the echo. In general, echo suppressers perform well when the round-trip delay of the echo is less than 100 ms. For a conversation routed via a geo-stationary satellite the round-trip delay may be as much as 600 ms. Such long delays change the pattern of conversation and result in a significant increase in the speech/echo classification

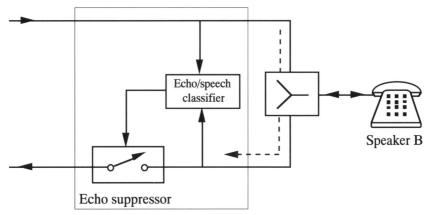

Figure 13.3 Block diagram illustration of an echo suppression system.

errors. When the delay is long, echo suppressers fail to perform satisfactorily. A system which is effective with both short and long time delays is the adaptive echo canceller introduced in the following section.

13.2 Adaptive Echo Cancellation

Figure 13.4 illustrates the operation of an adaptive line echo canceller. The speech on the line from speaker A to speaker B is input to the hybrid B and to the echo canceller. The echo canceller monitors the line from B to A and attempts to synthesis a replica of the echo of speaker A. This replica is used to subtract and cancel out the echo of speaker A on the line from B to A. The echo canceller is basically an adaptive linear filter. The coefficients of the filter are adapted so that the energy of the signal on the line is minimised. The echo canceller may be an infinite impulse response or a finite impulse response filter. The main advantage of an infinite impulse response filter is that a long-delay echo can be synthesised by a relatively small number of filter coefficients. In practice, echo cancellers are based on finite impulse response filters. This is mainly due to the practical difficulties associated with the adaptation and stable operation of adaptive IIR filters.

Assuming that the signal on the line from speaker B to speaker A, $y_B(m)$, is composed of the speech of speaker B $x_B(m)$, plus the echo of speaker A $x_A^{echo}(m)$ we have

$$y_B(m) = x_B(m) + x_A^{echo}(m) \tag{13.1}$$

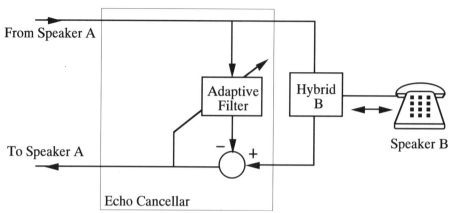

Figure 13.4 Block diagram illustration of an adaptive echo cancellation system.

In practice speech and echo signals are not simultaneously present on a line. This, as pointed out shortly, can be used to simplify the adaptation process. Assuming that the echo-synthesiser is a finite impulse response filter, the filter output can be expressed as

$$\hat{x}_A^{echo}(m) = \sum_{k=0}^{P-1} w_k(m)\, x_A(m-k) \qquad (13.2)$$

where $w_k(m)$ are the coefficients of an adaptive FIR filter, and $\hat{x}_A^{echo}(m)$ is an estimate of the echo of speaker A, on the line from speaker B to A. The residual echo, or the error signal, after subtraction is given by

$$
\begin{aligned}
e(m) &= x_A^{echo}(m) - \hat{x}_A^{echo}(m) \\
&= x_A^{echo}(m) - \sum_{k=0}^{P-1} w_k(m)\, x_A(m-k)
\end{aligned} \qquad (13.3)
$$

The filter coefficients $w_k(m)$ are adapted to minimise the energy of the signal on the line from B to A. Assuming that the speech signal $x_A(m)$ and $x_B(m)$ are uncorrelated the energy on the telephone line is minimised when the filter output is equal to the echo on the line so that the echo is cancelled out. The most widely used algorithm for adaptation of the coefficients of an echo canceller is the normalised least mean squared error (NLMS) method. The time-update equation describing the adaptation of the filter coefficient vector is given by

$$w(m) = w(m-1) + \mu\, \frac{e(m)}{x_A^T(m) x_A(m)} x_A(m) \qquad (13.4)$$

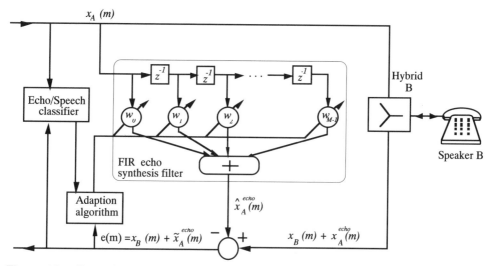

Figure 13.5 illustration of an echo canceller using an adaptive FIR filter and incorporation a echo/speech classifier.

where $x_A(m)=[x_A(m), ..., x_A(m-P)]$, and $w(m)=[w_0(m), ..., w_{P-1}(m)]$, are the input signal vector, and the coefficient vector of the echo synthesiser, and $e(m)$ is the difference between the signal on the echo line and the output of the echo synthesiser. Note that the normalising quantity $x_A^T(m)x_A(m)$ is the energy of the input speech to the adaptive filter. The scalar μ is the adaptation step size and controls the speed of convergence, the steady state error and the stability of the adaptation process. The filter coefficients may also be adapted using the block least squared error algorithm of Chapter 5 or the fast converging recursive least squared error method of Chapter 6.

13.2.1 Convergence of Line Echo Canceller

For satisfactory performance, the echo canceller should have a fast convergence rate, so that it can adequately track the changes in the telephone line and the signal, characteristics. The convergence of an echo canceller is affected by the following factors:

a) Non-stationary characteristics of telephone line and speech : The echo characteristics depends on the impedance mismatch between the subscribers loop and the hybrids. Any changes in the connecting paths affects the echo characteristics and the convergence process. Also as explained in Chapter 6, the nonstationary character and the eigenvalue spread of the input speech signal of an LMS adaptive filter, affects the convergence rates of the filter coefficients.

b) Simultaneous conversation : In a telephone conversation, usually the talkers do not speak simultaneously, and hence speech and echo are seldom present on a line at the same time. This observation simplifies the echo cancellation problem and substantially aids the correct functioning of adaptive echo cancellers. Problems arise during the periods when both speakers talk at the same time. This is because speech and its echo have similar characteristics and occupy basically the same frequency bandwidth. When the reference signal contains both echo and speech, the adaptation process can lose track and the echo cancellation process can attempt to cancel out and distort the speech signal. One method of avoiding this problem is to use a speech activity detector, and freeze the adaptation process during periods when speech and echo are simultaneously present on a line, as shown in Figure 13.5. In this system the effect of a speech/echo misclassification is that the echo may not be optimally cancelled out. This is more acceptable than is the case in echo suppressers where the effect of a misclassification is suppression of a part of the speech.

c) The adaptation Algorithm : Most echo cancellers use variants of the LMS adaptation algorithm. The attractions of the LMS are the relatively low memory and computational requirements and the ease of implementation and monitoring. The main drawback of LMS is that it can be sensitive to the eigenvalue spread of the input signal and is not particularly fast in its convergence rate. However, in practice, LMS adaptation has produced effective line echo cancellation systems. The recursive least square (RLS) error methods have a faster convergence rate and a better minimum mean squared error performance. With the increasing availability of low cost dedicated DSP processors, implementation of the higher performance, and computationally intensive echo cancellers based on RLS are feasible.

13.2.2   Echo Cancellation for Digital Data Transmission over Subscriber's Loop

The two-wire telephone lines of subscriber's loop can be used to provide telephone users with digital data links. This can be achieved by either using the entire usable bandwidth of the wire for data transmission or by transmitting the data on a bandwidth above that used to carry the speech. On an analog subscriber's loop, voice occupies the frequency spectrum between 300 Hz and 3.4 kHz, in this frequency rang the wire has almost a flat frequency response. Above 4 kHz the frequency response of the two-wire line drops to -50 dBm at 40 kHz. However, a data rate of up to 16 kbps can be supported by appropriate modulation of the digital signal onto a

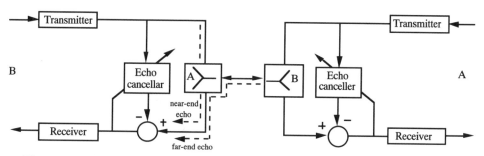

Figure 13.6 Echo cancellation in digital modems using two-wire subscriber's loop.

carrier at a band above 4 kHz. Figure 13.6 shows a system for providing full-duplex digital service over a two-wire subscriber's loop.

To provide simultaneous transmission of data on both directions within the same bandwidth over the subscriber's loop echo cancellation is needed. The echoes on a line consists of the near-end echo which results from the signal that loops back at the first or the near hybrid, and the far-end echoes which is the signal that loops back at a hybrid some distance away. The main purpose of the echo canceller is to cancel the near-end echo. Since the digital signal coming from a far-end may be attenuated by 40 to 50 dB, the near echo on a high speed data transmission line can be as much as 40 to 50 dB above the desired signal level. For reliable data communication the echo canceller must provide 50 to 60 dB attenuation of the echo signal so that the signal power remains at 10 dB above the echo.

13.3 Acoustic Feedback Coupling Between Loud-speaker and Microphone

A well known problem with hands-free telephones, teleconference systems, public address systems, and hearing aids, is the feedback coupling of the sound waves between the loudspeakers and the microphones. Acoustic feedbacks can be manifested in howling if a significant proportion of the sound energy, transmitted by the loudspeaker, is received back at the microphone and circulated in the feedback loop. The overall round-gain of an acoustic feedback loop depends on the frequency responses of the electrical and the acoustic signal paths. The undesirable effects of the electrical sections, on the acoustic feedback can be reduced by designing systems that have a flat frequency response. The main problem is in the acoustic feedback path, and the reverberating characteristics of the room. If the microphone-speaker-room system is excited at a frequency whose loop gain is greater than unity, then the signal

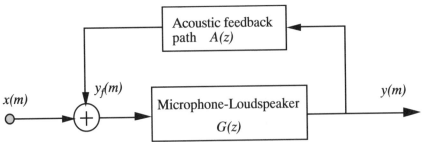

Figure 13.7 Configuration of a feedback model for a microphone-loudspeaker-room system.

is amplified each time it circulates round the loop, and a feedback howling results. In practice the howling is limited by the nonlinearities of the electronic system.

There are a number of methods for removing acoustic feedback. One method for alleviating the effects of acoustic feedback and the room reverberations is to place a frequency shifter (or a phase shifter) in the electrical path of the feedback loop. Each time a signal travels round the feedback loop it is shifted by a few hertz before being re-transmitted by the loudspeaker. This method has some effect in reducing the howling but is not effective for removal of the overall echo of the acoustic feedback. Another approach is to reduce the feedback loop-gain at those frequencies where the acoustic feedback energy is concentrated. This may be achieved by using adaptive notch filters to reduce the system gain at frequencies where acoustic oscillations occur. The drawback of this method is that in addition to reducing the feedback the notch filters also result in distortion of the desired signal frequencies. The most effective method of acoustic feedback removal is the use of an adaptive feedback cancellation system. Figure 13.7 illustrates a model of an acoustic feedback environment, comprising a microphone, a loudspeaker and the reverberating space of a room. The z-transfer function of a linear model of the acoustic feedback environment may be expressed as

$$H(z) = \frac{G(z)}{1 - G(z)A(z)} \tag{13.5}$$

where $G(z)$ is the z-transfer function model for the microphone-loudspeaker system and $A(z)$ is the z-transfer function model of the reverberating room environment. Assuming that the microphone-loudspeaker combination has a flat frequency response with a gain of G Eq. (13.5) can be simplified to

$$H(z) = \frac{G}{1 - G.A(z)} \tag{13.6}$$

Note that in Eq. (13.6) the acoustic feedback path is itself a feedback system due to the reverberating character of the room. The reverberating characteristics of the acoustic environment may be modelled by an all pole linear predictive model, or alternatively a relatively long finite impulse response model.

The equivalent time domain input/output relation for the linear filter model of Eq. (13.6) is given by the following difference equation

$$y(m) = G \sum_{k=0}^{P-1} a_k(m)y(m-k) + G x(m) \tag{13.7}$$

where $a_k(m)$ are the coefficients of a linear feedback model of the reverberating room environment and $x(m)$ and $y(m)$ are the time domain input and output signals of the microphone-loudspeaker system.

Figure 13.8 is an illustration of an acoustic feedback cancellation system. In an acoustic feedback environment the total input signal to the microphone is given as the sum of any new input $x(m)$ plus the unwanted feedback signal $y_f(m)$ as

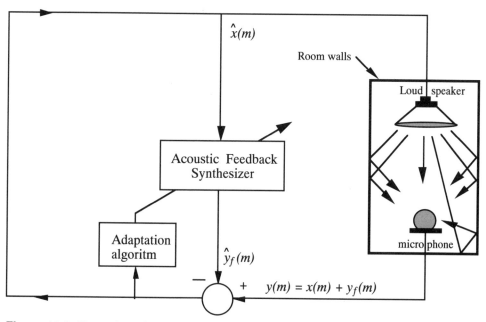

Figure 13.8 Illustration of adaptive acoustic feedback cancellation in a conference room environment.

$$y(m) = x(m) + y_f(m) \tag{13.8}$$

The most successful acoustic feedback control systems are based on adaptive estimation and cancellation of the feedback signal. As in a line echo canceller, an adaptive acoustic feedback canceller attempts to synthesis a replica of the acoustic feedback at its output as

$$\hat{y}_f(m) = \sum_{k=0}^{P-1} \hat{a}_k(m)y(m-k) \tag{13.9}$$

The filter coefficients are adapted to minimise the energy of an error signal defined as

$$e(m) = x(m) + y_f(m) - \hat{y}_f(m) \tag{13.10}$$

The adaptation criterion is usually the minimum mean squared error criterion and the adaptation algorithm is a variant of the LMS or the RLS method.

The problem of acoustic echo cancellation is more complex than that of the line echo cancellation for a number of reasons. Firstly, the acoustic echo is usually much longer (up to a second) that the terrestrial telephone line echoes. In fact, the delay of an acoustic echo is similar to a line echo routed via a geo-stationary satellite system.

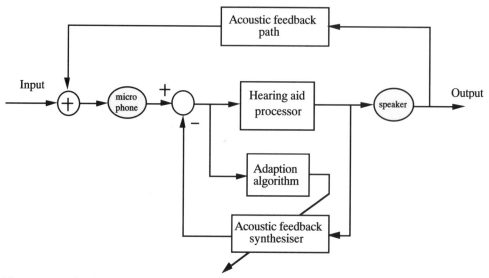

Figure 13.9 Configuration of an acoustic feedback canceller incorporated in a hearing aid system.

The large delay of an acoustic echo path implies that impractically large filters in the order of a few thousand coefficients may be required. The adaptation of filters of such length presents a difficult problem. Secondly, the characteristics of an acoustic echo path is more nonstationary compared with that of a telephone line echo. For example the opening or closing of a door, or a moving person, can suddenly change the acoustic character of a conference room. Thirdly, acoustic echoes are due to signals reflected back from a multitude of different paths, off the walls, the floors, the ceilings, the windows etc. Finally the propagation and diffusion characteristics of the acoustic space of a room is a nonlinear process, and is not well approximated by a lumped FIR (or IIR) linear filter. In comparison it is more reasonable to model the characteristics of a telephone line echo with a linear filter. In any case, for acoustic echo cancellation the filter must have a large impulse response and should be able to quickly track fast changes in echo path characteristics. An important application of acoustic feedback cancellation is in hearing aid systems. The hearing aid system can be modelled as a feedback system as shown in the Figure 13.9. The maximum usable gain of a hearing aid system is limited by the acoustic feedback between the microphone and the speaker. Figure 13.9 illustrates the configuration of a feedback canceller in a hearing aid system. The acoustic feedback synthesiser has the same input as the acoustic feedback path. An adaptation algorithm adjusts the coefficients of the synthesiser to cancel out the feedback signals picked up by the microphone, before the microphone output is fed into the speaker.

13.4 Sub-band Acoustic Echo Cancellation

In addition to the complex and varying nature of a room acoustics there are two main problems in acoustic echo cancellation. Firstly the echo delay is relatively long and therefore the FIR echo synthesiser must have a large number of coefficients, say 2000 or more. Secondly, the long impulse response of the FIR filter and the large eigenvalue spread of the speech signals result in a slow, and uneven, rate of convergence of the adaptation process.

A sub-band-based echo canceller alleviates the problems associated with the required filter length and the speed of convergence. The sub-band-based system is shown in Figure 13.10. The sub-band analyser splits the input signal into N subbands. Assuming that the sub-bands have equal bandwidth, each subband occupies only one N^{th} of the baseband frequency and can therefore be decimated (down sampled) without loss of information. For simplicity assume that all subbands are down sampled by the same factor R. The main advantages of a sub-band echo canceller are a reduction in filter length and a gain in the speed of convergence as explained below :

(a) Reduction in filter length : Assuming that the impulse response of each sub-band filter has the same duration, in seconds, as the impulse response of the full band FIR

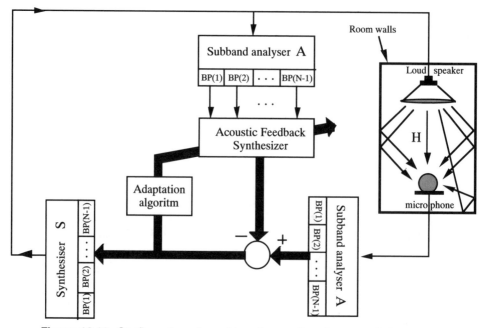

Figure 13.10 Configuration of a subband acoustic echo cancellation system.

filter, the length of the FIR filter for each down-sampled sub-band is $1/R$ of the full band filter.

(b) Reduction in computational complexity : The computational complexity of an LMS type adaptive filter depends directly on the product of the filter length and the sampling rate. As for each sub-band both the number of samples per second and the filter length decrease with $1/R$, it follows that the computational complexity of each subband filter is $1/R^2$ of that of the full band filter. Hence, the overall gain in computational complexity of a sub-band system is R^2/N of the full band system.

(c) Speed of convergence : The speed of convergence depends on both the filter length and the eigenvalue spread of the signal. The speed of convergence increases with the decrease in the length of the FIR filter for each sub-band. A more important factor affecting the convergence of adaptive filter is the eigenvalue spread of the autocorrelation matrix of the input signal. As the spectrum of a signal becomes more flat, the spread of its eigenvalues decreases, and the speed of convergence of the adaptive filter increases. In general, the signal within each subband is expected to have a flatter spectrum than the full band signal. This aids the speed of convergence.

However, it must be noted that the attenuation of subband filters at the edges of the spectrum of each band creates some very small eigenvalues.

Summary

Telephone line echoes and acoustic feedback echo effect the functioning of telecommunication and teleconferencing systems. In general, line echo cancellation is a relatively simpler problem than acoustic echo cancellation because acoustic cancellers need to model the more complex environment of the space of a room.

We began this chapter with a study of the telephone line echoes arising from the mismatch at the two-to-four wire hybrid bridge. In Section 13.2 line echo suppression and adaptive line echo cancellation were considered. For adaptation of an echo canceller the LMS or the RLS adaptation methods can be used. The RLS methods provides a faster convergence rate and a better overall performance at a higher computational complexity.

In Section 13.3 we considered the acoustic coupling between a loud speaker and a microphone system. Acoustic feedback echo can result in howling, and disrupt the performance of teleconference, hands-free telephones, and hearing aid systems. The main problems in implementation of acoustic echo cancellation systems are the requirement for a large filter to model the relatively long echo, and the adaptation problems associated with the eigenvalue spread of the signal. The subband echo canceller introduced in Section 13.4 alleviates these problems.

Bibliography

ALLEN J., BERKLEY D., BLAURET J. (1977), Multi-microphone Signal Processing Technique to Remove Room Reverberation from Speech Signals, J. Acoust. Soc. Am., Vol. 62, No. 4.

ARMBRUSTER W. (1992),Wideband Acoustic Echo Canceller with Two Filter Structure, Proc. EUSIPCO-92, Vol. 3, Pages 1611-17.

CARTER G. (1987),Coherence and Time Delay Estimation, Proc. IEEE, Vol. 75, No. 2 Pages 236-55.

FLANAGAN J. L.*et al.*, (1991), Autodirective Microphone systems, Acoustica Vol. 73, Pages 58-71.

FLANAGAN J. L.*et al.*, (1985), Computer-Steered Microphone Arrays for Sound Transduction in Large Rooms , J. Acoust. Soc. Amer., Vol. 78, Pages 1508-18.

GAO X. Y., SNELGROVE W. M. (1991), Adaptive Linearisation of a Loudspeaker, ICASSP-91 Vol. 3, Pages 3589-92.

GILLOIRE A., VETTERLI M. (1994), Adaptive Filtering in Sub-bands with Critical Sampling : Analysis, Experiments and Applications to Acoustic Echo Cancellation, IEEE. Trans. Signal Processing, Vol. 40, Pages 320-28.

GRITTON C. W., LIN D. W. (1984), Echo Cancellation Algorithms, IEEE ASSP Mag., Vol. 1, No. 2, Pages 30-37.

HANSLER E. (1992), The Hands-Free Telephone Problem An Annotated Bibliography, Signal Processing 27, Pages 259-71.

HART J. E., NAYLOR P. A., TANRIKULU O. (1993), Polyphase Allpass IIR Structures for Subband Acoustic Echo Cancellation, EuroSpeech-93, Vol. 3, Pages 1813-16.

KELLERMANN W. (1988), Analysis and Design of Multirate Systems for Cancellation of Acoustical Echoes, IEEE Proc. ICASSP-88, Pages 2570-73.

KNAPPE M. E. (1992), Acoustic Echo Cancellation : Performance and Structures, M. Eng. Thesis, Carleton University, Ottawa, Canada.

MARTIN R., ALTENHONER J. (1995), Coupled Adaptive Filters for Acoustic Echo Control and Noise Reduction, IEEE Proc. ICASSP-95, Vol. 5, Pages 3043-46.

OSLEN H. F. (1964), Acoustical Engineering, Toronto, D. Van Nostrand Inc.

SCHROEDER M. R. (1964),Improvement of Acoustic-Feedback Stability by Frequency Shifting, J. Acoust. Soc. Amer., Vol. 36, Pages 1718-24.

SILVERMAN H. F. et al., (1992), A Two-Stage Algorithm for Determining Talker Location from Linear Microphone Array Data, Computer Speech and Language, Vol. 6, Pages 129-52.

SONDHI M. M., BERKLEY D A.(1980), Silencing Echoes on the Telephone Network', Proc. IEEE Vol. 68, Pages 948-63.

SONDHI M. M., MORGAN D. R. (1991), Acoustic Echo Cancellation for Stereophonic Teleconferencing, IEEE Workshop on Applications of Signal Processing to Audio And Acoustics.

SONDHI M. M. (1967), An Adaptive Echo Canceller, Bell Syst. tech. J. Vol. 46, Pages 497-511.

TANRIKULU O., etal. (1995), Finite-Precision Design and Implementation of All-Pass Polyphase Networks for Echo Cancellation in subbands, IEEE Proc. ICASSP-95, Vol. 5, Pages 3039-42.

VAIDYANATHAN P. P. (1993), Multirate Systems and Filter Banks, Prentice-Hall.

WIDROW B., McCOOL J. M., LARIMORE M. G., JOHNSON C. R. (1976), Stationary and Nonstationary Learning Characteristics of the LMS Adaptive Filters, Proceedings of the IEEE, Vol. 64, No. 8, Pages 1151-62.

ZELINSKI R. (1988), A Microphone Array with Adaptive Post-Filtering for Noise Reduction in Reverberant Rooms, IEEE Proc. ICASSP-88, Pages 2578-81.

14

Blind Deconvolution and Channel Equalisation

14.1   Introduction
14.2   Blind-Equalisation Using Channel Input Power Spectrum
14.3   Equalisation Based on Linear Prediction Models
14.4   Bayesian Channel Equalisation
14.5   Blind Equalisation for Digital Communication Channels
14.6   Equalisation Based on Higher Order Statistics

Blind deconvolution is the process of unravelling two unknown signals that have been convolved. An important application of blind deconvolution is blind equalisation for the restoration of a signal distorted in transmission through a communication channel. Blind equalisation/deconvolution has a wide range of applications, for example in digital telecommunications for removal of inter-symbol interference, in speech recognition for removal of the effects of microphones and channels, in deblurring of distorted images, in dereverberation of acoustic recordings, in seismic data analysis etc.

In practice, blind equalisation is only feasible if some useful statistics of the channel input, and perhaps also of the channel itself, are available. The success of a blind equalisation method depends on how much is known about the statistics of the channel input, and how useful this knowledge is in the channel identification and equalisation process. This chapter begins with an introduction to the basics of deconvolution and channel equalisation. We study blind equalisation based on the channel input power spectrum, equalisation through model factorisation, Bayesian equalisation, nonlinear adaptive equalisation for digital communication channels, and equalisation of maximum phase channels using higher order statistics.

14.1 Introduction

In this chapter we consider the recovery of a signal distorted, in transmission through a channel, by a convolutional process and observed in additive noise. The process of recovery of a signal convolved with the impulse response of a communication channel or a recording medium is known as deconvolution or equalisation. Figure 14.1 illustrates a typical model for a distorted and noisy signal, followed by an equaliser. Let $x(m)$, $n(m)$ and $y(m)$ denote the channel input, the channel noise and the observed channel output respectively. The channel input/output relation can be expressed as

$$y(m) = h[x(m)] + n(m) \qquad (14.1)$$

where the function $h[]$ is the channel distortion. In general, the channel response may be time-varying and nonlinear. In this chapter it is assumed that the effects of a channel can be modelled using a stationary, or a slowly time-varying, linear transversal filter. For a linear transversal filter model of the channel, Eq. (14.1) becomes

$$y(m) = \sum_{k=0}^{P-1} h_k(m)x(m-k) + n(m) \qquad (14.2)$$

where $h_k(m)$ are the coefficients of a linear filter model of the channel. For a time-invariant channel model $h_k(m)=h_k$. In the frequency domain Eq. (14.2) becomes

$$Y(f) = X(f)H(f) + N(f) \qquad (14.3)$$

where $Y(f)$, $X(f)$, $H(f)$ and $N(f)$ are the frequency spectra of the channel output, the channel input, the channel and the additive noise respectively. Ignoring the noise term

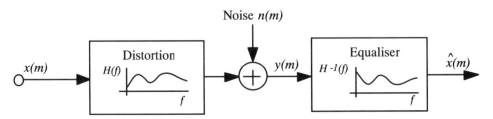

Figure 14.1 Illustration of a channel distortion model followed by an equaliser.

and taking the logarithm of Eq. (14.3) yields

$$\ln|Y(f)| = \ln|X(f)| + \ln|H(f)| \tag{14.4}$$

From Eq. (14.4), in the log-frequency domain the effect of channel distortion is the addition of a 'tilt' term $\ln|H(f)|$ to the signal spectrum.

14.1.1   The Ideal Inverse Channel Filter

The ideal inverse-channel filter, or the ideal equaliser, recovers the original channel input from the channel output signal. In the frequency domain the ideal inverse channel filter can be expressed as

$$H(f)H^{inv}(f) = 1 \tag{14.5}$$

In Eq. (14.5) $H^{inv}(f)$ is used to denote the inverse channel filter. For the ideal equaliser $H^{inv}(f)=H^{-1}(f)$, or equivalently in the log-frequency domain $\ln H^{inv}(f)=-\ln H(f)$. The general form of Eq. (14.5) is given by the z-transform relation

$$H(z)H^{inv}(z) = z^{-N} \tag{14.6}$$

for some delay N that makes the channel inversion causal. Taking the inverse Fourier transform of Eq. (14.5) we have the following convolutional relation between the impulse responses of the channel $\{h_k\}$ and the ideal inverse channel response $\{h_k^{inv}\}$

$$\sum_k h_k^{inv} h_{i-k} = \delta(i) \tag{14.7}$$

where $\delta(i)$ is the Kronecker delta function. Assuming the channel output is noise-free and the channel invertible, the ideal inverse channel filter can reproduce the channel input signal with zero error, as follows. The inverse filter output $\hat{x}(m)$, with the distorted signal $y(m)$ as the input, is given as

$$\begin{aligned}
\hat{x}(m) &= \sum_k h_k^{inv} y(m-k) \\
&= \sum_k h_k^{inv} \sum_j h_j x(m-k-j) \tag{14.8} \\
&= \sum_i x(m-i) \sum_k h_k^{inv} h_{i-k}
\end{aligned}$$

The last line of Eq. (14.8) is derived by a change of variables $i=k+j$ in the second line and rearrangement of the terms. For the ideal inverse channel filter, substitution of Eq. (14.7) in Eq. (14.8) yields

$$\hat{x}(m) = \sum_i \delta(i)x(m-i) = x(m) \tag{14.9}$$

which is the desired result. In practice $H^{inv}(f)$ is not implemented simply as $H^{-1}(f)$ because in general a channel may be non-invertible. Even for invertible channels, a straight forward implementation of the inverse channel filter $H^{-1}(f)$ can cause problems. For example, at frequencies where $H(f)$ is small, its inverse $H^{-1}(f)$ is large, and this can lead to noise amplification if the signal to noise ratio is low.

14.1.2 Equalisation Error, Convolutional Noise

The equalisation error signal, also called the convolutional noise, is defined as the difference between the channel equaliser output and the desired signal :

$$
\begin{aligned}
v(m) &= x(m) - \hat{x}(m) \\
&= x(m) - \sum_{k=0}^{P-1} \hat{h}_k^{inv} y(m-k)
\end{aligned} \tag{14.10}
$$

where $\hat{h}_k^{inv}$ is an estimate of the inverse channel filter. Assuming that there is an ideal equaliser h_k^{inv} that can recover the channel input signal $x(m)$ from the channel output $y(m)$ we have

$$x(m) = \sum_{k=0}^{P-1} h_k^{inv} y(m-k) \tag{14.11}$$

Substitution of Eq. (14.11) in Eq. (14.10) yields

$$
\begin{aligned}
v(m) &= \sum_{k=0}^{P-1} h_k^{inv} y(m-k) - \sum_{k=0}^{P-1} \hat{h}_k^{inv} y(m-k) \\
&= \sum_{k=0}^{P-1} \tilde{h}_k^{inv} y(m-k)
\end{aligned} \tag{14.12}
$$

where $\tilde{h}_k^{inv} = h_k^{inv} - \hat{h}_k^{inv}$. The equalisation error signal $v(m)$ may be viewed as the output of an error filter $\tilde{h}_k^{inv}$ with input $y(m-k)$ and hence the name convolutional noise. When the equalisation process is proceeding well, so that $\hat{x}(m)$ is a good estimate of the channel input $x(m)$, then the convolutional noise is relatively small and decorrelated and can be modelled as a zero mean Gaussian random process.

14.1.3 Blind Equalisation

The equalisation problem is relatively simple when the channel response is known and invertible, and when the channel output is not noisy. However, in most practical cases the channel is unknown, time-varying, and may also be non-invertible, and the channel output is observed in additive noise.

Digital communication systems provide equaliser training periods, during which a *training* pseudo-noise (PN) sequence, also available at the receiver, is transmitted. A synchronised version of the PN sequence is generated at the receiver, where the channel input and output signals are used for the identification of the channel equaliser as illustrated in Figure 14.2(a). The obvious drawback of using training periods for channel equalisation is that power and bandwidth are consumed for the equalisation process. It is preferable to have a "blind" equalisation scheme that can operate without access to the channel input, as illustrated in Figure 14.2(b). Furthermore, in some applications, such as the restoration of acoustic recordings, or blurred images, all that is available is the distorted signal and the only restoration method applicable is blind equalisation.

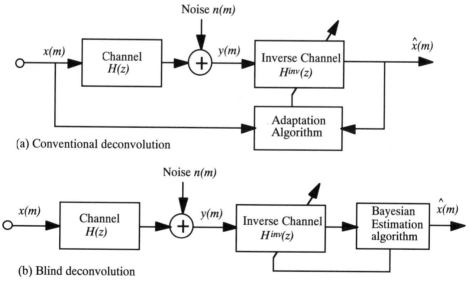

Figure 14.2 A comparative illustration of (a) a conventional equaliser which has access to channel input and output, and (b) a blind equaliser.

Blind equalisation is feasible only if some statistical knowledge of the channel input, and perhaps also the channel, are available. Blind equalisation involves two stages of channel identification and deconvolution of the channel response as follows :

(a) Channel Identification. The general form of a channel estimator can be expressed as

$$\hat{h} = \psi(y, \mathcal{M}_x, \mathcal{M}_h) \tag{14.13}$$

Where ψ is the channel estimator, the vector $\hat{h}$ is an estimate of the channel response, y is the channel output, and M_x and M_h are statistical models of the channel input and the channel respectively. Channel identification methods rely on utilisation of a knowledge of the following characteristics of the input signal and the channel :

(i) The distribution of the channel input signal. For example in decision-directed channel equalisation, described in Section 14.5, the knowledge that the input is a binary signal is used in a binary decision device to estimate the channel input and to direct the equaliser adaptation process.

(ii) The relative durations of the channel input and the channel impulse response. The duration of a channel impulse response is usually orders of magnitude smaller than that of the channel input. This observation is used in Section 14.3.1 to estimate a stationary channel from the long-time averages of the channel output .

(iii) The stationary, or time-varying characteristics of the input signal process and the channel. In Section 14.3.1 a method is described for the recovery of a nonstationary signal convolved with the impulse response of a stationary channel.

(b) Channel Equalisation. Assuming that the channel is invertible, the channel input signal $x(m)$ can be recovered using an inverse channel filter as

$$\hat{x}(m) = \sum_{k=0}^{P-1} \hat{h}_k^{inv} y(m-k) \tag{14.14}$$

In the frequency domain Eq. (14.14) becomes

$$\hat{X}(f) = \hat{H}^{inv}(f)Y(f) \tag{14.15}$$

In practice, perfect recovery of the channel input may not be possible either because the channel is non-invertible or because the output is observed in noise. A channel is non-invertible if :

(i) The channel transfer function is maximum phase. The transfer function of a maximum phase channel has zeros outside the unit circle, and hence the inverse channel has unstable poles. Maximum phase channels are considered in the following section.

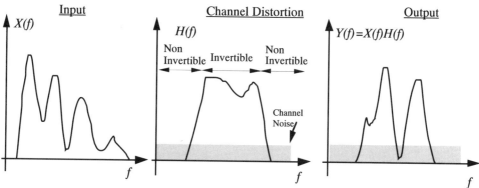

Figure 14.3 Illustration of the invertible and noninvertible regions of a channel.

(ii) The channel transfer function maps many inputs to the same output. In these situations a stable closed-form equation for the inverse channel does not exist, and instead an iterative deconvolution method is used. Figure 14.4 illustrates the frequency response of a channel which has one invertible and two non-invertible regions. In the non-invertible regions the signal frequencies are heavily attenuated and lost to channel noise. In the invertible region the signal is distorted but recoverable. This example illustrates that the inverse filter must be implemented with care in order to avoid undesirable results such as noise amplification at frequencies with low SNR.

14.1.4 Minimum and Maximum Phase Channels

For stability, all the poles of the transfer function of a channel must lie inside the unit circle. If all the zeros of the transfer function are also inside the unit circle then the channel is said to be a minimum phase channel. If some of the zeros are outside the unit circle then the channel is said to be a maximum phase channel. The inverse of a minimum phase channel has all its poles inside the unit circle, and is therefore stable. The inverse of a maximum phase channel has some of its poles outside the unit circle, therefore it has an exponentially growing impulse response and is unstable. However, a stable approximation of the inverse of a maximum phase channel may be obtained by truncating the impulse response of the inverse filter. Figure 14.3 illustrates an example of a maximum phase and a minimum phase fourth order FIR filters.

When both the channel input and output signals are available, in the correct synchrony, it is possible to estimate the channel magnitude and phase response using the conventional least squared error criterion. In blind deconvolution there is no access to the exact instantaneous value or the timing of the channel input signal. The

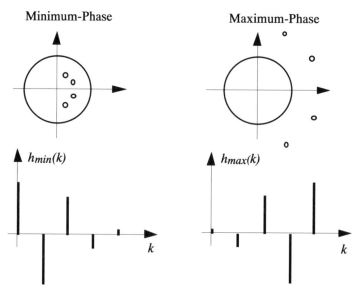

Figure 14.4 Illustration of the zero diagram and impulse response of fourth order maximum phase and minimum phase FIR filters.

only information available are the channel output and some statistics of the channel input. The second order statistics of a signal (i.e. the correlation or the power spectrum) do not include the phase information, hence it is not possible to estimate the channel phase from the second order statistics. Furthermore, the channel phase can not be recovered if the input signal is Gaussian, because a Gaussian process of a known mean is entirely specified by the autocovariance matrix, and autocovariance matrices do not include any phase information. For estimation of the phase of a channel we can either use a non-linear estimate of the desired signal to direct the adaptation of a channel equaliser as in Section 14.5, or we can use the higher order statistics as in Section 14.6.

14.1.5   Wiener Equaliser

In this section we consider the least squared error Wiener equalisation. Note that in its conventional form, the Wiener equalisation is not a form of blind equalisation, because the implementation of a Wiener equaliser requires the cross correlation of the channel input and output signals which are not available in a blind equalisation application. The Wiener filter estimate of the channel input signal is given by

$$\hat{x}(m) = \sum_{k=0}^{P-1} \hat{h}_k^{inv} y(m-k) \tag{14.16}$$

where $\hat{h}_k^{inv}$ is a transversal Wiener filter estimate of the inverse channel impulse response. The equalisation error signal $v(m)$ is defined as

$$v(m) = x(m) - \sum_{k=0}^{P-1} \hat{h}_k^{inv} y(m-k) \qquad (14.17)$$

The Wiener equaliser with input $y(m)$ and desired output $x(m)$ is obtained from Eq. (5.10) in Chapter 5 as

$$\hat{h}^{inv} = R_{yy}^{-1} r_{xy} \qquad (14.18)$$

where R_{yy} is the $P \times P$ autocorrelation matrix of the channel output, and r_{xy} is the P-dimensional cross correlation vector of the channel input and output signals. A more expressive form of Eq. (14.18) can be obtained by writing the noisy channel output signal in a vector equation as

$$y = Hx + n \qquad (4.19)$$

where y is an N-sample channel output vector, x is an $N+P$ sample channel input vector including the P initial samples, H is an $N \times (N + P)$ channel distortion matrix whose elements are composed of the coefficients of the channel filter, and n is a noise vector. The autocorrelation matrix of the channel output can be obtained from Eq. (14.19) as

$$R_{yy} = \mathcal{E}[yy^T] = HR_{xx}H^T + R_{nn} \qquad (14.20)$$

where $E[\,]$ is the expectation operator. The cross correlation vector r_{xy} of the channel input and output signals becomes

$$r_{xy} = \mathcal{E}[xy] = Hr_{xx} \qquad (14.21)$$

substitution of Eq. (14.20) and (14.21) in (14.18) yields the Wiener equaliser as

$$\hat{h}^{inv} = [HR_{xx}H^T + R_{nn}]^{-1}Hr_{xx} \qquad (14.22)$$

The derivation of the Wiener equaliser in the frequency domain is as follows. The Fourier transform of the equaliser output is given by

$$\hat{X}(f) = \hat{H}^{inv}(f)Y(f) \qquad (14.23)$$

where $Y(f)$, the channel output and $\hat{H}^{inv}(f)$ is the frequency response of the Wiener equaliser. The error signal $V(f)$ is defined as

$$V(f) = X(f) - \hat{X}(f)$$
$$= X(f) - \hat{H}^{inv}(f)Y(f)$$
(14.24)

As in Section 5.5 the minimisation of expectation of the squared magnitude of $V(f)$ results in the frequency Wiener equaliser given by

$$\hat{H}^{inv}(f) = \frac{P_{XY}(f)}{P_{YY}(f)}$$
$$= \frac{P_{XX}(f)H^*(f)}{P_{XX}(f)|H(f)|^2 + P_{NN}(f)}$$
(14.25)

Where $P_{XX}(f)$ is the channel input power spectrum, $P_{NN}(f)$ is the noise power spectrum, $P_{XY}(f)$ is the cross power spectrum of the channel input and output signals, and $H(f)$ is the frequency response of the channel. Note that in the absence of noise $P_{NN}(f)=0$ and the Wiener inverse filter becomes $H^{inv}(f) = H^{-1}(f)$.

14.2   Blind Equalisation Using Channel Input Power Spectrum

One of the early papers on blind deconvolution was by Stockham *et al* (1975) on dereverberation of old acoustic recordings. Acoustic recorders, illustrated in Figure (14.5) had a bandwidth of about 200 Hz to 4 kHz. However, this limited bandwidth, or even the additive noise or scratch noise pulses, are not considered the major causes of distortions of acoustic recordings. The main distortion on acoustic recordings is due to the reverberations of the recording horn instrument. An acoustic recording can be modelled as the convolution of the input signal $x(m)$ and the impulse response of a linear filter model of the recording instrument $\{h_k\}$, as in Eq. (14.2) reproduced here for convenience

$$y(m) = \sum_{k=0}^{P-1} h_k x(m-k) + n(m)$$
(14.26)

or in the frequency domain as

$$Y(f) = X(f)\,H(f) + N(f)$$
(14.27)

where $H(f)$ is the frequency response of a linear time-invariant model of the acoustic recording instrument, and $N(f)$ is an additive noise. Multiplying both sides of Eq. (14.27) with their complex conjugates, and taking the expectation, we obtain

$$\mathcal{E}[Y(f)Y^*(f)] = \mathcal{E}\left[(X(f)H(f) + N(f))(X(f)H(f) + N(f))^*\right] \qquad (14.28)$$

Assuming the signal $X(f)$ and the noise $N(f)$ are uncorrelated Eq. (14.28) becomes

$$P_{YY}(f) = P_{XX}(f)|H(f)|^2 + P_{NN}(f) \qquad (14.29)$$

where $P_{YY}(f)$, $P_{XX}(f)$, $P_{NN}(f)$ are the power spectra of the distorted signal, the original signal, and the noise respectively. From Eq. (14.29) an estimate of the spectrum of the channel response can be obtained:

$$|H(f)|^2 = \frac{P_{YY}(f) - P_{NN}(f)}{P_{XX}(f)} \qquad (14.30)$$

In practice, Eq. (14.30) is implemented using time-averaged estimates of the power spectra.

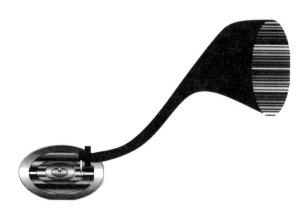

Figure 14.5 Illustration of the early acoustic recording process on a wax disc. Acoustic recordings were made by focusing the sound energy, through a horn via a sound box, diaphragm, and stylus mechanism, onto a wax disc. The sound was distorted by the reverberations of the horn.

14.2.1   Homomorphic Equalisation

In homomorphic equalisation, the convolutional distortion is transformed, first through a Fourier transform of the distorted signal into a multiplicative distortion, and then into an additive distortion by taking the logarithm of the spectrum of the distorted signal. A further inverse Fourier transform operation converts the log-frequency variables into cepstral variables as illustrated in Figure 14.6. Through homomorphic transformation convolution becomes addition, and equalisation becomes subtraction. Ignoring the additive noise term and transforming both sides of Eq. (14.27) into log spectral variables yields

$$\ln Y(f) = \ln X(f) + \ln H(f) \tag{14.31}$$

Note that in the log-frequency domain the effect of channel distortion is the addition of a tilt to the spectrum of the channel input. Taking the expectation of Eq. (14.31) yields

$$\mathcal{E}[\ln Y(f)] = \mathcal{E}[\ln X(f)] + \ln H(f) \tag{14.32}$$

In Eq. (14.32) it is assumed that the channel is time-invariant hence $\mathcal{E}[\ln H(f)] = \ln H(f)$. Using the relation $\ln z = \ln|z| + j \angle z$ the term $\mathcal{E}[\ln X(f)]$ can be expressed as

$$\mathcal{E}[\ln X(f)] = \mathcal{E}[\ln|X(f)|] + j\mathcal{E}[\angle X(f)] \tag{14.33}$$

The first term in the right hand side of Eq. (14.33), $\mathcal{E}[\ln|X(f)|]$ is non-zero and represents the frequency distribution of the signal power in decibels, whereas the second term $\mathcal{E}[\angle X(f)]$ is the expectation of the phase and can be assumed to be zero. From Eq. (14.32) the log frequency spectrum of the channel can be estimated as

$$\ln H(f) = \mathcal{E}[\ln Y(f)] - \mathcal{E}[\ln X(f)] \tag{14.34}$$

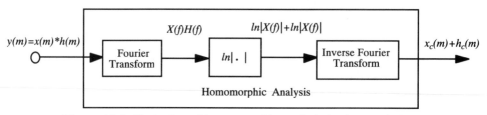

Figure 14.6 Illustration of homomorphic analysis in deconvolution.

In practice, when only a single record is available, the signal is divided into a number of segments, and the average signal spectra is obtained over time across the segments. Assuming that the length of each segment is long compared to the duration of the channel impulse response, we can write an approximate convolutional relation for the ith signal segment as

$$y_i(m) \approx x_i(m) * h_i(m) \tag{14.35}$$

The segments are windowed, using a Hamming or a Hanning window, to reduce the spectral leakage due to end-effects at the edges of the segment. Taking the complex logarithm of the Fourier transform of Eq. (14.35) yields

$$\ln Y_i(f) = \ln X_i(f) + \ln H_i(f) \tag{14.36}$$

Taking the time-averages over N segments of the distorted signal record yields

$$\frac{1}{N}\sum_{i=0}^{N-1}\ln Y_i(f) = \frac{1}{N}\sum_{i=0}^{N-1}\ln X_i(f) + \frac{1}{N}\sum_{i=0}^{N-1}\ln H_i(f) \tag{14.37}$$

Estimation of the channel response from Eq. (14.37) requires the average log spectrum of the undistorted signal. In Stockham's method for restoration of acoustic records, the expectation of the signal spectrum was obtained from a modern recording of the same musical material as that of the acoustic recording. From Eq. (14.37) the estimate of the logarithm of the channel is given by

$$\ln \hat{H}(f) = \frac{1}{N}\sum_{i=0}^{N-1}\ln Y_i(f) - \frac{1}{N}\sum_{i=0}^{N-1}\ln X_i^M(f) \tag{14.38}$$

where $X^M(f)$ is the spectrum of a modern recording. The equaliser can then be defined as

$$\ln H^{inv}(f) = \begin{cases} -\ln \hat{H}(f) & 200 \le f \le 4000 \ Hz \\ -40 \ dB & otherwise \end{cases} \tag{14.39}$$

In Eq. (14.39) the inverse acoustic channel is implemented in the range between 200 and 4000 Hz where the channel is assumed to be invertible. Outside this range, the signal is dominated by noise, and the inverse filter is designed to attenuate the noisy signal.

14.2.2   Homomorphic Equalisation Using a Bank of High- Pass Filters

In the log-frequency domain channel distortion can be eliminated using a bank of highpass filters. Consider a time sequence of log-spectra of the output of a channel described as

$$\ln Y_t(f) = \ln X_t(f) + \ln H_t(f) \qquad (14.40)$$

Where $Y_t(f)$ and $X_t(f)$ are the channel input and output derived from a Fourier transform of the tth signal segment. From Eq. (14.40) the effect of a time invariant channel is to add a constant term $\ln H(f)$ to each coefficient of the channel input $X_t(f)$, and the overall result is a time-invariant tilt of the log frequency spectrum of the original signal. This observation suggests the use of a bank of narrow band highpass notch filters for the removal of the additive distortion term $lnH(f)$. A simple first order recursive digital filter with its notch at the zero frequency is given by

$$\ln \hat{X}_t(f) = \alpha \ln \hat{X}_{t-1}(f) + \ln Y_t(f) - \ln Y_{t-1}(f) \qquad (14.41)$$

where the parameter α controls the bandwidth of the notch at the zero hertz frequency.

14.3   Equalisation Based on Linear Prediction Models

Linear prediction models, described in Chapter 6, are routinely used in applications, such as seismic signal analysis and speech processing, for the modelling and identification of a minimum-phase channel. Linear prediction theory is based on two basic assumptions : that the channel is minimum phase, and that the channel input is a random signal. The standard linear prediction analysis can be viewed as a blind deconvolution method, because both the channel response and the channel input is unknown, and the only information are the channel output and the assumption that the channel input is random and hence has a flat power spectrum. In this section we consider blind deconvolution using linear predictive models for the channel and its input. The channel input signal is modelled as

$$X(z) = E(z)A(z) \qquad (14.42)$$

where $X(z)$ is the z-transform of the channel input signal, $A(z)$ is the z-transfer function of a linear predictive model of the channel input, and $E(z)$ is the z-transform

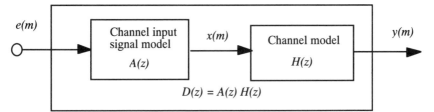

Figure 14.7 A distorted signal modelled as combination of a signal model and a channel model.

of a random excitation signal. Similarly the channel output can be modelled by a linear predictive model $H(z)$ with input $X(z)$ and output $Y(z)$ as

$$Y(z) = X(z)H(z) \qquad (14.43)$$

Figure 14.7 illustrates a cascade linear prediction model for a channel input process $X(z)$ and a channel $H(z)$. The channel output can be expressed as

$$\begin{aligned} Y(z) &= E(z)A(z)H(z) \\ &= E(z)D(z) \end{aligned} \qquad (14.44)$$

where

$$D(z)=A(z)H(z) \qquad (14.45)$$

The z-transfer function of the linear prediction models of the channel input signal and the channel can be expanded as

$$A(z) = \frac{G_1}{1-\displaystyle\sum_{k=1}^{P} a_k z^{-k}} = \frac{G_1}{\displaystyle\prod_{k=1}^{P}(1-\alpha_k z^{-1})} \qquad (14.46)$$

and

$$H(z) = \frac{G_2}{1-\displaystyle\sum_{k=1}^{Q} b_k z^{-k}} = \frac{G_2}{\displaystyle\prod_{k=1}^{Q}(1-\beta_k z^{-1})} \qquad (14.47)$$

where $\{a_k, \alpha_k\}$ and $\{b_k, \beta_k\}$ are the coefficients and the poles of the models for the channel input signal and the channel respectively. Substitution of Eqs. (14.46) and (14.47) in Eq. (14.45) yields the combined input-channel model as

$$D(z) = \frac{G}{1 - \sum_{k=1}^{P+Q} d_k z^{-k}} = \frac{G}{\prod_{k=1}^{P+Q}(1 - \gamma_k z^{-1})} \tag{14.48}$$

The total number of poles of the combined model for the input signal and the channel is the sum of the poles of the input signal model and the channel model.

14.3.1 Blind Equalisation Through Model Factorisation

A model-based approach to blind equalisation, is to factorise the channel output model, $D(z)=A(z)H(z)$, into a channel input signal model $A(z)$, and a channel model $H(z)$. If the channel input model $A(z)$ and the channel model $H(z)$ are non-factorable, then the only factors of $D(z)$ are $A(z)$ and $H(z)$. However, z-transfer functions are factorable into the roots, the so called poles and zeros, of the models. One approach to model based deconvolution, is to factorise the model for the convolved signal into its poles and zeros, and classify the poles and zeros as either those of the signal or those of the channel.

Spencer and Rayner (1990) developed a method for blind deconvolution through factorisation of linear prediction models, based on the assumptions that the channel is stationary with time-invariant poles, whereas the input signal is nonstationary with time-varying poles. As an application they considered the restoration of old acoustic recordings where a time-varying audio signal is distorted by the time-invariant frequency response of the recording equipment. For a simple example, consider the case when the signal and the channel are each modelled by a second order linear predictive model. Let the time-varying second order linear predictive model for the channel input signal $x(m)$ be

$$x(m) = a_1(m)x(m-1) + a_2(m)x(m-2) + G_1(m)e(m) \tag{14.49}$$

where $a_1(m)$ and $a_2(m)$ are the time-varying coefficients of the linear predictor model, $G_1(m)$ is the input gain factor, and $e(m)$ is a zero mean, unit variance, random signal. Now let $\alpha_1(m)$ and $\alpha_2(m)$ denote the time-varying poles of the predictor model of Eq. (14.49), these poles are the roots of the following polynomial

$$1 - a_1(m)z^{-1} - a_2(m)z^{-2} = (1 - z^{-1}\alpha_1(m))(1 - z^{-1}\alpha_2(m)) = 0 \tag{14.50}$$

Similarly, assume that the channel can be modelled by a second order stationary linear predictive model as

$$y(m) = h_1 y(m-1) + h_2 y(m-2) + G_2 x(m) \tag{14.51}$$

where h_1 and h_2 are the time-invariant predictor coefficients and G_2 is the channel gain. Let β_1 and β_2 denote the poles of the channel model, these are the roots of the polynomial

$$1 - h_1 z^{-1} - h_2 z^{-2} = (1 - z^{-1}\beta_1)(1 - z^{-1}\beta_2) = 0 \tag{14.52}$$

The combined cascade of the two second order models of Eq. (14.49) and (14.51) can be written as a fourth order predictive model with input $e(m)$ and output $y(m)$ as

$$y(m) = d_1(m)y(m-1) + d_2(m)y(m-2) + d_3(m)y(m-3) + d_4(m)y(m-4) + Ge(m) \tag{14.53}$$

Where the combined gain $G=G_1 G_2$. The poles of the fourth order predictor model of Eq. (14.53) are the roots of the following polynomial:

$$1 - d_1(m)z^{-1} - d_2(m)z^{-2} - d_3(m)z^{-3} - d_4(m)z^{-4} =$$
$$(1 - z^{-1}\alpha_1(m))(1 - z^{-1}\alpha_2(m))(1 - z^{-1}\beta_1)(1 - z^{-1}\beta_2) = 0 \tag{14.54}$$

In Eq. (14.54) the poles of the fourth order predictor are $\alpha_1(m)$, $\alpha_2(m)$, β_1 and β_2. The above argument on factorisation of the poles of time-varying and stationary models can be generalised to a signal model of order P and a channel model of order Q.

In Spencer and Rayner, the separation of the stationary poles of the channel, from the time-varying poles of the channel input, is achieved through a clustering process. The signal record is divided into N segments and each segment is modelled by an all-pole model of order $P+Q$, where P and Q are the model orders for the channel input and the channel respectively. In all, there are $N(P+Q)$ values which are clustered to form $P+Q$ clusters. Even if both the signal and the channel were stationary, the poles extracted from different segments would have variations due to the random character of the signal. Assuming that the variance of the estimates of the stationary poles are small compared to the variations of the time-varying poles, it is expected that, for each stationary pole of the channel, the N values extracted from N segments would form an N-point cluster of a relatively small variance. These clusters can be identified and the centre of each cluster taken as a pole of the channel model This method assumes that the poles of the time-varying signal are well separated in space from the poles of the time-invariant signal.

14.4 Bayesian Blind Deconvolution and Equalisation

The Bayesian inference method, described in Chapter 3, provides a general framework for inclusion of statistical models of the channel input and the channel response. In this section we consider the Bayesian equalisation method, and study the case where the channel input is modelled by a set of hidden Markov models. The Bayesian risk for a channel estimate $\hat{h}$ is defined as

$$
\begin{aligned}
\mathcal{R}(\hat{h}|y) &= \iint_{HX} C(\hat{h}, h) f_{X,H|Y}(x,h|y) \, dx \, dh \\
&= \frac{1}{f_Y(y)} \int_H C(\hat{h}, h) f_{Y|H}(y|h) f_H(h) \, dh
\end{aligned}
\tag{14.55}
$$

where $C(\hat{h}, h)$ is the cost of estimating the channel h as $\hat{h}$, $f_{X,H|Y}(x,h|y)$ is the joint posterior density of the channel and the channel input, $f_{Y|H}(y|h)$ is the observation likelihood, and $f_H(h)$ is the prior pdf of the channel. The Bayesian estimate is obtained by minimisation of the risk function $\mathcal{R}(\hat{h}|y)$. There are a variety of Bayesian-type solutions depending on the choice of the cost function and the prior knowledge as described in Chapter 3.

In this section it is assumed that the convolutional channel distortion is transformed into an additive distortion through transformation of the channel output into log spectral or cepstral variables. Ignoring the channel noise, the relation between the cepstra of the channel input and output signals is

$$
y(m) = x(m) + h
\tag{14.56}
$$

where the cepstral vectors $x(m)$, $y(m)$ and h are the channel input, the channel output and the channel respectively.

14.4.1 Conditional Mean Channel Estimation

A commonly used cost function in the Bayesian risk of Eq. (14.55) is the mean square error $C(h - \hat{h}) = |h - \hat{h}|^2$ which results in the conditional mean (CM) estimate defined as

$$
\hat{h}^{CM} = \int_H h \, f_{H|Y}(h|y) \, dh
\tag{14.57}
$$

The posterior density of the channel input signal may be conditioned on an estimate of the channel vector $\hat{h}$ and expressed as $f_{X|Y,H}(x|y,\hat{h})$. The conditional mean of the channel input signal given the channel output y and an estimate of the channel $\hat{h}$ is

$$\hat{x}^{CM} = \mathcal{E}[x|y,\hat{h}]$$
$$= \int_X x\, f_{X|Y,H}(x|y,\hat{h})\, dx \tag{14.58}$$

Equations (14.57) and (14.58) suggest a two stage method for channel estimation and the recovery of the channel input signal.

14.4.2   Maximum Likelihood Channel Estimation

The ML channel estimate is equivalent to the case when the Bayes cost function and the channel prior are uniform. Assuming that the channel input signal has a Gaussian distribution with a mean vector μ_x and a covariance matrix Σ_{xx}, the likelihood of a sequence of N P-dimensional channel output vectors $\{y(m)\}$ given a channel input vector h is

$$f_{Y|H}(y(0),\ ...,\ y(N-1)|\,h) = \prod_{m=0}^{N-1} f_X(y(m)-h)$$
$$= \prod_{m=0}^{N-1} \frac{1}{(2\pi)^{P/2}|\Sigma_{xx}|^{1/2}} \exp\big((y(m)-h-\mu_x)^T \Sigma_{xx}^{-1}(y(m)-h-\mu_x)\big) \tag{14.59}$$

To obtain the maximum likelihood estimate of the channel h, the derivative of the log likelihood function $\ln(f_Y(y|h))$ is set to zero to yield

$$\hat{h}^{ML} = \frac{1}{N}\sum_{m=0}^{N-1}(y(m)-\mu_x) \tag{14.60}$$

14.4.3   Maximum a Posterior Channel Estimation

The MAP estimate, like the ML estimate, is equivalent to a Bayesian estimator with a uniform cost function. However, the MAP estimate includes the prior pdf of the channel. The prior can be used to confine the channel estimate within a desired region of the parameter space. Assuming that the channel input vectors are statistically independent, the posterior pdf of the channel given the observation sequence $Y=\{y(0), ..., y(N-1)\}$ is

$$f_{H|Y}(h|y(0), ..., y(N-1)) = \prod_{m=0}^{N-1} \frac{1}{f_Y(y(m))} f_{Y|H}(y(m)|h) f_H(h)$$

$$= \prod_{m=0}^{N-1} \frac{1}{f_Y(y(m))} f_X(y(m)-h) f_H(h)$$

(14.61)

Assuming the channel input $x(m)$ is Gaussian, $f_X(x(m))=N(x, \mu_x, \Sigma_{xx})$, with a mean vector μ_x, and a covariance matrix Σ_{xx}, and the channel h is also Gaussian, $f_H(h)=N(h, \mu_h, \Sigma_{hh})$, with a mean vector μ_h and a covariance matrix Σ_{hh}, the logarithm of the posterior pdf is

$$\ln f_{H|Y}(h|y(0), ..., y(N-1)) = \left(-\sum_{m=0}^{N-1} \ln f(y(m))\right) - NP\ln(2\pi) - \frac{1}{2}\ln(|\Sigma_{xx}||\Sigma_{hh}|) -$$

$$\sum_{m=0}^{N-1} \frac{1}{2}\left((y(m)-h-\mu_x)^T \Sigma_{xx}^{-1}(y(m)-h-\mu_x) + (h-\mu_h)^T \Sigma_{hh}^{-1}(h-\mu_h)\right)$$

(14.62)

The MAP channel estimate, obtained by setting the derivative of the log posterior function $\ln f_{H|Y}(h|y)$ to zero, is

$$\hat{h}^{MAP} = (\Sigma_{xx} + \Sigma_{hh})^{-1} \Sigma_{hh} (\bar{y} - \mu_x) + (\Sigma_{xx} + \Sigma_{hh})^{-1} \Sigma_{xx} \mu_h \qquad (14.63)$$

where

$$\bar{y} = \frac{1}{N}\sum_{m=0}^{N-1} y(m) \qquad (14.64)$$

is the time-averaged estimate of the mean of observation vector. Note that for a Gaussian process the MAP and conditional mean estimates are identical.

14.4.4 Channel Equalisation Based on Hidden Markov Models

This section considers blind deconvolution in applications where the statistics of the channel input are modelled by a set of hidden Markov models. An application of this method, illustrated in Figure 14.8, is in recognition of speech distorted by a communication channel or a microphone. A hidden Markov model (HMM) is a finite-

HMMs of the channel input

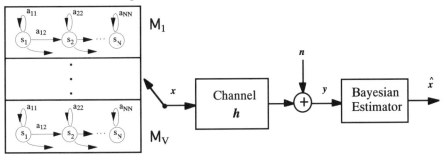

Figure 14.8 Illustration of a channel with the input alphabet modelled by a set of HMMs.

state Bayesian model, with a Markovian state prior and a Gaussian observation likelihood (Chapter 4). An N-state HMM can be used to model a nonstationary process, such as speech, as a chain of N stationary states connected with a set of Markovian state transitions. The likelihood of an HMM M_i and a sequence of N P-dimensional channel input vectors $X=[x(0), ..., x(N-1)]$ can be expressed in terms of the state transition and the observation pdfs of M_i as

$$f_{X|M}(X|M_i) = \sum_s f_{X|M,S}(X|M_i,s)P_{S|M}(s|M_i) \tag{14.65}$$

where $f_{X|M,S}(X|M_i,s)$ is the likelihood that the sequence $X=[x(0), ..., x(N-1)]$ was generated by the state sequence $s=[s(0), ..., s(N-1)]$ of model M_i, and $P_{S|M}(s|M_i)$ is the Markovian prior pmf of the state sequence s. The Markovian prior entails that the probability of a transition to state i at time m depends only on the state at time m-1 and is independent of the previous states. The transition probability of a Markov process is defined as

$$a_{ij} = P\big(s(m) = i \mid s(m-1) = j\big) \tag{14.66}$$

where a_{ij} is the probability of making a transition from state j to state i. The HMM state observation probability is often modelled by a multi-variate Gaussian pdf as

$$f_{X|M,S}(x|M_i,s) = \frac{1}{(2\pi)^{P/2}|\Sigma_{xx,s}|^{1/2}} \exp\left(-\frac{1}{2}(x - \mu_{x,s})^T \Sigma_{xx,s}^{-1}(x - \mu_{x,s})\right) \tag{14.67}$$

where $\mu_{x,s}$ and $\Sigma_{xx,s}$ are the mean vector and the covariance matrix of the Gaussian observation pdf of the HMM state s.

HMM-based channel equalisation can be stated as follows : Given a sequence of N P-dimensional channel output vectors $Y=[y(0), ..., y(N-1)]$, and the prior knowledge that the channel input sequence is drawn from a set of V HMMs $M=\{M_i \ i=1, \ldots, V\}$ estimate the channel response and the channel input. The joint posterior pdf of an input word M_i and the channel vector h can be expressed as

$$f_{\mathcal{M},H|Y}(\mathcal{M}_i,h|Y) = P_{\mathcal{M}|H,Y}(\mathcal{M}_i|h,Y) f_{H|Y}(h|Y) \tag{14.68}$$

Simultaneous joint estimation of the channel vector h and classification of the unknown input word M_i is a nontrivial exercise. The problem is usually approached iteratively by making an estimate of the channel response, and then using this estimate to obtain the channel input as follows. From Bayes rule, the posterior pdf of the channel h conditioned on the assumption that the input model is M_i and given the observation sequence Y can be expressed as

$$f_{H|\mathcal{M},Y}(h|\mathcal{M}_i,Y) = \frac{1}{f_{Y|\mathcal{M}}(Y|\mathcal{M}_i)} f_{Y|\mathcal{M},H}(Y|\mathcal{M}_i,h) f_{H|\mathcal{M}}(h|\mathcal{M}_i) \tag{14.69}$$

The likelihood of the observation sequence, given the channel and the input word model can be expressed as

$$f_{Y|\mathcal{M},H}(Y|\mathcal{M}_i,h) = f_{X|\mathcal{M}}(Y - h|\mathcal{M}_i) \tag{14.70}$$

For a given input model M_i, and state sequence $s=[s(0), s(1), \ldots, s(N-1)]$, the pdf of a sequence of N independent observation vectors $Y=[y(0), y(1), \ldots, y(N-1)]$ is

$$
\begin{aligned}
f_{Y|H,S,\mathcal{M}}(Y|h,s,\mathcal{M}_i) &= \prod_{m=0}^{N-1} f_{X|S,\mathcal{M}}(y(m) - h|s(m),\mathcal{M}_i) \\
&= \prod_{m=0}^{N-1} \frac{1}{(2\pi)^{P/2}|\Sigma_{xx,s(m)}|^{1/2}} \exp\left(-\frac{1}{2}(y(m) - h - \mu_{x,s(m)})^T \Sigma_{xx,s(m)}^{-1}(y(m) - h - \mu_{x,s(m)})\right)
\end{aligned}
\tag{14.71}
$$

Taking the derivative of the log-likelihood of Eq. (14.71) with respect to the channel vector h yields a maximum likelihood channel estimate as

$$\hat{h}_{ML}(Y,s) = \sum_{m=0}^{N-1} \left(\sum_{k=0}^{N-1} \Sigma_{xx,s(k)}^{-1}\right)^{-1} \Sigma_{xx,s(m)}^{-1} \left(y(m) - \mu_{x,s(m)}\right) \tag{14.72}$$

Note that when all state observation covariance matrices are identical the channel estimate becomes

$$\hat{h}_{ML}(Y,s) = \frac{1}{N} \sum_{m=0}^{N-1} \left(y(m) - \mu_{x,s(m)}\right) \tag{14.73}$$

The ML estimate of Eq. (14.73) is based on the ML state sequence s of M_i. In the following section we consider the conditional mean estimate over all state sequences of a model.

14.4.5 MAP Channel Estimate Based on HMMs

The conditional pdf of the channel averaged over all HMMs can be expressed as

$$
f_{H|Y}(h|Y) = \sum_{i=1}^{V} \sum_{s} f_{H|Y,S,\mathcal{M}}(h|Y,s,\mathcal{M}_i) P_{S|\mathcal{M}_i}(s|\mathcal{M}_i) P_{\mathcal{M}}(\mathcal{M}_i) \tag{14.74}
$$

where $P_{\mathcal{M}}(\mathcal{M}_i)$ is the prior pmf of the input words. Given a sequence of N P-dimensional observation vectors $Y=[y(0), ..., y(N-1)]$ the posterior pdf of the channel h along a state sequence s of an HMM M_i, is defined as

$$
f_{Y|H,S,\mathcal{M}}(h|Y,s,\mathcal{M}_i) = \frac{1}{f_Y(Y)} f_{Y|H,S,\mathcal{M}}(Y|h,s,\mathcal{M}_i) f_H(h)
$$

$$
= \frac{1}{f_Y(Y)} \prod_{m=0}^{N-1} \frac{1}{(2\pi)^P |\Sigma_{xx,s(m)}|^{1/2} |\Sigma_{hh}|^{1/2}} \exp\left(-\frac{1}{2}\left(y(m)-h-\mu_{x,s(m)}\right)^T \Sigma_{xx,s(m)}^{-1}\left(y(m)-h-\mu_{x,s(m)}\right)\right) \times
$$

$$
\exp\left(-\frac{1}{2}(h-\mu_h)^T \Sigma_{hh}^{-1}(h-\mu_h)\right) \tag{14.75}
$$

where it is assumed that each state of the HMM has a Gaussian distribution with a mean vector $\mu_{x,s(m)}$ and a covariance matrix $\Sigma_{xx,s(m)}$, and that the channel h is also Gaussian distributed with a mean vector μ_h and a covariance matrix Σ_{hh}. The MAP estimate along state s, in the left hand side of Eq. (14.75), can be obtained as

$$
\hat{h}^{MAP}(Y,s,\mathcal{M}_i) = \sum_{m=0}^{N-1} \left(\sum_{k=0}^{N-1}\left(\Sigma_{xx,s(k)}^{-1}+\Sigma_{hh}^{-1}\right)\right)^{-1} \Sigma_{xx,s(m)}^{-1}\left(y(m)-\mu_{x,s(m)}\right) +
$$

$$
\left(\sum_{k=0}^{N-1}\left(\Sigma_{xx,s(k)}^{-1}+\Sigma_{hh}^{-1}\right)\right)^{-1} \Sigma_{hh}^{-1} \mu_h \tag{14.76}
$$

The MAP estimate of the channel over all state sequences of all HMMs can be obtained as

$$
\hat{h}(Y) = \sum_{i=1}^{V} \sum_{S} \hat{h}^{MAP}(Y,s,\mathcal{M}_i) P_{S|\mathcal{M}}(s|\mathcal{M}_i) P_{\mathcal{M}}(\mathcal{M}_i) \tag{14.77}
$$

14.4.6 Implementations of HMM-based Deconvolution

In this section we consider three implementation methods for HMM-based channel equalisation.

I Use of the Statistical Averages Over All HMMs

A simple approach to blind equalisation, similar to that proposed by Stockham, is to use as the channel input statistics the average of the mean vectors and the covariance matrices, taken over all the states of all the HMMs as

$$
\mu_{\mathbf{x}} = \frac{1}{V N_s} \sum_{i=1}^{V} \sum_{j=1}^{N_s} \mu_{\mathcal{M}_i, j}
$$

$$
\Sigma_{\mathbf{xx}} = \frac{1}{V N_s} \sum_{i=1}^{V} \sum_{j=1}^{N_s} \Sigma_{\mathcal{M}_i, j}
$$

(14.78)

where $\mu_{\mathcal{M}_i, j}$ and $\Sigma_{\mathcal{M}_i, j}$ are the mean and the covariance of the jth state of the ith HMM, V and N_s denote the number of models and number of states per model respectively. The maximum likelihood estimate of the channel, $\hat{h}^{ML}$, is defined as

$$
\hat{h}^{ML} = \left(\bar{y} - \mu_{\mathbf{x}} \right)
$$

(14.79)

where $\bar{y}$ is the time-averaged channel output. The estimate of the channel input is

$$
\hat{x}(m) = y(m) - \hat{h}^{ML}
$$

(14.80)

Using the averages over all states and models, the MAP channel estimate becomes

$$
\hat{h}^{MAP}(Y) = \sum_{m=0}^{N-1} \left(\Sigma_{xx} + \Sigma_{hh} \right)^{-1} \Sigma_{hh} \left(y(m) - \mu_x \right) + \left(\Sigma_{xx} + \Sigma_{hh} \right)^{-1} \Sigma_{xx} \, \mu_h
$$

(14.81)

II Hypothesised-Input HMM Equalisation

In this method, for each candidate HMM in the input vocabulary, a channel estimate is obtained and used to equalise the channel output, prior to computation of a likelihood score for the HMM. Thus a channel estimate $\hat{h}_w$ is based on the hypothesis that the input word is w. It is expected that a better channel estimate is obtained from

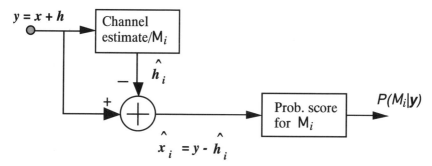

$y = x + h$

Figure 14.9 Hypothesised channel estimation procedure.

the correctly hypothesised HMM, and a poorer estimate from an incorrectly hypothesised HMM. The hypothesised-input HMM algorithm is as follows :

For $i = 1$ to number of words V {

   step 1 Using each HMM, M_i, make an estimate of the channel, $\hat{h}_i$,

   step 2 Using the channel estimate, $\hat{h}_i$, estimate the channel input $\hat{x}(m) = y(m) - \hat{h}_i$

   step 3 Compute a probability score for model M_i, given the estimate $[\hat{x}(m)]$. }

Select the channel estimate associated with the most probable word.

Figure 14.10 shows the ML channel estimates of two channels using unweighted average and hypothesised input methods.

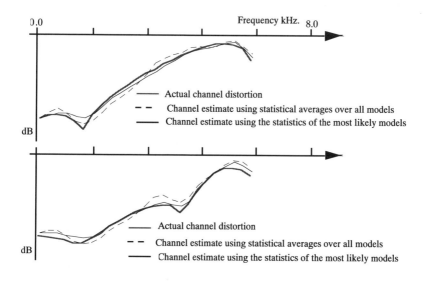

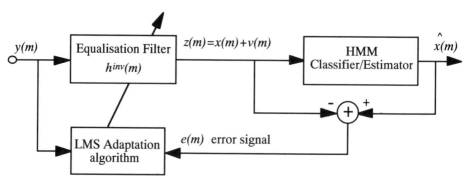

Figure 14.11 A decision-directed equaliser.

III   Decision-Directed Equalisation

Blind adaptive equalisers are composed of two distinct sections : an adaptive linear equaliser followed by a nonlinear estimator to improve the equaliser output. The output of the nonlinear estimator is the final estimate of the channel input, and is used as the desired signal *to direct* the equaliser adaptation. The use of the output of the nonlinear estimator as the desired signal, assumes that the linear equalisation filter removes a large part of the channel distortion, thereby enabling the nonlinear estimator to produce an accurate estimate of the channel input. A method of ensuring that the equaliser locks into, and cancels a large part of the channel distortion is to use a start up, equaliser training, period during which a known signal is transmitted.

Figure 14.11 illustrates a blind equaliser incorporating an adaptive linear filter followed by a hidden Markov model classifier/estimator. The HMM classifies the output of the filter as one of a number of likely signals and provides an enhanced output which is also used for adaptation of the linear filter. The output of the equaliser $z(m)$ is expressed as sum of the input to the channel $x(m)$ and a so called convolutional noise term $v(m)$ as

$$z(m) = x(m) + v(m) \tag{14.82}$$

The HMM may incorporate state based Wiener filters for suppression of the convolutional noise $v(m)$ as described in Section 4.5. Assuming that the LMS adaptation method is employed, the adaptation of the equaliser coefficient vector is governed by the following recursive equation:

$$\hat{h}^{inv}(m) = \hat{h}^{inv}(m-1) + \mu\, e(m) y(m) \tag{14.83}$$

where $\hat{h}^{inv}(m)$ is an estimate of the optimal inverse channel filter, μ is the adaptation step size and the error signal $e(m)$ is defined as

$$e(m) = \hat{x}^{HMM}(m) - z(m) \qquad (14.84)$$

where $\hat{x}^{HMM}(m)$ is the output of the HMM based estimator and is used as correct estimate of the desired signal to direct the adaptation.

14.5 Blind Equalisation for Digital Communication Channels

High speed transmission of digital data over analog channels, such as a telephone line or a radio channel, requires adaptive equalisation to reduce decoding errors caused by channel distortions. In telephone lines the channel distortions are due to the nonflat magnitude response and the nonlinear phase response of the lines. In radio channels environments the distortions are due to the effects of muti-path propagation of the radio waves via a multitude of different routes with different attenuations and delays. In general, the main types of distortions suffered by transmitted symbols are amplitude distortion, time dispersion, and fading. Of these, time dispersion is perhaps the most important and has received a great deal of attention. Time dispersion has the effect of smearing and elongating the duration of each symbol. In high speed communication systems, where the data symbols closely follow each other, time dispersion results in an overlap of successive symbols, an effect known as inter-symbol-interference (ISI) illustrated in Figure 14.12.

In a digital communication system, the transmitter modem takes N bits of binary data at a time, and encodes them into one of 2^N analog symbols for transmission, at the signalling rate, over an analog channel. At the receiver the analog signal is sampled and decoded into the required digital format. Most digital modems are based on multi-level phase shift keying, or combined amplitude and phase shift keying schemes. In this section we consider multi-level pulse amplitude modulation (M-ary PAM) as a convenient scheme for the study of adaptive equalisation.

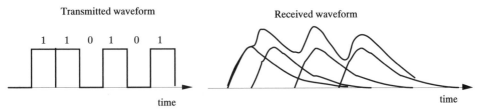

Figure 14.12 Illustration of intersymbol interference in a binary pulse amplitude modulation system.

Assume that at the transmitter modem, the kth set of N binary digits is mapped into a pulse of duration T_s seconds and an amplitude $a(k)$. Thus the modulator output signal, which is the input to the communication channel, is given as

$$x(t) = \sum_k a(k)r(t - kT_s) \tag{14.85}$$

where $r(t)$ is a pulse of duration T_s seconds and the amplitude $a(k)$ can assume one of $M=2^N$ distinct levels. Assuming that the channel is linear, the channel output can be modelled as the convolutional of the input signal and channel response as

$$y(t) = \int_{-\infty}^{\infty} h(\tau)x(t - \tau)d\tau \tag{14.86}$$

where $h(t)$ is the channel impulse response. The sampled channel output is given by

$$y(m) = \sum_k h_k x(m - k) \tag{14.87}$$

To remove the channel distortion, the sampled channel output $y(m)$ is passed to an equaliser with impulse response $\hat{h}_k^{inv}$. The equaliser output $z(m)$ is given as

$$\begin{aligned} z(m) &= \sum_k \hat{h}_k^{inv} y(m - k) \\ &= \sum_j x(m - j) \sum_k \hat{h}_k^{inv} h_{j-k} \end{aligned} \tag{14.88}$$

where Eq. (14.87) is used to obtain the second line of Eq. (14.88). The ideal equaliser output is $z(m)=x(m-D)=a(m-D)$ for some delay D that depends on the channel and the length of the equaliser. From Eq. (14.88), the channel distortion would be cancelled if

$$h_m^c = h_m * \hat{h}_m^{inv} = \delta(m - D) \tag{14.89}$$

where h_m^c is the combined impulse response of the cascade of the channel and the equaliser. A particular form of channel equaliser, for the elimination of ISI, is the Nyquist's *zero-forcing* filter where the impulse response of the combined channel and equaliser is defined as

$$h^c(kT_s + D) = \begin{cases} 1 & k = 0 \\ 0 & k \neq 0 \end{cases} \tag{14.90}$$

Note that in Eq. (14.90), at the sampling instances the channel distortion is cancelled, and hence there is no ISI at the sampling instances. A function that satisfies Eq. (14.90) is the sinc function $h^c(t) = \sin(\pi f_s t)/\pi f_s t.$, where $f_s = 1/T_s$. Zero forcing methods are sensitive to deviations of $h_e(t)$ from the requirement of Eq. (14.90), and also to the jitters in the synchronisation and the sampling process.

In this section we consider the more general form of the LMS based adaptive equaliser followed by a nonlinear estimator. In a conventional sample-adaptive filter, the filter coefficients are adjusted to minimise the mean squared distance between the filter output and the desired signal. In blind equalisation the desired signal (which is the channel input) is not available. The use of an adaptive filter, for blind equalisation, requires an internally generated desired signal as illustrated in Figure 14.13. Digital blind equalisers are composed of two distinct sections : (a) an adaptive equaliser that removes a large part of the channel distortion, followed by (b) a nonlinear estimator for an improved estimate of the channel input. The output of the nonlinear estimator is the final estimate of the channel input, and is used as the desired signal *to direct* the equaliser adaptation. A method of ensuring that the equaliser removes a large part of the channel distortion is to use a start up, equaliser training, period during which a known signal is transmitted.

Assuming that the LMS adaptation method is employed, the adaptation of the equaliser coefficient vector is governed by the following recursive equation

$$\hat{h}^{inv}(m) = \hat{h}^{inv}(m-1) + \mu e(m) y(m) \tag{14.91}$$

where $\hat{h}^{inv}(m)$ is an estimate of the optimal inverse channel filter h^{inv}, the scalar μ is the adaptation step size, and the error signal $e(m)$ is defined as

$$\begin{aligned} e(m) &= \psi(z(m)) - z(m) \\ &= \hat{x}(m) - z(m) \end{aligned} \tag{14.92}$$

where $\hat{x}(m) = \psi(z(m))$ is a nonlinear estimate of the channel input. For example in a binary communication system with an input alphabet $\{\pm a\}$ we can use a signum non-linearity such that $\hat{x}(m) = a.sgn(z(m))$ where the function $sgn()$ gives the sign of the argument. In the followings we use a Bayesian framework to formulate the nonlinear estimator $\psi()$.

Assuming that the channel input is an uncorrelated process and the equaliser removes a large part of the channel distortion, the equaliser output can be expressed as the sum of the desired signal (the channel input) plus an uncorrelated additive noise term as

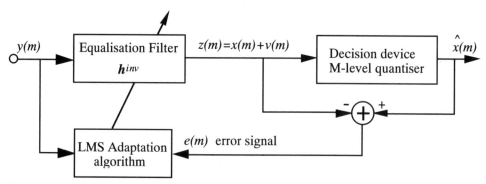

Figure 14.13 Configuration of an adaptive channel equaliser with an estimate of the channel input used as an 'internally' generated desired signal.

$$z(m) = x(m) + v(m) \tag{14.93}$$

where $v(m)$ is the so called convolutional noise defined as

$$
\begin{aligned}
v(m) &= x(m) - \sum_k \hat{h}_k^{inv} y(m-k) \\
&= \sum_k (h_k^{inv} - \hat{h}_k^{inv}) y(m-k)
\end{aligned}
\tag{14.94}
$$

In the followings we assume that the nonlinear estimates of the channel input are correct, and hence the error signals $e(m)$ and $v(m)$ are identical. Due to the averaging effect of the channel and the equaliser each sample of convolutional noise is affected by many samples of the input process. From the central limit theorem the convolutional noise $e(m)$ can be modelled by a zero mean Gaussian process as

$$f_E(e(m)) = \frac{1}{\sqrt{2\pi}\,\sigma_e} \exp\left(-\frac{e^2(m)}{2\sigma_e^2}\right) \tag{14.95}$$

where σ_e^2, the noise variance, can be estimated using the recursive time-update equation

$$\sigma_e^2(m) = \rho\,\sigma_e^2(m-1) + (1-\rho)e^2(m) \tag{14.96}$$

where $\rho < 1$ is the adaptation factor. The Bayesian estimate of the channel input given the equaliser output can be expressed in a general form as

$$\hat{x}(m) = \underset{\hat{x}(m)}{\arg\min} \int_X C(x(m), \hat{x}(m)) f_{X|Z}(x(m)|z(m)) \, dx(m) \qquad (14.97)$$

where $C(x(m), \hat{x}(m))$ is a cost function and $f_{X|Z}(x(m)|z(m))$ is the posterior pdf of the channel input signal. The choice of the cost function determines the type of the estimator as described in Chapter 3. Using a uniform cost function in Eq. (14.97) yields the maximum a posterior (MAP) estimate

$$
\begin{aligned}
\hat{x}^{MAP}(m) &= \underset{x(m)}{\arg\max} \; f_{X|Z}(x(m)|z(m)) \\
&= \underset{x(m)}{\arg\max} \; f_E(z(m) - x(m)) P_X(x(m))
\end{aligned}
\qquad (14.98)
$$

Now consider an M-ary pulse amplitude modulation system, and let $\{a_i \; i=1, ..., M\}$ denote the set of M pulse amplitudes with a probability mass function

$$P_X(x(m)) = \sum_{i=1}^{M} P_i \delta(x(m) - a_i) \qquad (14.99)$$

The pdf of the equaliser output $z(m)$ can be expressed as the mixture pdf

$$f_Z(z(m)) = \sum_{i=1}^{M} P_i \, f_E(x(m) - a_i) \qquad (14.100)$$

The posterior density of the channel input is

$$P_{X|Z}(x(m) = a_i|z(m)) = \frac{1}{f_Z(z(m))} f_E(z(m) - a_i) P_X(x(m) = a_i) \qquad (14.101)$$

and the MAP estimate is obtained from

$$\hat{x}^{MAP}(m) = \underset{a_i}{\arg\max}(f_E(z(m) - a_i) P_X(x(m) = a_i)) \qquad (14.102)$$

Note that the classification of the continuous-valued equaliser output $z(m)$ into one of M discrete channel input symbols is basically a nonlinear process. Substitution of the zero mean Gaussian model for the convolutional noise $e(m)$ in Eq. (102) yields

$$\hat{x}^{MAP}(m) = \underset{a_i}{\arg\max}\left(P_X(x(m) = a_i)\exp\left(-\frac{(z(m) - a_i)^2}{2\sigma_e^2}\right)\right) \qquad (14.103)$$

Note that when the symbols are equi-probable, the MAP estimate reduces to a simple threshold decision device.

Figure 14.13 shows a channel equaliser followed by an M-level quantiser. In this system , the output of the equaliser filter is passed to an M-ary decision circuit. The decision device, which is essentially an M-level quantiser, classifies the channel output into one of M valid symbols. The output of the decision device is taken as an internally generated desired signal to direct the equaliser adaptation.

Equalisation of a binary digital channel

Consider a binary PAM communication system with an input symbol alphabet $\{a_0, a_1\}$ and symbol probability $P(a_0)=P_0$ and $P(a_1)=P_1=1-P_0$. The pmf of the amplitude of the channel input signal can be expressed as

$$P(x(m)) = P_0\delta(x(m) - a_0) + P_1\delta(x(m) - a_1) \tag{14.104}$$

Assume that at the output of the linear adaptive equaliser Figure 14.13, the convolutional noise $v(m)$ is a zero-mean Gaussian process with variance σ_v^2. Therefore the pdf of the equaliser output $z(m)=x(m)+v(m)$ is a mixture of two Gaussian pdfs and can be described as

$$f_Z(z(m)) = \frac{P_0}{\sqrt{2\pi}\,\sigma_v}\exp\left(-\frac{(z(m) - a_0)^2}{2\sigma_v^2}\right) + \frac{P_1}{\sqrt{2\pi}\,\sigma_v}\exp\left(-\frac{(z(m) - a_1)^2}{2\sigma_v^2}\right) \tag{14.105}$$

The MAP estimate of the channel input signal is

$$\hat{x}(m) = \begin{cases} a_0 & if \;\; \dfrac{P_0}{\sqrt{2\pi}\sigma_v}\exp\left(-\dfrac{(z(m)-a_0)^2}{2\sigma_v^2}\right) < \dfrac{P_1}{\sqrt{2\pi}\sigma_v}\exp\left(-\dfrac{(z(m)-a_1)^2}{2\sigma_v^2}\right) \\ a_1 & otherwise \end{cases} \tag{14.106}$$

For the case when the channel alphabet consists of $a_0=-a$, $a_1=a$ and $P_0=P_1$, the MAP estimator is identical to the function $sgn(x(m))$, and the error signal is given by

$$e(m) = z(m) - sgn(z(m))a \tag{14.107}$$

Figure 14.12 shows the error signal as a function of $z(m)$. An undesirable property of a hard nonlinearity, such as the $sgn()$ function, is that it produces a large error signal at those instances when $z(m)$ is around zero, and a decision based on the sign of $z(m)$ is most likely to be incorrect. A large error signal based on an incorrect decision would have an unsettling effect on the convergence of the adaptive equaliser.

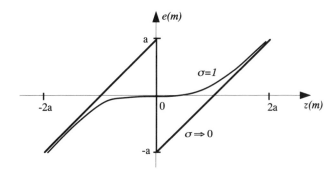

Figure 14.14 Comparison of the error functions produced by the hard nonlinearity of a sign function Eq. (14.107) and the soft nonlinearity of Eq. (14.108).

It is desirable to have an error function that produces small error signals when $z(m)$ is around zero. Nowlan and Hinton proposed a soft non-linearity of the following form

$$e(m) = z(m) - \frac{e^{2az(m)/\sigma^2} - 1}{e^{2az(m)/\sigma^2} + 1} a \qquad (14.108)$$

As shown in Figure 14.14 the error $e(m)$ is small when magnitude of $z(m)$ is small and large when magnitude of $z(m)$ is large.

14.6 Equalisation Based on Higher Order Statistics

The second order statistics of a random process, namely the autocorrelation or its Fourier transform the power spectrum, are central to linear estimation theory and form the basis of most statistical signal processing methods such as Wiener filters and linear predictive models. An attraction of the correlation function is that a Gaussian process, of a known mean vector, can be completely described in terms of the covariance matrix, and many random processes can be well characterised by Gaussian or mixture Gaussian models. A shortcoming of second order statistics is that they do not include the phase characteristics of the process. Therefore, given the channel output, it is not possible to estimate the channel phase from the second order statistics. Furthermore, as a Gaussian process of a known mean depends entirely on the autocovariance function, it follows that blind deconvolution, based on a Gaussian model of the channel input, can not estimate the channel phase.

Higher order statistics, and the probability models based on them, can model both the magnitude and the phase characteristics of a random process. In this section we consider blind deconvolution based on higher order statistics and their Fourier transforms known as the higher order spectra. The prime motivation in using the higher order statistics is their ability to model the phase characteristics. Further motivations are the potential of the higher order statistics to model channel nonlinearities, and to estimate a non-Gaussian signal in a high level of Gaussian noise.

14.6.1 Higher Order Moments, Cumulants and Spectra

The kth order moment of a random variable X is defined as

$$
\begin{aligned}
m_k &= \mathcal{E}[x^k] \\
&= (-j)^k \left. \frac{\partial^k \Phi_X(\omega)}{\partial \omega^k} \right|_{\omega=0}
\end{aligned}
\tag{14.109}
$$

where $\Phi_X(\omega)$ is the characteristic function of the random variable X defined as

$$
\begin{aligned}
\Phi_X(\omega) &= \mathcal{E}[\exp(j\omega x)] \\
&= \int_{-\infty}^{\infty} f_X(x) e^{j\omega x} \, dx
\end{aligned}
\tag{14.110}
$$

Where $f_X(x)$ is the pdf of X. Note from Eq. (14.110) that $\Phi_X(\omega)$ is the complex conjugate Fourier transform of the pdf $f_X(x)$. From Eqs. (14.109) and (14.110) the first moment of X is $m_1=E[x]$, the second moment of X is $m_2=E[x^2]$ and so on. The joint k^{th} $(k=k_1+k_2)$ order moment of two random variables X_1 and X_2 is defined as

$$
\mathcal{E}[x_1^{k_1} x_2^{k_2}] = (-j)^{k_1+k_2} \left. \frac{\partial^{k_1} \partial^{k_2} \Phi_{X_1 X_2}(\omega_1, \omega_2)}{\partial \omega_1^{k_1} \partial \omega_2^{k_2}} \right|_{\omega_1=\omega_2=0}
\tag{14.111}
$$

And in general the joint k^{th} order moment of N random variables is defined as

$$
\begin{aligned}
m_k &= \mathcal{E}[x_1^{k_1} x_2^{k_2} \ldots x_N^{k_N}] \\
&= (-j)^k \left. \frac{\partial^k \Phi(\omega_1, \omega_2, \ldots, \omega_N)}{\partial \omega_1^{k_1} \partial \omega_2^{k_2} \cdots \partial \omega_N^{k_N}} \right|_{\omega_1=\omega_2=\cdots=\omega_N=0}
\end{aligned}
\tag{14.112}
$$

where $k=k_1+k_2+\ldots + k_N$ and the joint characteristic function is

$$
\Phi(\omega_1, \omega_2, \ldots, \omega_N) = \mathcal{E}\left[\exp\left(j\omega_1 x_1 + \omega_2 x_2 + \cdots + \omega_N x_N\right)\right]
\tag{14.113}
$$

A discrete-time random process generates a time series of random variables. Hence, the kth order moment of a random process $x(m)$ is defined as

$$m_x(\tau_1, \tau_2, \ldots, \tau_{K-1}) = \mathcal{E}[x(m), x(m + \tau_1)x(m+, \tau_2) \cdots x(m + \tau_{k-1})] \quad (14.114)$$

Note that the second order moment $E[x(m)x(m+\tau)]$ is the autocorrelation function.

Cumulants

Cumulants are similar to moments, the difference is that the moments of a random process are derived from the characteristic function $\Phi_X(\omega)$, whereas the cumulant generating function $C_X(\omega)$ is defined as the logarithm of the characteristic function as

$$C_X(\omega) = \ln \Phi_X(\omega) = \ln \mathcal{E}[\exp(j\omega x)] \quad (14.115)$$

Using a Taylor's series expansion of the term $E[\exp(j\omega x)]$ in Eq. (14.115) the cumulant generating function can be expanded as

$$C_X(\omega) = \ln\left(1 + m_1(j\omega) + \frac{m_2}{2!}(j\omega)^2 + \frac{m_3}{3!}(j\omega)^3 + \cdots + \frac{m_n}{n!}(j\omega)^n + \cdots\right) \quad (14.116)$$

where $m_k = E[x^k]$ is the kth moment of the random variable x. The kth order cumulants of a random variable is defined as

$$c_k = (-j)^k \left. \frac{\partial^k C_X(\omega)}{\partial \omega^k} \right|_{\omega=0} \quad (14.117)$$

from Eq. (14.116) and (14.117) we have

$$c_1 = m_1 \quad (14.118)$$
$$c_2 = m_2 - m_1^2 \quad (14.119)$$
$$c_3 = m_3 - 3m_1 m_2 + 2m_1^2 \quad (14.120)$$

and so on. The general form of the kth order $(k=k_1+k_2+\ldots+k_N)$ joint cumulant generating function is

$$c_{k_1 \cdots k_N} = (-j)^{k_1 + \cdots + k_N} \left. \frac{\partial^{k_1 + \cdots + k_N} \ln \Phi_X(\omega_1, \cdots, \omega_N)}{\partial \omega_1^{k_1} \cdots \partial \omega_N^{k_N}} \right|_{\omega_1 = \omega_2 = \cdots = \omega_N = 0} \quad (14.121)$$

The cumulants of a zero mean random process $x(m)$ are given as

$$c_x = \mathcal{E}[x(k)] = m_x = 0 \qquad \text{mean} \qquad (14.122)$$

$$c_x(k) = \mathcal{E}[x(m)x(m+k)] - \mathcal{E}[x(m)]^2 = m_x(k) - m_x^2 = m_x(k) \qquad \text{covariance} (14.123)$$

$$\begin{aligned} c_x(k_1,k_2) &= m_x(k_1,k_2) - m_x[m_x(k_1) + m_x(k_2) + m_x(k_2 - k_1)] + 2(m_x)^3 \\ &= m_x(k_1,k_2) \end{aligned} \qquad \text{skewness}$$

$$(14.124)$$

$$c_x(k_1,k_2,k_3) = m_x(k_1,k_2,k_3) - m_x(k_1)m_x(k_3 - k_2) - m_x(k_2)m_x(k_3 - k_1) - m_x(k_3)m_x(k_2 - k_1) \qquad (14.125)$$

and so on. Note that $m_x(k_1, k_2, ..., k_N) = E\ [x(m)x(m+k_1), x(m+k_2), ..., x(m+k_N)]$. The general formulation of the kth order cumulant of a random process $x(m)$, [Rosenblatt], is defined as

$$c_x(k_1,k_2,...,k_n) = m_x(k_1,k_2,...,k_n) - m_x^G(k_1,k_2,...,k_n) \qquad (14.126)$$
$$\text{for } n = 3, 4, ...$$

where $m_x^G(k_1,k_2,...,k_n)$ is the kth order moment of a Gaussian process having the same mean and autocorrelation as the random process $x(m)$. From Eq. (14.126) it follows that for a Gaussian process the cumulants of order greater than 2 are identically zero.

Higher Order Spectra

The kth order spectrum of a signal $x(m)$ is defined as the $(k-1)$-dimensional Fourier transform of the kth order cumulant sequence as

$$C_X(\omega_1,...,\omega_{k-1}) = \frac{1}{(2\pi)^{k-1}} \sum_{\tau_1=-\infty}^{\infty} \cdots \sum_{\tau_{k-1}=-\infty}^{\infty} c_x(\tau_1, ..., \tau_{k-1}) e^{-j(\omega_1\tau_1 + \cdots + \omega_{k-1}\tau_{k-1})}$$

$$(14.127)$$

for the case when $k=2$, the second order spectrum is the power spectrum given as

$$C_X(\omega) = \frac{1}{2\pi} \sum_{\tau=-\infty}^{\infty} c_x(\tau)\, e^{-j\omega\tau} \qquad (14.128)$$

The *bi-spectrum* is defined as

$$C_X(\omega_1, \omega_2) = \frac{1}{(2\pi)^2} \sum_{\tau_1 = -\infty}^{\infty} \sum_{\tau_2 = -\infty}^{\infty} c_x(\tau_1, \tau_2) e^{-j(\omega_1 \tau_1 + \omega_2 \tau_2)} \qquad (14.129)$$

and the *tri-spectrum* is

$$C_X(\omega_1, \omega_2, \omega_3) = \frac{1}{(2\pi)^3} \sum_{\tau_1 = -\infty}^{\infty} \sum_{\tau_2 = -\infty}^{\infty} \sum_{\tau_3 = -\infty}^{\infty} c_x(\tau_1, \tau_2, \tau_3) e^{-j(\omega_1 \tau_1 + \omega_2 \tau_2 + \omega_3 \tau_3)} \quad (14.130)$$

Since the term $e^{j\omega t}$ is periodic with a period of 2π, it follows that higher order spectra are periodic in each ω_k with a period of 2π.

14.6.2 Higher Order Spectra of Linear Time-Invariant Systems

Consider a linear time-invariant system with an impulse response sequence $\{h_k\}$, input signal $x(m)$ and output $y(m)$. The relation between the kth order cumulant spectra of the input and output signals is given by

$$C_Y(\omega_1, \ldots, \omega_{k-1}) = H(\omega_1) H(\omega_{k-1}) H^*(\omega_1 + \cdots + \omega_{k-1}) \, C_X(\omega_1, \ldots, \omega_{k-1}) \quad (14.131)$$

where $H(\omega)$ is the frequency response of the linear system $\{h_k\}$. The magnitude of the kth order spectrum of the output signal is given as

$$|C_Y(\omega_1, \ldots, \omega_{k-1})| = |H(\omega_1)| \cdots |H(\omega_{k-1})| |H(\omega_1 + \cdots + \omega_{k-1})| \, |C_X(\omega_1, \ldots, \omega_{k-1})|$$
$$(14.132)$$

and the phase of the kth order spectrum is

$$\Phi_Y(\omega_1, \ldots, \omega_{k-1}) = \Phi_H(\omega_1) + \cdots + \Phi_H(\omega_{k-1}) - \Phi_H(\omega_1 + \cdots + \omega_{k-1}) + \Phi_X(\omega_1, \ldots, \omega_{k-1})$$
$$(14.133)$$

14.6.3 Blind Equalisation Based on Higher Order Cepstrum

In this section we consider blind equalisation of a maximum phase channel, based on higher order cepstrum. Assume that the channel can be modelled by an all-zero filter, and that its z-transfer function $H(z)$ can be expressed as the product of a maximum phase polynomial factor and a minimum phase factor as

$$H(z) = G H_{min}(z) H_{max}(z^{-1}) z^{-D} \tag{14.134}$$

$$H_{min}(z) = \prod_{i=1}^{P_1}(1 - \alpha_i z^{-1}) \qquad |\alpha_i| < 1 \tag{14.135}$$

$$H_{max}(z^{-1}) = \prod_{i=1}^{P_2}(1 - \beta_i z) \qquad |\beta_i| < 1 \tag{14.136}$$

where G is a gain factor, $H_{min}(z)$ is a minimum phase polynomial with all its zeros inside the unit circle, $H_{max}(z^{-1})$ is a maximum phase polynomial with all its zeros outside the unit circle, and z^{-D} inserts D unit delays in order to make Eq. (14.134) causal. The complex cepstrum of $H(z)$ is defined as

$$h_c(m) = Z^{-1}(\ln H(z)) \tag{14.137}$$

where Z^{-1} denotes the inverse z-transform. At $z=e^{j\omega}$, the z-transform is the discrete Fourier transform (DFT), and the cepstrum of a signal is obtained by taking the inverse DFT of the logarithm of the signal spectrum. In the followings we consider cepstrum based on the power spectrum and the higher order spectra and show that the higher order cepstra have the ability to retain maximum phase information. Assuming that the channel input $x(m)$ is a zero-mean uncorrelated process with variance σ_x^2, the power spectrum of the channel output can be expressed as

$$P_Y(\omega) = \frac{\sigma_x^2}{2\pi} H(\omega) H^*(\omega) \tag{14.138}$$

The cepstrum of the power spectrum of $y(m)$ is defined as

$$
\begin{aligned}
y_c(m) &= IDFT(\ln P_Y(\omega)) \\
&= IDFT\left(\ln\left(\sigma_x^2 G^2 / 2\pi\right) + \ln H_{min}(\omega) + H_{max}(-\omega) + \ln H^*_{min}(\omega) + H^*_{max}(-\omega)\right)
\end{aligned}
\tag{14.139}
$$

where IDFT is the inverse Discrete Fourier transform. Substituting Eq. (14.135-36) in (14.139), the cepstrum can be expressed as

$$
y_c(m) = \begin{cases}
\ln\left(G^2 \sigma_x^2 / 2\pi\right) & m = 0 \\
-(A(m) + B(m))/m & m > 0 \\
(A(-m) + B(-m))/m & m < 0
\end{cases}
\tag{14.140}
$$

where $A(m)$ and $B(m)$ are defined as

$$A(m) = \sum_{i=1}^{P_1} \alpha_i^m \qquad (14.141)$$

$$B(m) = \sum_{i=1}^{P_2} \beta_i^m \qquad (14.142)$$

Note from Eq. (14.140) that the along the index m, the maximum phase information $B(m)$ and the minimum phase information $A(m)$ overlap and can not be separated.

Bi-cepstrum

The bi-cepstrum of a signal is defined as the inverse Fourier transform of the logarithm of the bi-spectrum as

$$y_c(m_1, m_2) = IDFT_2[\log C_Y(\omega_1, \omega_2)] \qquad (14.143)$$

where $IDFT_2[.]$ denotes the 2-dimensional inverse discrete Fourier transform. The relationship between the bi-spectra of the input and output of a linear system is

$$C_Y(\omega_1, \omega_2) = H(\omega_1)H(\omega_2)H^*(\omega_1 + \omega_2)\, C_X(\omega_1, \omega_2) \qquad (14.144)$$

Assuming that the input $x(m)$ of the linear time invariant system $\{h_k\}$ is an uncorrelated non-Gaussian process, the bi-spectrum of the output can be written as

$$C_Y(\omega_1, \omega_2) = \frac{\gamma_x^{(3)}G^3}{(2\pi)^2} H_{min}(\omega_1)H_{max}(-\omega_1)H_{min}(\omega_2)H_{max}(-\omega_2) \times$$
$$H_{min}^*(\omega_1 + \omega_2)H_{max}^*(-\omega_1 - \omega_2) \qquad (14.145)$$

where $\gamma_x^{(3)}/(2\pi)^2$ is the third-order cumulant of the uncorrelated random input process $x(m)$. Taking the logarithm of Eq. (14.145) yields

$$\ln C_y(\omega_1, \omega_2) = \ln|A| + \ln H_{min}(\omega_1) + \ln H_{max}(-\omega_1) + \ln H_{min}(\omega_2) + \ln H_{max}(-\omega_2)$$
$$\ln H_{min}^*(\omega_1 + \omega_2) + \ln H_{max}^*(-\omega_1 - \omega_2)$$
$$(14.146)$$

where $A = \gamma_x^{(3)}G^3/(2\pi)^2$. The bi-cepstrum is obtained through the inverse Discrete Fourier transform of Eq. (14.146) as

$$
y_c(m_1, m_2) = \begin{cases}
\ln|A| & m_1 = m_2 = 0 \\
-A(m_1)/m_1 & m_1 > 0, \ m_2 = 0 \\
-A(m_2)/m_2 & m_2 > 0, \ m_1 = 0 \\
-B(-m_1)/m_1 & m_1 < 0, \ m_2 = 0 \\
B(-m_2)/m_2 & m_2 < 0, \ m_1 = 0 \\
-B(m_2)/m_2 & m_1 = m_2 > 0 \\
A(-m_2)/m_2 & m_1 = m_2 < 0 \\
0 & otherwise
\end{cases}
\tag{14.147}
$$

Note from Eq. (14.147) that the maximum phase information $B(m)$ and the minimum phase information $A(m)$ are separated and appear in different regions of the bi-cepstrum indices m_1 and m_2.

The higher order cepstral coefficients can be obtained either from the IDFT of higher order spectra as in Eq. (14.147) or using parametric methods as follows. In general the cepstral and cumulant coefficients can be related by a convolutional equation. Pan and Nikias (1988) have shown that the recursive relation between the bi-cepstrum coefficients and the third order cumulants of a random process are

$$
y_c(m_1, m_2) * \left[-m_1 c_y(m_1, m_2) \right] = -m_1 c_y(m_1, m_2)
\tag{14.148}
$$

Substituting Eq. (14.147) in Eq. (14.148) yields

$$
\sum_{i=1}^{\infty} A^{(i)}[c_x(m_1 - i, m_2) - c_x(m_1 + i, m_2 + i)] + B^{(i)}[c_x(m_1 - i, m_2 - i) - c_x(m_1 + i, m_2)]
$$
$$
= -m_1 c_x(m_1, m_2)
$$

$$(14.149)$$

The truncation of the infinite summation in Eq. (14.149) provides an approximate equation as

$$
\sum_{i=1}^{P} A^{(i)}[c_x(m_1 - i, m_2) - c_x(m_1 + i, m_2 + i)] \ +
$$

$$(14.150)$$

$$
\sum_{i=1}^{Q} B^{(i)}[c_x(m_1 - i, m_2 - i) - c_x(m_1 + i, m_2)] \approx -m_1 c_x(m_1, m_2)
$$

Eq. (14.150) can be used to solve for the cepstral parameters $A(m)$ and $B(m)$.
Tri-cepstrum

The tri-cepstrum of a signal $y(m)$ is defined as the inverse Fourier transform of the tri-spectrum as

$$y_c(m_1, m_2, m_3) = IDFT_3[\ln C_Y(\omega_1, \omega_2, \omega_3)] \qquad (14.151)$$

where $IDFT_3[.]$ denotes the 3-dimensional inverse discrete Fourier transform. The tri-spectra of the input and output of the linear system is

$$C_Y(\omega_1, \omega_2, \omega_3) = H(\omega_1)H(\omega_2)H(\omega_3)H^*(\omega_1 + \omega_2 + \omega_3)\, C_X(\omega_1, \omega_2, \omega_3) \quad (14.152)$$

Assuming that the channel input $x(m)$ is uncorrelated Eq. (14.152) becomes

$$C_Y(\omega_1, \omega_2, \omega_3) = \frac{\gamma_x^{(4)} G^4}{(2\pi)^3} H(\omega_1)H(\omega_2)H(\omega_3)H^*(\omega_1 + \omega_2 + \omega_3) \qquad (14.153)$$

where $\gamma_x^{(4)}/(2\pi)^3$ is the fourth-order cumulant of the input signal. Taking the logarithm of the tri-spectrum gives

$$\ln C_Y(\omega_1, \omega_2, \omega_3) = \frac{\gamma_x^{(4)} G^4}{(2\pi)^3} + \ln H_{\min}(\omega_1) + \ln H_{\max}(-\omega_1) + \ln H_{\min}(\omega_2) + \ln H_{\max}(-\omega_2)$$

$$+ \ln H_{\min}(\omega_3) + \ln H_{\max}(-\omega_3) + \ln H^*_{\min}(\omega_1 + \omega_2 + \omega_3) + \ln H^*_{\max}(-\omega_1 - \omega_2 - \omega_3)$$
$$(14.154)$$

From Eqs. (14.151) and (14.154) we have

$$y_c(m_1, m_2, m_3) = \begin{cases} \ln A & m_1 = m_2 = m_3 = 0 \\ -A(m_1)/m_1 & m_1 > 0,\, m_2 = m_3 = 0 \\ -A(m_2)/m_2 & m_2 > 0,\, m_1 = m_3 = 0 \\ -A(m_3)/m_3 & m_3 > 0,\, m_1 = m_2 = 0 \\ B(-m_1)/m_1 & m_1 < 0,\, m_2 = m_3 = 0 \\ B(-m_2)/m_2 & m_2 < 0,\, m_1 = m_3 = 0 \\ B(-m_3)/m_3 & m_3 < 0,\, m_1 = m_2 = 0 \\ -B(m_2)/m_2 & m_1 = m_2 = m_3 > 0 \\ A(m_2)/m_2 & m_1 = m_2 = m_3 < 0 \\ 0 & otherwise \end{cases} \qquad (14.155)$$

where $A = \gamma_x^{(4)} G^4 / (2\pi)^3$. Note from Eq. (14.155) that the maximum phase information $B^{(m)}$ and the minimum phase information $A^{(m)}$ are separated and appear in different regions of the tri-cepstrum indices m_1, m_2, and m_3.

Calculation of Equaliser Coefficients from Tri-cepstrum

Assuming that the channel z-transfer function can be described by Eq. (14.134), the inverse channel can be written as

$$H^{inv}(z) = \frac{1}{H(z)} = \frac{1}{H_{min}(z) H_{max}(z^{-1})} = H_{min}^{inv}(z) H_{max}^{inv}(z^{-1}) \qquad (14.156)$$

where it is assumed that the channel gain G is unity. In the time domain Eq. (14.156) becomes

$$h^{inv}(m) = h_{min}^{inv}(m) * h_{max}^{inv}(m) \qquad (14.157)$$

Pan and Nikias (1988) describe an iterative algorithm for estimation of the truncated impulse response of the maximum phase and the minimum phase factors of the inverse channel transfer function. Let $\hat{h}_{min}^{inv}(i,m)$, $\hat{h}_{max}^{inv}(i,m)$ denote the estimates of the mth coefficients of the maximum phase and minimum phase parts of the inverse channel at the ith iteration . The Pan and Nikias algorithm is the following:
(a) Initialisation

$$\hat{h}_{min}^{inv}(i,0) = \hat{h}_{max}^{inv}(i,0) = 1 \qquad (14.158)$$

(b) Calculation of the minimum phase polynomial

$$\hat{h}_{min}^{inv}(i,m) = \frac{1}{m} \sum_{k=2}^{m+1} \hat{A}(k-1) \hat{h}_{min}^{inv}(i,m-k+1) \qquad i=1, ..., P_1 \qquad (14.159)$$

(c) Calculation of the maximum-phase polynomial

$$\hat{h}_{max}^{inv}(i,m) = \frac{1}{m} \sum_{k=m+1}^{0} \hat{B}(1-k) \hat{h}_{max}^{inv}(i,m-k+1) \qquad i=-1, ..., -P_2 \qquad (14.160)$$

The maximum-phase and the minimum-phase components of the inverse channel response are combined in Eq. (14.157) to give the inverse channel equaliser.

Summary

In this chapter we considered a number of different approaches to channel equalisation. The chapter began with an introduction to models for channel distortions, the definition of an ideal channel equaliser, and the problems that arise in channel equalisation due to noise and possible non-invertibility of the channel. In some problems such as speech recognition or restoration of distorted audio signals we are mainly interested in restoring the magnitude spectrum of the signal and phase restoration is not a primary objective. Whereas in other applications such as digital telecommunication the restoration of both the amplitude and the timing of the transmitted symbols are of interest, and hence we need to equalise for both the magnitude and the phase distortions.

In section 14.1 we considered the least squared error Wiener equaliser. The Wiener equaliser can only be used if we have access to the channel input or the cross correlation of the channel input and output signals.

For cases when a training signal can not be employed to identify the channel response, the channel input is recovered through a Blind equalisation method. Blind equalisation is feasible only if some statistics of the channel input signal is available. In Section 14.2 we considered blind equalisation using the power spectrum of the input signal. This method was introduced by Stockham for restoration of the magnitude spectrum of distorted acoustic recordings. In Section 14.3 we considered a blind deconvolution method based on the factorisation of a linear predictive model of the convolved signals.

The Bayesian inference provides a frame work for inclusion of the statistics of the channel input and perhaps also that of the channel environment. In Section 14.4 we considered the Bayesian equalisation methods and studied the case when the channel input are modelled by a set of hidden Markov models. Section 14.5 introduced channel equalisation methods for removal of intersymbol interference in digital telecommunication systems, and finally in Section 14.6 we considered the use of higher order spectra for equalisation of non-minimum phase channels.

Bibliography

BENVENISTE A., GOURSAT M., RUGET G. (1980), Robust Identification of a Nonminimum Phase System : Blind Adjustment of Linear Equaliser in Data Communications, IEEE Trans, Automatic Control, Vol AC-25, Pages 385-99.

BELLINI S. (1986), Bussgang Techniques for Blind Equalisation, IEEE GLOBECOM Conf. Rec., Pages 1634-40.

BELLINI S., ROCCA F (1988), Near Optimal Blind Deconvolution, IEEE Proc. Int. Conf. Acoustics, Speech, and Signal Processing., ICASSP-88, Pages 2236-39.

BELFIORE C. A., PARK J. H. (1979), Decision Feedback Equalisation, Proc. IEEE, Vol. 67, Pages 1143-56.

COWAN C. F. N., GIBSON G. J., SIU S., (1989), Data Equalisation using Highly Non-linear Adaptive Architectures, SPIE, San Diego, California.

GERSHO A (1969), Adaptive Equalisation of Highly Dispersive Channels for Data Transmission, Bell System Technical Journal , Vol. 48, Pages 55-70.

GODARD, D. N. (1974), Channel Equalisation using a Kallman Filter for Fast Data Transmission, IBM J. Res. Dev., Vol. 18, Pages 267-73.

GODARD, D. N. (1980), Self-recovering Equalisation and Carrier Tracking in a Two-Dimensional Data Communication System, IEEE Trans. Comm. , Vol. COM-28, Pages 1867-75.

HANSON B. A., APPLEBAUM T. H. (1993), Subband or Cepstral Domain Filtering for Recognition of Lombard and Channel-Distorted Speech, IEEE Int. Conf. Acoustics, Speech and Signal Processing Vol. II, Pages 79-82.

HARIHARAN S., CLARK A. P. (1990), HF Channel Estimation using a Fast Transversal Filter Algorithm, IEEE Trans. Acoustics, Speech and Signal Processing, Vol. 38, Pages 1353-62.

HATZINAKO S. D.(1990), Blind Equalisation Based on Polyspectra, Ph.D. Thesis, Northeastern University, Boston, Mass.

HERMANSKY H, MORGAN N (1992), Towards Handling the Acoustic Environment in Spoken Language Processing, Int. Conf. on Spoken Language Processing Tu.fPM.1.1, Pages 85-88.

LUCKY R. W. (1965), Automatic Equalisation of Digital Communications, Bell System technical Journal, Vol, 44, Pages 547-88.

LUCKY R. W. (1965), Techniques for Adaptive Equalisation of Digital Communication Systems, Bell System Technical Journal, Vol. 45, Pages 255-86.

MENDEL J. M. (1990), Maximum Likelihood Deconvolution : A Journey into Model Based Signal Processing, Springer-Verlag, New York.

MENDEL J. M. (1991), Tutorial on Higher Order Statistics (Spectra) in Signal Processing and System Theory : Theoretical results and Some Applications, Proc. IEEE, Vol. 79, Pages 278-305.

MOKBEL C., MONNE J. JOUVET D. (1993), On-Line Adaptation of A Speech Recogniser to Variations in Telephone Line Conditions, Proc. 3rd European Conf. On Speech Communication and Technoplogy, EuroSpeech-93, Vol. 2, Pages 1247-50.

MONSEN P. (1971), Feedback Equalisation for Fading Dispersive Channels, IEEE Trans. Information Theory, Vol. IT-17, Pages 56-64.

NIKIAS C. L., CHIANG H. H. (1991), Higher-Order Spectrum Estimation via Non-Causal Autoregressive Modeling and Deconvolution, IEEE Trans. Acoustics, Speech and Signal Processing Vol. ASSP-36, Pages 1911-13.

NOWLAN S. J., HINTON G. E. (1993), A Soft Decision-Directed Algorithm for Blind Equalisation IEEE Transactions on Communications, Vol. 41, No. 2, Pages 275-79.

PAN R., NIKIAS C. L. (1988), Complex Cepstrum of Higher Order Cumulants and Nonminimum Phase identification, IEEE Trans. Acoustics, Speech and Signal Processing, Vol. ASSP-36, Pages 186-205.

PICCHI G., PRATI G. (1987), Blind Equalisation and Carrier Recovery using a Stop-and-Go Decision-Directd Algorithm:, IEEE Trans. Commun, Vol. COM-35, Pages 877-87.

RAGHUVEER M.R., NIKIAS C.L. (1985), Bispectrum Estimation : A Parameteric Approach, IEEE Trans. Acoustics, Speech, and Signal Processing, Vol. ASSP-33, No. 5, Pages 35-48.

ROSENBLATT M. (1985), Stationary Sequences and Random Fields, Birkhauser, Boston, Mass.

SPENCER P.S , RAYNER P.J.W (1990) ,Separation of Stationary and Time-Varying Systems and Its Applications to the Restoration of Gramophone Recordings, Ph.D. Thesis Cambridge Universoty Engineering Department, Cambridge UK.

STOCKHAM T. G., CANNON T. M., INGEBRETSEN R.B (1975), Blind Deconvolutiopn Through Digital Signal Processing, IEEE Proc. Vol.63, No 4, Pages 678-92.

QURESHI S. U. (1985), Adaptive Equalisation Proceedings of the IEEE, Vol 73, No. 9, Pages 1349-87.

UNGERBOECK G. (1972), Theory on the Speed of Convergence in Adaptive Equalisers for Digital Communication, IBM J. Res. Dev., Vol. 16, Pages 546-55.

Frequently used Symbols and Abbreviations

$AWGN$	Additive White Gaussian Noise
$ARMA$	AutoRegressive MovingAverage Process
AR	AutoRegressive Process
A	Matrix of predictor coefficients
a_k	Linear predictor coefficients
a	Linear predictor coefficients vector
a_{ij}	Probability of transition from state i to state j in a Markov model
$\alpha_i(t)$	Forward probability in an HMM
$b(m)$	Backward prediction error
$b(m)$	Binary state signal
$\beta_i(t)$	Backward probability in an HMM
$c_{xx}(m)$	Covariance of signal $x(m)$
$c_{XX}(k_1,k_2,\cdots,k_N)$	K^{th} order cumulant of $x(m)$
$C_{XX}(\omega_1,\omega_2,\cdots,\omega_{K-1})$	K^{th} order cumulant spectra of $x(m)$
D	Diagonal matrix
$e(m)$	Estimation error
$E[]$	Expectation operator
f	Frequency variable
$f_X(x)$	Probability density function for process X
$f_{X,Y}(x,y)$	Joint probability density function of X and Y
$f_{X|Y}(x|y)$	Probability density function of X conditioned on Y
$f_{X;\Theta}(x;\theta)$	Probability density function of X with θ as a parameter
$f_{X|S,\mathcal{M}}(x|s,\mathcal{M})$	Probability density function of X given a state sequence s of an HMM M of the process X
$\Phi(m,m-1)$	State transition matrix in Kalman filter
G	Filter gain factor
h	Filter coefficient vector, Channel response
h_{max}	Maximum phase channel response
h_{min}	Minimum phase channel response
h^{inv}	Inverse channel response
$H(f)$	Channel frequency response
$H^{inv}(f)$	Inverse channel frequency response
H	Observation matrix, Distortion matrix

I	Identity matrix
J	Fishers information matrix
J	Jacobian of a transformation
$LSAR$	Least squares AR interpolation
$K(m)$	Kalman gain matrix
λ	Eigenvalue
Λ	Diagonal matrix of eigenvalues
MAP	Maximum a Posterior Estimate
MA	Moving Average process
ML	Maximum Likelihood Estimate
$MMSE$	Minimum Mean Squared Error Estimate
m	Discrete time index
m_k	K^{th} order moment
M	A model, e.g. an HMM
μ	Adaptation Convergence factor
μ_x	Expected mean of vector x
$n(m)$	Noise
$n(m)$	A noise vector of N samples
$n_i(m)$	Impulsive noise
$N(f)$	Noise spectrum
$N^*(f)$	Complex conjugate of $N(f)$
$\overline{N(f)}$	Time-averaged noise spectrum
$\mathcal{N}(x, \mu_{xx}, \Sigma_{xx})$	A Gaussian pdf with mean vector μ_{xx} and covariance matrix Σ_{xx}
P	Filter order (length)
pdf	Probability density function
pmf	Probability mass function
$P_X(x_i)$	Probability mass function of x_i
$P_{X,Y}(x_i, y_j)$	Joint probability mass function of x_i and y_j
$P_{X|Y}(x_i|y_j)$	Conditional probability mass function of x_i given y_j
$P_{NN}(f)$	Power spectrum of noise $n(m)$
$P_{XX}(f)$	Power spectrum of the signal $x(m)$
$P_{XY}(f)$	Cross power spectrum of signals $x(m)$ and $y(m)$
θ	Parameter vector
$\hat{\theta}$	Estimate of the parameter vector θ
r_k	Reflection coefficients
$r_{xx}(m)$	Autocorrelation function
$r_{xx}(m)$	Autocorrelation vector
R_{xx}	Autocorrelation matrix of signal $x(m)$

R_{xy}	Cross correlation matrix
s	State sequence
s^{ML}	Maximum likelihood state sequence
SNR	Signal to Noise Ratio
$SINR$	Signal to Impulsive Noise Ratio
σ_n^2	Variance of noise $n(m)$
Σ_{nn}	Covariance matrix of noise $n(m)$
Σ_{xx}	Covariance matrix of signal $x(m)$
σ_x^2	Variance of signal $x(m)$
σ_n^2	Variance of noise $n(m)$
$x(m)$	Clean signal
$\hat{x}(m)$	Estimate of clean signal
$x(m)$	Clean signal vector
$X(f)$	Frequency spectrum of the signal $x(m)$
$X^*(f)$	Complex conjugate of $X(f)$
$\overline{X(f)}$	Time-averaged frequency spectrum of the signal $x(m)$
$X(f,t)$	Time-Frequency spectrum of the signal $x(m)$
X	Clean signal matrix
X^H	Hermitian transpose of X
$y(m)$	Noisy signal
$y(m)$	Noisy signal vector
$\hat{y}(m \mid m-i)$	Prediction of $y(m)$ based on observations up to time m-i
Y	Noisy signal matrix
Y^H	Hermitian transpose of Y
Var	Variance
w_k	Wiener filter coefficients
$w(m)$	Wiener filter coefficients vector
$W(f)$	Wiener filter frequency response
z	z-transform variable

INDEX

A

Absolute value of error, 309
Acoustic feedbacks, 335
Adaptation formula, 171
Adaptation step size, 179, 333
Adaptive echo cancellation, 331
Adaptive filter, 371, 171
Adaptive noise cancellation and noise reduction, 5
Adaptive subtraction of noise pulses, 323
Aliasing, 19
Analog signals, 17
Analog to digital conversion, 16
AR-based restoration, 324
Auto-regressive process, 54
Auto-regressive-moving-average (ARMA) model, 226
Autocorrelation, 36, 219, 297
Autocorrelation of impulsive noise, 39
Autocorrelation of the output of a Linear time-invariant (LTI) system, 37
Autocorrelation of white noise, 38
Autocovariance, 37
Autoregressive, 88, 226
Autoregressive (AR) model, 25, 112, 279, 317
Autoregressive model of transient noise, 317
Autoregressive moving average (ARMA) power spectral estimation, 231
Autoregressive power spectrum estimation, 229

B

AWGN, 83
Backward predictor, 194
Backward probability, 123
Bartlett periodogram, 221
Baum-Welch model re-estimation, 124
Bayes rule, 28, 133, 204
Bayesian blind deconvolution and equalisation, 360

Bayesian estimation, 74
Bayesian MMSE, 79
Bayesian risk function, 74
Bayesian statistical signal processing, 4
Beamforming, 14
Bernoulli-Gaussian model, 299
Bi-cepstrum, 381
Bi-spectrum, 379
Bi-variate pdf, 30
Bias, 69
Binary-state model of impulsive noise, 300
Binary-state classifier, 8
Binary-state Gaussian process, 49
Blackman-Tukey method, 224
Blind deconvolution, 343
Blind equalisation, 347
Blind equalisation through model factorisation, 358
Block least squared error estimation, 146
Brownian motion, 25
Burg's method, 200

C

Central limit theorem, 43, 45, 372
Channel equalisation, 7, 343
Channel equalisation based on hidden
    Markov models, 362
Channel impulse response, 296
Channel response, 344
Characteristic function, 376
Classification, 96
Classification of discrete-valued
    parameters, 96
Clutters, 54
Coding of audio signals, 11
Coherence, 42
Complete data, 90
Conditional mean channel estimation,
    360
Conditional multi-variate Gaussian
    probability, 47
Conditional probability density, 30
Consistent estimator, 70
Continuous density HMMs, 118
Continuous density HMMs, 127
Continuous observation pdfs, 127
Continuous-valued random variables,
    29
Continuously variable state process,
    112
Convergence of line echo canceller,
    333
Convergence rate, 180
Convolutional noise, 372
Correlation subtraction, 207
Correlation-ergodic, 44, 145
Correlator, 13
Cost function, 309
Cost of error function, 74, 309
Cramer-Rao lower bound, 93
Cross correlation, 40
Cross covariance, 40
Cross power spectral density, 42
Cross-correlation function, 324
Cubic spline interpolation, 273
Cumulants, 377
Cumulative distribution function, 29

D

Decision-directed equalisation, 368
Decoding of signals, 129
Deconvolution, 344
Decorrelation filter, 193
Detection, 304
Detection of signals in noise, 13
Deterministic signals, 24
DFT, 256
Digital signal, 16
Directional reception of waves, 14
Discrete Fourier transform, 12, 217
Discrete observation density HMMs,
    125
Discrete state observation HMM, 118
Discrete-time stochastic process, 26
Discrete-valued random variable, 28
Distortion matrix, 165
Distribution function, 46
Divided differences, 272
Durbins algorithm, 198

E

Echo cancellation, 328
Echo canceller, 331
Echo suppresser, 330
Echo synthesiser, 339
Efficient estimator, 69
Eigenvalue, 179
Eigenvalue spread, 180
Eigenvalues, 179
EM algorithm, 91
Energy density spectrum, 216
Energy spectral density, 218
Entropy, 227
Equalisation, 344
Equalisation based on higher-order
    statistics, 375
Ergodic HMM, 115
Ergodic processes, 27
Ergodic random processes, 42
ESPIRIT algorithm, 238

Estimate-maximise (EM) method, 90
Estimation, 66
Estimation of the mean and variance of a Gaussian process, 76
Expected values, 35

F

Factorisation of linear prediction models, 358
Feedback coupling, 335
Finite state process, 112, 99
Fisher's information matrix, 96
Forgetting factor, 174
Forward predictor model, 193
Forward probability, 122
Forward-backward probability computation, 122
Fourier transform, 215
Fourier transform of a sampled signal, 216
Frequency domain signal restoration, 209
Frequency resolution, 217

G

Gauss-Markov process, 55
Gaussian pdf, 118
Gaussian process, 45
Gaussian-AR process, 88

H

Hard nonlinearity, 374
Hermite polynomials, 273
Hermitian transpose, 236
Hidden Markov models, 50, 111-136, 301, 318, 362
High resolution spectral estimation, 232
Higher-order moments, 376
Higher-order spectra, 378

HMM-based estimation of signals in noise, 133
HMM-based Wiener filters, 135
HMMs with mixture Gaussian pdfs, 127
Homogeneous Poisson process, 51
Homogenous Markov chain, 57
Homomorphic equalisation, 354, 356
Howling, 335
Huber's function, 310
Hypothesised-input HMM equalisation, 366

I

Ideal equaliser, 345
Ideal interpolation, 262
Impulsive noise, 294, 295
Impulsive noise detection, 304
Impulsive noise removal using linear prediction models, 304
Incomplete data, 90
Influence function, 309
Information, 2
Inhomgeneous Markov chains, 57
Innovation signal, 166, 187, 207
Inter-symbol-interference, 369
Interpolation, 261
Interpolation error, 285
Interpolation in frequency-time domain, 287
Interpolation through signal substitution, 291
Interpolation using adaptive code books, 289
Inverse discrete Fourier transform, 217
Inverse Fourier transform, 216
Inverse linear predictor, 192
Inverse-channel filter, 345

J

Jacobian, 61
Joint characteristic function, 376

Joint statistical averages of two random processes, 40

K

K-means algorithm, 103
Kalman filter, 165
Kalman filtering algorithm, 169
Kalman gain, 167
Kronecker delta function, 297

L

Lagrange interpolation, 269
Lattice predictors, 198
Leaky LMS algorithm, 182
Least squared ar (LSAR) interpolation, 282
Least squared error estimation, 141
Left-right hmm, 115
Levinson-Durbin algorithm, 196
Levinson-Durbins algorithm, 195
Line echoes, 329
Line-interpolator, 269
Linear array, 14
Linear least squared error filters, 140
Linear prediction, 186
Linear prediction models, 185, 356
Linear predictive models, 9
Linear time invariant channel, 157
Linear transformation, 62
Linear transformation of a Gaussian process, 62
LMS adaptation algorithm, 334
LMS filter, 181
Log-normal process, 59

M

M-ary PAM, 369
M-ary pulse amplitude modulation, 373
M-variate pdf, 30

Magnitude spectral subtraction, 245
Many-to-one mapping of random signals, 60
MAP estimation, 87
MAP estimation of predictor coefficients, 204
Marginal density, 54
Marginal probabilities, 49
Marginal probability mass functions, 28
Markov chain, 55
Markov process, 54
Markovian prior, 363
Markovian state transition prior, 117
Matched filter, 13, 320
Matrix inversion lemma, 174
Maximum a posterior (MAP) estimate, 75
Maximum a posterior channel estimation, 361
Maximum a posterior classification, 98
Maximum a posterior estimate, 206
Maximum a posterior interpolation, 277
Maximum entropy correlation, 227
Maximum entropy spectral estimation, 227
Maximum likelihood channel estimation, 361
Maximum likelihood classification, 98
Maximum likelihood estimation, 76
Maximum phase channel, 349, 379
Maximum phase information, 382
Mean squared error signal, 150
Mean value of a process, 35
Mean-ergodic, 43
Median, 81
Median filters, 302
Minimisation of backward and forward prediction error, 201
Minimum mean absolute value of error, 81
Minimum mean squared error, 143
Minimum mean squared error classification, 99
Minimum mean squared error estimation, 79
Minimum phase channel, 349
Minimum phase information, 382

Mixture Gaussian densities, 49
Mixture Gaussian density, 118
Mixture Gaussian model, 104
Mixture pdf, 373
Model order selection, 201
Model-based power spectrum
  estimation, 225
Modelling Noise, 136
Monotonic transformation, 58
Moving-Average, 226
Multi-variate Gaussian pdf, 47
Multi-variate probability mass
  functions, 30
MUSIC algorithm, 235
Musical noise, 249, 252

N

Neural networks, 4
Newton polynomials, 270
Noise detectability, 306
Noise detection based on HMM, 322
Noise detection based on inverse
  filtering, 321
Noise pulse templates, 317
Noisy speech model, 305
Non-linear spectral subtraction, 252
Non-parametric power spectrum
  estimation, 220
Non-parametric signal processing, 3
Non-stationary process, 35
Nonstationary process, 31, 112
Normal process, 45
Normalised least mean squared error,
  332
Nyquist sampling theorem, 19, 262

O

Observation equation, 165
Outlier, 303
Over-subtraction, 253

P

Parameter estimation, 68
Parameter space, 67
Parsevals theorem, 152, 216
Partial correlation, 198
Partial correlation (PACOR)
  coefficients, 200
Pattern recognition, 8
Performance measures, 69
Periodogram, 220
Pisarenko harmonic decomposition,
  232
Poisson process, 50
Poisson-Gaussian model, 300
Poles and zeros, 358
Polynomial interpolation, 268
Posterior pdf, 71
Posterior signal space 71
Power, 33
Power spectra, 353
Power spectral density, 38, 219
Power spectral subtraction, 245
Power spectrum, 152, 215, 220, 297
Power spectrum estimation, 214
Power spectrum of a white noise, 39
Power spectrum of impulsive noise.,
  39
Power spectrum subtraction, 246
Prediction error filter, 192
Prediction error signal, 193
Predictive model, 66
Principal eigenvectors., 237
Prior pdf, 72
Prior pdf of the predictor coefficients,
  206
Prior space of a signal, 71
Probability density function, 29
Probability mass function, 28
Probability models, 27
Processing distortions, 249, 251

Q

QR decomposition, 146
Quantisation, 17, 20

Quantisation noise, 20

R

Random signals, 24
Random variable, 27
Rayner, 358
Rearrangement matrices, 277
Recursive least squared error (RLS)
        filter, 172
Reducing the noise variance, 251
Reflection coefficient, 196, 198
RLS adaptation algorithm, 176
Robust estimator, 309
Rotation matrix, 239

S

Sample-and-hold, 19
Sampling, 16, 17
Sampling frequency, 19
Scalar Gaussian random variable, 46
Second order statistics, 38
Separability of signal and noise, 155
Short time Fourier transform (STFT),
        288
Short-term and long-term predictors,
        202
Shot noise, 52
Signal, 2
Signal classification, 8
Signal processing methods, 3
Signal restoration, 68, 207
Signal space, 67
Signal to impulsive noise ratio, 302
Signal to noise ratio, 153, 154
Signal to quantisation noise ratio, 20
Signum non-linearity, 371
SINR, 302
Sinusoidal signal, 24
Soft non-linearity, 375
Source-filter model, 186
Spectral coherence, 42
Spectral subtraction, 243, 244
Spectral whitening, 193

Spectral-time representation, 288
Speech processing, 9
Speech recognition, 9
Squared root Wiener filter, 156
State observation models, 118
State transition probabilities, 119
State transition probability, 125
State transition-probability matrix, 117
State-dependent Wiener filters, 136
State-equation model, 165
State-space Kalman filters, 165
State-time diagram, 120
Statistical interpolation, 276
Statistical models, 66
Statistics, 23
Steepest descent method, 177
Stochastic models for impulsive noise,
        298
Stochastic processes, 23, 25
Stockham, 352
Strict sense stationary process, 33
Sub-band acoustic echo cancellation,
        339
Subspace eigen analysis, 232

T

Time delay estimation, 41
Time delay of arrival, 159
Time-alignment of signals, 158
Time-averaged correlations, 145
Time-varying processes, 35
Toeplitz matrix, 145, 191
Training hidden markov models, 121
Transform-based coder, 12
Transformation of a random process,
        57
Transient noise pulse models, 316
Transient noise pulses, 315
Trellis, 120
Tri-cepstrum, 383
Tri-spectrum, 379
Tukeys bi-weight function, 310
Two-sided predictor, 308
Two-wire to four-wire hybrid, 330

U

Unbiased estimator, 69
Uni-variate pdf, 30
Uniform cost function, 75
Vandermonde matrix, 269

V

Vector quantisation, 103
Vector space, 148
Vector space interpretation of Wiener
    filters, 148
Viterbi decoding algorithm, 131

W

Welch power spectrum, 223
White noise, 38, 215
Wide sense stationary processes, 34
Wiener channel equaliser, 157
Wiener equalisation, 350
Wiener filters, 7, 140, 141, 247
Wiener filter in frequency domain, 151
Wiener-Kinchin, 38

Z

Zero-forcing filter, 370
Zero-inserted signal, 264
Zero-padding, 217